CELL WALL BIOCHEMISTRY
RELATED TO
SPECIFICITY IN HOST - PLANT
PATHOGEN INTERACTIONS

SCANDINAVIAN UNIVERSITY BOOKS

Universitetsforlaget Oslo, Bergen, Tromsø
Munksgaard København
Esselte Studium Stockholm, Göteborg, Lund

CELL WALL BIOCHEMISTRY

RELATED TO
SPECIFICITY IN HOST - PLANT PATHOGEN
INTERACTIONS

Proceedings of a Symposium held at
the University of Tromsø, Tromsø, Norway
2 -6 August 1976

Edited by

B. SOLHEIM and J. RAA
University of Tromsø

UNIVERSITETSFORLAGET
TROMSØ - OSLO - BERGEN

© Universitetsforlaget 1977
ISBN 82-00-05141-2

Distribution offices:

NORWAY
Universitetsforlaget
Box 6589, Rodeløkka
Oslo 5

UNITED KINGDOM
Global Book Resources Ltd.
37, Queen Street
Henley on Thames
Oxon RG9 1AJ

UNITED STATES and CANADA
Columbia University Press
136 South Broadway
Irvington-on-Hudson
New York 10533

Cover design: Kåre Andersen

Printed in Norway by
Lorentzens offsettrykkeri, Oslo

PREFACE

One of the most challenging tasks for all biologists today is to reveal the molecular mechanism behind the distinct specificity in cell-cell interactions. In some cell-cell recognitions surface carbohydrates and proteins are involved as shown for mating cells of the yeast Hansenula, aggregation of sponges and attachment of phages to the bacterial cell wall. Similar mechanisms may be involved in specificity of parasitic and symbiotic associations between plants and microorganisms. The present symposium focused attention mainly on the role of cell wall components either in the plant or in the microorganism as determinants for this specificity.

This is not a field of only academic interest. Further advances in the understanding of the interactions between plants and pathogens is a prerequisite for developing better methods to control plant diseases.

The Editors wish to acknowledge the economic support given to this venture by The University of Tromsø and by The Norwegian Research Council for Science and the Humanities.

The Editors thank the Chairmen for their help with editing the discussions. We would also like to thank Miss Lisen Møller Enoksen for her secretarial work before the start of the Symposium and her excellent help during the meetings. We are grateful to Mrs. Kari Andersen for preparing the final typescripts from which this book was copied. We would also like to thank all our staff colleagues and students for help and support in arranging the Symposium and the production of this volume.

December 1976.   Bjørn Solheim
                                                    Jan Raa

In cases where there is more than one author on a paper the name of the person who read the paper at the Symposium is underlined.

CONTENTS

- 9 -

CELL SURFACE BIOCHEMISTRY RELATED TO SPECIFICITY OF PATHOGENESIS AND VIRULENCE OF MICROORGANISMS

Jan Raa, Børre Robertsen, Bjørn Solheim and Arne Tronsmo, University of Tromsø, Norway

Pathogenic microorganisms have highly specific host requirements, as a rule. Pathogenesis of Cladosporium fulvum is, for example, restricted to tomato, pathogenesis of Cladosporium cucumerinum to cucumber.

Exceptions to this rule exist. However, less specialized pathogens have a rather "crude" mode of pathogenicity which is caused by toxins or enzymes, which destroy plant tissues in an un-specific manner. Such pathogens are "opportunists" which attack when the host is weakened.

The major concern of this symposium will be the interactions between host plants and pathogenic, or symbiotic, microorganisms with a narrow host range.

We can not approach the fundamental aspects within this field without raising the question why saprophytic microorganisms are unable to damage a living plant, even though they produce the necessary enzymes to degrade a dead plant. In other words: What distinguishes a saprophyte from a pathogen, in biochemical terms?

The basic molecular mechanisms underlying specificity in host-pathogen systems may be related to those involved in other cellular recognition processes (Callow, 1975). Examples of such recognition processes are:
1) Binding of bacteriophages to bacterial cell walls (Lindberg, 1973);
2) Conjugation of bacteria (Anderson et al., 1957; Sermonti, 1969);
3) Sexual agglutination of mating types of the yeast Hansenula wingei (Crandall & Brock, 1968) and of the green algae Chlamydomonas (Wiese, 1973);
4) Acceptance or rejection of the pollen grain on the stigma surface (Linskens, 1969; Knox et al., 1972) and of sperm cells on the egg cell of animals (Metz, 1969); 5) Specific aggregation of cells of slime molds (Rosen et al., 1974) and of sponges (Henkart et al., 1973).

As far as the molecular basis for such specific recognition reactions has been revealed it appears that surface-localised complementary "lock-to-key" macromolecular reactions are involved. The molecules involved in the reaction are proteins, glycoproteins and carbohydrates. Such molecules have a sufficiently high structural variability to account for the high degree of specificity of the recognition reactions.

It has been claimed that any macromolecule complex which is exposed to the surface of a bacterium may be a specific receptor of a bacteriophage (Lindberg, 1973). In most cases, however, carbohydrates in the bacterial cell wall are the receptor molecules for proteins on the surface of the virus. The specific topography of the receptor site is in such cases determined by end group sugar molecules. The specificity of such "lock-to-key" reactions may be so high that the virus recognition factor can discriminate between α- and β-glycosidic linkages, H-, NH_2- and OH-groups on the sugar molecule, acetylated or non-acetylated OH-groups, equatorial or axial position of OH-groups, etc.

In several cases (cf. Lindberg, 1973) it has not been possible to isolate any receptor-molecule from the bacterial cell wall. This may be because the receptor consists of more than one molecule which make up a specific arrangement in the intact cell wall. Thus, a specific receptor site may exist, although it is impossible to isolate any single molecule with receptor properties.

Also higher plants contain molecules with the ability to recognize specific configurations of sugar residues in glycoproteins. These molecules are the lectins (formerly called phytohemagglutinins). These peculiar molecules, which are proteins or glycoproteins, can bind to cell surfaces through specific carbohydrate containing receptor sites. When bound to cell surfaces the lectins can exert toxic effects, induce mitosis in lymphocytes and cause agglutination of red blood cells (Lis & Sharon, 1973). The effect on animal cells has been studied most extensively, although such effects not necessarily are informative of their function in the plant. The ability of one lectin, the wheat germ agglutinin, to inhibit the germination of fungal spores may represent one of the essensial functions of the plant lectins (Mirelman, Galun, Sharon & Lotan, 1975).

It seems that lectins provide surface receptor sites for Rhizobium-infections (Bohlool & Schmidt, 1974; Hamblin & Kent, 1973). This receptor site may be identical to the site of binding of the root hair curling factors which are produced by Rhizobium (Solheim & Raa, 1973). According to

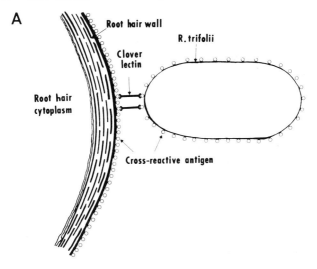

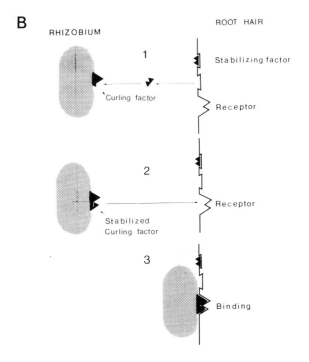

Figure 1. Models of the binding of Rhizobium trifolii to clover root hairs according to (A) Dazzo & Hubbel (1975) and (B) Solheim (1975).

Dazzo & Hubbel (1975, a, b), infective <u>Rhizobium</u> bacteria adsorb on the
surface of clover roots by a cross-bridging of common surface antigens via
a multivalent lectin from clover (Figure 1 A). A corresponding model
(Figure 1 B) has been proposed with basis in studies of the curling reaction of clover root hairs which are exposed to extracts of <u>Rhizobium</u>
(Solheim, 1975; Solheim & Raa, 1973).

The binding of <u>Rhizobium</u> bacteria to their host plant through the
lectin may be the necessary first step in the sequence of events which
finally result in the symbiotic association. Both partners in the symbiosis will accordingly benefit from a discriminatory interaction which
secures that only the symbiotic bacterium can establish that proper intimate contact which is a prerequisite for further expansion into the plant.

Common antigenic substances between pathogens and their plant hosts
have also been found, for example between <u>Melampsora lini</u> and flax (Doubly,
Flor & Clagett, 1960), <u>Xanthomonas malvacearum</u> and cotton, <u>Ceratocystis
fimbriata</u> and sweet potatoe (DeVay, Schnathorst & Foda, 1966), to mention
some. Also metazoa parasitic to animals do produce host-like antigens
(Brown, 1975). It is still uncertain whether these antigenically similar
cell constituents underlie host-pathogen specificity. However, according to
one theory a strong similarity between a host and a pathogen will secure the
least disruptive effect on the host and thus be important for success in
disease development. In other words, the host will not recognize the
presence of "non-self", which otherwise invokes a defence reaction. It may
also be that the common antigens function as they do in the clover-<u>Rhizobium</u>
association, as complementary anchor sites which adhere the pathogen to the
host via a lectin. This theory has not been proved, however, and there are
strong arguments against it.

One should for instance expect heavy selection against any molecule
that can anchor a pathogen to the plant surface. If specific anchor sites
for pathogenic bacteria do exist at the plant surface, it is necessary to
postulate that such receptor sites serve some other function which is beneficial for the plant.

Figure 2 shows the interface between cucumber and <u>Cladosporium cucume-
rinum</u>. The conidia which adhere so firmly to the host surface that they
are not washed away during the preparation procedure have some "adhesive
material" in the interface. The germ tube coalesces with the host cell
wall as it grows into the middle lamella.

We have never succeeded in showing a similar firm attachment of a

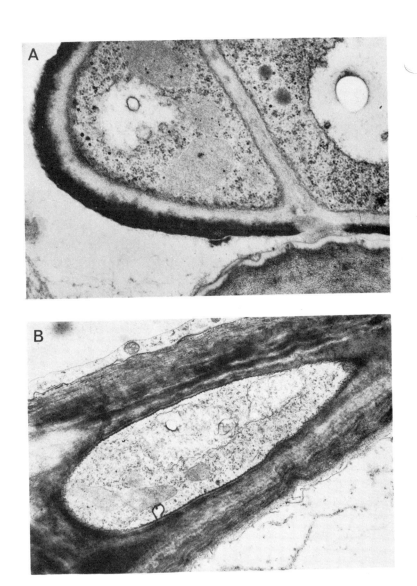

Figure 2.   A: Conidium of Cladosporium cucumerinum attached to the surface of
of a cucumber hypocotyl. X (3.9 x 1200).   B: Hyphae growing in the
middle lamellae of the cucumber hypocotyl tissue. X (3.0 x 5000).

Raa et al. - Cell surface biochemistry

saprophyte (<u>Aspergillus niger</u>) at the surface of a cucumber plant. It is
our experience that non-host pathogens and saprophytes are unable to coalesce so firmly to the plant surface that they remain there after the preparation procedure for electron microscopy.

This is noteworthy since saprophytes may adhere very firmly to inor-
ganic surfaces and to plastic. The mechanism of adhesion to a living plant
may accordingly be another than that to a dead surface. With animal cells
this is shown to be the case (Snere & Milam, 1974), since sulfhydryl-
reagents, chelating agents, hormones and enzymes affect cell-to-surface
adhesions and cell-to-cell adhesions in different ways.

The adhesion characteristics of lymphocytes is apparently controlled
by surface glycoproteins, since glycosidases and trypsin profoundly affect
their circulation and ability to "home" in the lymph nodes, without affecting their viability (Cook & Stoddart, 1973). It is not known whether plant
surface enzymes can modify the surface structure of microorganisms to that
they are rendered unable to adhere to the plant.

The biochemical mechanism of adhesion and coalescence of a pathogen to
its host plant is not known. The initial steps in the infection process
may include complementary lock-to-key reactions which cause binding of the
pathogen to its host. It is also possible that the complementary molecules are enzymes and substrates on opposing cell surfaces, in particular
the glycosyltransferases and the cell surface carbohydrates.

In this mechanism, a glycosyl transferase on the surface of one cell
may not only be involved in synthesis of surface saccharides, but may also
be able to bind to a sugar molecule on an opposing cell. This process
would result in the adherence of the two different cells.

The glycosyltransferases are evidently involved in the adhesion of
platelets to collagen, because the same substances which inhibit adhesion
also inhibit the glycosyltransferase activity in the membrane of the platelet (Barber & Jamieson, 1971).

The glycosyltransferase model for cell-to-cell adhesions has been
worked out mainly after experiments with animal cells. It may as well
turn out to be a mechanism in the adhesion and coalescence process of
pathogens to their hosts. It seems reasonable to postulate that a combined
action of glycosyltransferases, which may cause coalescence with the host
cell wall, and a sublethal glycosidase activity, which causes a corrosion
of the host cell wall, can explain why a pathogen can remain coalesced with

the host cell wall, even though the pathogen actually degrades the wall when it expands.

The typical coalesced mode of growth of plant pathogens is obviously the result of a more subtle process than a crude hydrolysis of the host cell wall by enzymes secreted by the pathogen.

The ability to adhere firmly to a plant surface is rare among microorganisms. This may be because the cuticle covers the carbohydrates of the cell wall so that very few possible anchor sites are exposed. However, this does not hold for the root. Root cells carry a layer of microorganisms in rather intimate association. This microflora is still as a rule unable to grow into the living cells or beneath the very outer surface layers. In this case it seems necessary to postulate a continuously active defence by the host.

However, it may be that the rhizosphere microflora of the healthy plant is unable to degrade or damage the outer structures of the root cells, and thus lives in a commensalistic association with the plant. If that is the case, it is not necessary to have an active defensive system in the host to regulate the commensalistic balance between the partners. The rhizosphere microflora may on the contrary play an important role in the plant's defensive system against potentially harmful microorganisms, thus similar to the function of the lactobacilli in the mucous layers of animals (Speck, 1976). The lactobacilli are attached securely to the intestinal epithelium but they do not penetrate the cells (cf. Savage, 1972).

If it really holds true that it is only the pathogenic microorganisms which have succeeded to develop mechanisms for penetration into and growth in the living plant, it follows that a given plant should have developed active defence systems only against its specific pathogens, and not against the enormous number of saprophytes in the environment. We will come back to the question of how this assumption stands in relation to the present knowledge in biochemical plant pathology.

There are two levels of specific interactions between a plant and a pathogenic microorganism. These are the interactions which condition specificity of the pathogenic property, and the interactions which determine specificity of virulence.

Pathogenicity is an inherent and stable quality of a given genus or species of a microorganism. The expression of this quality is usually restricted to certain genera within a plant family.

Both virulent and avirulent strains of a pathogenic species possess the pathogenic quality. But the expression of this quality is interrupted if the pathogenic strain carries a gene for avirulence which has a specific counterpart gene for resistance in the host (gene-for-gene relationship). This is the highest degree of specificity in the interactions between a plant and a pathogenic microorganism.

Pathogenicity is interrupted in an avirulent combination of pathogen and host, because the avirulent strain provokes a defence reaction in the host. The virulent strain can grow inside the host tissue without provoking such a reaction (Paus & Raa, 1973). Cytological studies have conformed that both virulent and avirulent strains of a pathogenic microorganism are able to penetrate the cell walls before the defence reaction is provoked (Skipp & Deverall, 1972). This is different from saprophytes, and from pathogens which are exposed to non-hosts. These are usually unable to adsorb firmly to the outer surface structures and they do not coalesce with the host cell wall (own experiments).

Pathogenicity is governed by a series of genes which are specifically directed towards a certain plant. This statement is supported by the facts that saprophytes do not readily mutate to become pathogenic and that pathogens not frequently get their host range extended.

Many host-pathogen systems have evolved a gene-for-gene interrelationship between virulence/avirulence genes on one side and susceptibility/resistance genes on the other. This reflects the ecological competition between the two organisms. Mutation and selection of resistance genes in the host population will create a subsequent selection pressure in the pathogen population of genes, which specifically extinguish the effect of the resistance genes.

The biochemical explanation of the gene-for-gene interrelationship is still discussed. According to one model an avirulence gene in the pathogen directs the synthesis of a compound (Albersheim & Anderson-Prouty, 1975) (elicitor) which specifically elicits a defence reaction in those host varieties which carry the corresponding gene for resistance. This defence reaction manifests itself as visible, either macroscopically or microscopically, alterations of host cells.

Often do the host cells succumb hypersensitively as the avirulent pathogen approaches the cell membrane.

Hypersensitive death of host cells would straight away restrict further growth of obligate pathogens. It seems necessary, however, to postulate that the hypersensitive host cells offer an active defence against facultative pathogens. Production of transient antimicrobial principles during hypersensitive death of host cells has been suggested as a defence reaction against such pathogens (Raa, 1968; Raa & Overeem, 1968). The elicitor may also upset the metabolism of host cells, so that secondary metabolic products with antimicrobial activity (phytoalexins) are produced, before the host cells succumb (Frank & Paxton, 1970; Bailey & Deverall, 1971).

In gene-for-gene relationships the avirulence genes and their corresponding resistance genes usually are dominant characters. Such gene pairs may code for the elicitor in the pathogen and its receptor in the host. It seems not necessary to ascribe any function to the recessive alleles for virulence and susceptability. If either of these genes is not transcribed the defence reaction is not provoked and the pathogen becomes virulent again.

The gene-for-gene resistance is very efficient. However, it is ecologically vulnerable, because it can be extinguised by single gene mutations in the pathogen. It seems therefore reasonable to expect additional defence mechanisms which are less specific than the gene-for-gene resistance and more stable towards mutation and selection in the pathogen population.

The socalled "horizontal resistance" or "field resistance" or "polygenic resistance" may be due to such a more general defence. This resistance protects against all races of a pathogen, but it is usually less efficient than that given by gene-for-gene resistance. It seems that tissues of horizontally resistant plants are less suitable than susceptible plants for growth of the pathogen (Tarr, 1972). This may be due to antagonistic actions of cell wall components which become exposed to those microorganisms which are able to make intimate contact with the cell wall. It may also be that the pathogen has been rendered less able to adhere to and coalesce with its host (due to structural circumstances of the host cell wall). Such defence mechanisms would not necessarily manifest themselves as visible alterations of the host cytoplams, as is the case in gene-for-gene resistance.

Plant cell walls do contain enzymes which can degrade cell walls of bacteria and fungi, such as lysozyme (Howard & Glazer, 1967), chitinase (Abeles et al., 1970; Pegg & Vessey, 1973), β-1,4-N-acetylglucosaminidase (Li & Li, 1970) and β-1,3-glucanase (Clarke & Stone, 1962). Since some of these enzymes do not find their substrates in the plant itself, but in cell walls of microorganisms, it seems reasonable that they play a role in the defence against microorganisms. In fact it has been shown that tomato chitinase is involved in the lysis of Verticillium which infects tomato (Pegg & Vessey, 1973). The presence in pathogens of components which can negate the action of such enzymes would offer a teleological evidence for their role in the defence. As yet this has been shown in one case; the inhibition of the endo-β-1,3-glucanase of bean by Colletotrichum lindemuthianum (Albersheim & Valent, 1974).

Many pathogens secrete enzymes which can degrade cell walls of higher plants. Among such enzymes is the polygalacturonidase most certainly involved in pathogenesis, at least of pathogens which grow in the middle lamellae of their host (Paus & Raa, 1973), and have polygalacturonidase as a constitutive enzyme (Skare, Paus & Raa, 1975).

The significance of this enzyme activity in pathogenesis has moreover a teleological support from the fact that dicotyledoneous plants contain compounds which inhibit the polygalacturonidase produced by their pathogens (Albersheim & Anderson, 1971; Anderson & Albersheim, 1972; Fisher, Anderson & Albersheim, 1973; Skare, Paus & Raa, 1975). Such inhibitors may have been selected as a response to pathogens which use this enzyme as one of their offensive weapons. This assumption implies that the host's inhibitors should act selectively against the polygalacturonidase produced by its pathogens, and not against saprophytes and non-host pathogens in general. Our experimental results are in accordance with this.

We have shown that cell walls of cucumber seedlings contain a high molecular weight substance which inhibits the polygalacturonidase produced by Cladosporium cucumerinum, but not that produced by Aspergillus niger and a variety of Botrytis cinerea which causes the dry eye rot disease in apple fruit (Figure 3). Moreover, the polygalacturonidase produced by Cladosporium cucumerinum binds to isolated cell walls of cucumber, whereas those of Aspergillus and Botrytis do not (Figure 4). Apple fruit tissue contains, on the other hand, an inhibitor of the polygalacturonidase produced by B. cinerea, but not of that produced by the two other fungi (Tronsmo & Raa, unpublished). Other workers (Fisher, Anderson & Aldersheim, 1973) have found that the inhibitor of Red Kidney bean is more effective against the polygalacturonidase from one of its pathogens than against non-host pathogens, but the polygalacturonidase from A. niger was inhibited. In spite of the fact that the A. niger polygalacturonidase was inhibited, the experimental results may support our suggestion that

INHIBITION OF POLYGALACTURONIDASE BY AN EXTRACT OF CUCUMBER SEEDLINGS

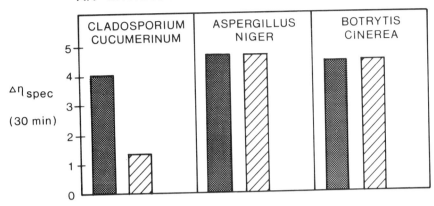

Figure 3.   Polygalacturonidase activity of a dialyzed culture filtrate of Cladosporium cucumerinum, Aspergillus niger and Botrytis cinerea after 4 days at 22°C in a shake culture on mineral medium with 0.5% pectin (apple) as carbon source. The left columns show control activity, the right ones the activity when a dialyzed 1 M NaCl extract of isolated cell walls from cucumber seedlings were added in the enzyme assay mixture (cf. Skare, Paus & Raa, 1975).

BINDING OF POLYGALACTURONIDASES TO ISOLATED CELL WALLS OF CUCUMBER SEEDLING

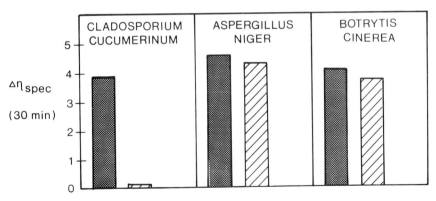

Figure 4.   Corresponds to Figure 3, but shows binding of the polygalacturo-nidases to isolated cell walls (extracted with 1 M NaCl) from cucumber seedlings. Left columns show control activities, the right ones the activity in the supernatant after exposure to isolated cell walls.

these inhibitors have been generated as a defence against those microorganisms which can advance so far in a plant tissue that their polygalacturonidases represent a threat to the plant; i.e. against pathogens but not against saprophytes or non-host pathogens.

It has been pointed out by the Albersheim group (Anderson & Albersheim, 1971; Albersheim & Anderson, 1972; Jones, Anderson & Albersheim, 1972; Albersheim & Anderson-Prouty, 1975) that the polygalacturonidase inhibitors are not sufficiently specific to account for specific virulence in host pathogen systems.

Although we claim that polygalacturonidase plays a predominant role in the pathogenesis of C. cucumerinum in cucumber, we have never succeeded in detecting this enzyme activity in homogenates or in any extracts of cucumber tissue with disease symptoms. This may be because the amount of inhibitor overcomes the amount of enzyme in an extract (Skare, Paus & Raa, 1975). In the infected tissue, however, the pathogen may still be able to utilize pectine because the polygalacturonidase activity may overcome the inhibition at the very vicinity of the hyphae.

Restriction of pectine degradation to the very vicinity of the infecting hypha, by binding and inhibition of the polygalacturonidase, may be beneficial to the pathogen. In this way the production capacity of the plant can be maintained while the pathogen expands in the cell wall in a slinking way, without provoking cell death or a hypersensitive defence.

Plant cell walls contain peroxidase activity (Raa, 1971; Raa, 1973). This activity resides in the same moiety as indolacetic acid (IAA) oxidase activity. The physiological significance of the latter enzyme may be to degrade exogenously produced IAA, which otherwise might interfere with hormone controlled regulation mechanisms in the host cytoplasm (Raa, 1971; Raa, 1973).

The biological significance of the peroxidase activity in the cell wall is still unknown. It has been claimed, however, that peroxidase activity must play some role in disease defence, since infection of plant tissues is associated with increase of this enzyme activity and changes in peroxidase isozyme patterns. This may of course be a noncausal correlation. However, it has been shown that peroxidase exerts toxic effects on germinating spores if included in the germination medium together with hydrogen peroxide (Stahmann & Demorest, 1973). Moreover, it may be bactericidal in mixture with a suitable hydrogen donor and hydrogen peroxide

Raa et al. - Cell surface biochemistry

(Rama Raje Uhrs & Dunleavy, 1975).

It should be borne in mind that antimicrobial activity may result from cooperative effects of several compounds. Bactericidal effect of serum depends for instance on a cooperation of antibody, complement and lysozyme (Wardlaw, 1963; Schwab & Reeves, 1966). It has also been shown that antibody and complement together with transferrin has bactericidal and bacteristatic effects (Griffiths, 1974).

The systemic chemoterapeutant phenylthiourea may be an important tool in the studies of the role of peroxidase in disease resistance. This compound protects cucumber plants effectively against scab at concentrations which are non-inhibiting to Cladosporium cucumerinum in vitro (Sijpesteijn & Pluijgers, 1962). Phenylthiourea inhibits the polyphenoloxidase activity in the plant, increases the level of peroxidase, and causes enhanced lignification of the cell walls near the penetrating hyphae (Sijpesteijn, 1969). Recently we (unpublished) have shown that phenylthiourea also causes increased level of the β-1,3-glucanase of cucumber seedlings.

What is the role of lectins in the plant? This is a question which is a challenge not only for plant pathologists, but for plant physiologists in general. Are the lectins indeed the plant's "antibodies"? Are they structural elements which control cell wall deposition and extension? Do they control cell division in plants as they do when exposed to animal cells? Are they simply storage proteins? It is of course not necessary to ascribe only one function to all compounds with lectin properties, and one compound may combine more than one function.

Lectins are possibly involved in the specific adsorption of Rhizobium on the clover roots. In host-pathogen interactions their role is still obscure.

The most essential finding which relates to the plant lectins' role in disease resistance may be that of Sharon's group (Mirelman, Galun, Sharon & Lotan, 1975). They have shown that the wheat germ agglutinin inhibits conidial formation, spore germination and growth of Trichoderma viride and incorporation of acetate into cell walls of Fusarium solani. This lectin binds to chitin of the emerging hyphal tube and then it exerts the antagonistic action. This may be an unspecific action against all fungi with the chitin wall exposed when advancing into a host.

We have the impression that this field now is pregnant with regard to ideas which may grope the way for fundamental advances in plant immunology.

References

Abeles, F.B., Bosshart, R.P., Forrence, L.E. & Habig, W.H. (1970). Preparation and purification of glucanase and chitinase from bean leaves. Plant Physiology 47, 129-134.

Albersheim, P. & Anderson, A.J. (1971). Proteins from plant cell walls inhibit polygalacturonases secreted by plant pathogens. Proceedings of the National Academy of Science (USA) 68 (8), 1815-1819.

Albersheim, P. & Anderson-Prouty, A.J. (1975). Carbohydrates, proteins, cell surfaces, and the biochemistry of pathogenesis. Annual Review of Plant Physiology 26, 31-52.

Albersheim, P. & Valent, B.S. (1974). Plant pathogens secrete proteins which inhibit enzymes of the host capable of attacking the pathogen. Plant Physiology 53, 684-687.

Anderson, A.J. & Albersheim, P. (1972). Host pathogen interactions V. Comparison of the abilities of proteins isolated from 3 varieties of Phaseolus vulgaris to inhibit the polygalacturonases secreted by three races of Colletotrichum lindemuthianum. Physiological Plant Pathology 2, 339-346.

Anderson, T.F., Wollman, E.L. & Jacob, F. (1957). Sur les processus de conjugaison et de recombinaision chez E. coli. Aspects morphologiques en microscopie electronique. Annales de l'Institut Pasteur 93, 450-455.

Bailey, J.A. & Deverall, B.J. (1971). Formation and activity of phaseollin in the interaction between bean hypocotyls (Phaseolus vulgaris) and physiological races of Colletotrichum lindemuthianum. Physiological Plant Pathology 1, 435-449.

Barber, A.J. & Jamieson, G.A. (1971). Platelet collagen adhesion characterization of collagen glycosyltransferase of plasma membranes of human blood platelets. Biochimica et Biophysica Acta 252, 533-545.

Bohlool, B.B. & Schmidt, E.L. (1974). A possible basis for specificity in the Rhizobium-legume root nodule symbiosis. Science 185, 269-271.

Brown, K.N. (1975). Specificity in host-parasite interactions. p. 119-175 in Receptors and recognition. Ed. by P. Cuatrecases and M.F. Greaves. Chapman and Hall, London.

Callow, J.A. (1975). Plant lectins. Current Advances in Plant Science 18, 181-193.

Raa et al. - Cell surface biochemistry

Clarke, A.E. & Stone, B.A. (1962). β-1,3-glucan hydrolases from the grape vine (Vitis vinifera) and other plants. Phytochemistry 1, 175-188.

Cook, G.M.W. & Stoddart, R.W. (1973). Surface carbohydrates of the eucaryotic cell. Academic Press, London and New York.

Crandall, M.A. & Brock, T.D. (1968). Molecular basis of mating in the yeast Hansenula. Bacteriological Review 32, 139-163.

Dazzo, F.B. & Hubbell, D.H. (1975a). Antigenic differences between infective and noninfective strains of Rhizobium trifolii. Applied Microbiology 30, 172-177.

Dazzo, F.B. & Hubbell, D.H. (1975b). Cross-reactive antigens and lectin as determinants of symbiotic specificity in the Rhizobium-clover association. Applied Microbiology 30, 1017-1033.

DeVay, J.E., Schnathorst, W.C. & Foda, M.S. (1966). Common antigens and host-parasite interactions. p. 313-328 in The dynamic role of molecular constituents in plant-parasitic interactions. Ed. by C. Mirocha and I. Urithani. Bruce Publishing Co., St. Paul, Minn.

Doubly, J.A., Flor, H.H. & Clagett, C.O. (1960). Relation of antigens of Melampsora lini and Linum usitatissimum to resistance and susceptibility. Science 131, 229.

Fisher, M.L., Anderson, A.J. & Albersheim, P. (1973). A single plant protein efficiently inhibits endopolygalacturonases secreted by Colletotrichum lindemuthianum and Aspergillus niger. Plant Physiology 51, 489-491.

Frank, J.A. & Paxton, J.D. (1970). Time sequence for phytoalexin production in Harosoy and Harosoy 63 Soy beans. Phytopathology 60, 315-318.

Griffiths, E. (1974). Rapid degradation of ribosomal RNA in Pasteurella septica induced by specific antiserum. Biochimica et Biophysica Acta 340, 400-412.

Hamblin, J. & Kent, S.P. (1973). Possible role of phytohaemagglutinin in Phaseolus vulgaris L. Nature New Biol 245, 28-30.

Henkart, P., Humphreys, S. & Humphreys, T. (1973). Characterization of sponge aggregation factor. A unique proteoglycan complex. Biochemistry 12, 3045-3050.

Howard, J.B. & Glazer, A.N. (1967). Studies of the physiochemical and enzymatic properties of papaya lysozyme. Journal of Biological Chemistry 242, 5715-5723.

Jones, T.M., Anderson, A.J. & Albersheim, P.(1972). Host-pathogen interactions IV. Studies on the polysaccharide-degrading enzymes secreted by Fusarium oxysporum f.sp. lycopersici. Physiological Plant Pathology 2, 153-166.

Knox, R.B., Willing, R.R. & Ashford, A.E. (1972). Role of pollen-wall proteins as a recognition substance in interspecific incompatibility in poplars. Nature 237, 381-383.

Li, S. & Li, Y. (1970). Studies on the glycosidases of Jack bean meal. III. Crystallization and properties of β-N-acetylhexosaminidase. Journal of Biological Chemistry 246 (19), 5153-5160.

Lindberg, A.A. (1973). Bacteriophage receptors. Annual Review of Microbiology 27, 205-241.

Linskens, H.F. (1969). Fertilization mechanisms in higher plants. Vol. 2, p. 189-253, in Fertilization. Ed. by C.B. Metz & A. Monroy. Academic Press, New York and London.

Lis, H. & Sharon, N. (1973). The biochemistry of plant lectins (phytohaemagglutinins). Annual Review of Biochemistry 42, 541-574.

Matsson, O., Knox, R.B., Heslop-Harrison, J. & Heslop-Harrison, Y. (1974). Protein pellicle of stigmatic papillae as a probable recognition site in incompatibility reactions. Nature 247, 298-300.

Metz, C.B. (1969). Gamete surface components and their role in fertilization. Vol. 1, p. 163-236 in Fertilization. Ed. by C.B. Metz & A. Monroy. Academic Press, New York and London.

Mirelman, D., Galun, E., Sharon, N. & Lotan, R. (1975). Inhibition of fungal growth by wheat germ agglutinin. Nature 256, 414-416.

Nowell, P.C. (1960). Phytohaemagglutinin: An initiator of mitosis in cultures of normal human leukocytes. Cancer Research 20, 462-466.

Paus, F. & Raa, J. (1973). An electron microscope study of infection and disease development in cucumber hypocotyls inoculated with Cladosporium cucumerinum. Physiological Plant Pathology 3, 461-464.

Pegg, G.F. & Vessey, J.C. (1973). Chitinase activity in Lycopersicon esculentum and its relationship to the in vivo lysis of Verticillium albo-atrum mycelium. Physiological Plant Pathology 3, 207-222.

Raa, J. (1968). Polyphenols and natural resistance of apple leaves against Venturia inaequalis. Netherland Journal of Plant Pathology 74 (1968 Suppl. 1), 37-45.

Raa, J. (1971). Indole-3-acetic acid levels and the role of indole-3-acetic acid oxidase in the normal root and club-root of cabbage. Physiologia plantarum 25, 130-134.

Raa, J. (1973). Cytochemical localization of peroxidase in plant cells. Physiologia plantarum 28, 132-133.

Raa, J. & Overeem, J.C. (1968). Transformation reactions of phloridzin in the presence of apple leaf enzymes. Phytochemistry 7, 721-731.

Rama Raje Uhrs, N.V. & Dunleavy, J.M. (1975). Enhancement of bactericidal activity of a peroxidase system by phenolic compounds. Phytopathology 65, 686-690.

Rosen, S.D., Simpson, D.L., Rose, J.E. & Barondes, S.H. (1974). Carbohydrate-binding protein from Polyspondylium pallidum implicated in intercellular adhesion. Nature 252, 128, 149, 150.

Savage, D.C. (1972). Associations and physiological interactions of idigenous microorganisms and gastrointestinal epithelia. American Journal of Clinical Nutrition 25, 1372-1379.

Schwab, G.E. & Reeves, P.R. (1966). Comparison of the bactericidal activity of different vertebrate sera. Journal of Bacteriology 91, 106-110.

Sermonti, G. (1969). Bacteria. Vol. 2, p. 47-94 in Fertilization. Ed. by C.B. Metz & A. Monroy. Academic Press, New York and London.

Sharon, N. & Lis, H. (1972). Cell-agglutinating and sugar-specific proteins. Science 177, 949-959.

Sijpesteijn, A.K.S. (1969). Mode of action of phenylthiourea, a therapeutic agent for cucumber scab. Journal of the Science of Food and Agriculture 20, 403-405.

Sijpesteijn, A.K.S. & Pluijgers, C.W. (1962). On the action of phenylthioureas as systemic compounds against fungal disease of plants. Mededelingen Landbouwhogescholl Gent 17, 1199-1203.

Skare, N., Paus, F. & Raa, J. (1975). Production of pectinase and cellulase by Cladosporium cucumerinum with dissolved carbohydrates and isolated cell walls of cucumber as carbon sources. Physiologia plantarum 33, 229-233.

Skipp, R.A. & Deverall, B.J. (1972). Relationships between fungal growth and host changes visible by light microscopy during infection of bean hypocotyls (Phaseolus vulgaris) susceptible and resistant to physiological races of Colletotrichum lindemuthianum. Physiological Plant Pathology 2, 357-374.

Snere, P.A. & Milam, M. (1974). Stochastic studies on cell surface stickiness. Vol. 7, p. 21-47 in Biology and chemistry of eucaryote cell surfaces (Miami Winter Symposia). Ed. by E.Y.C. Lee & E.E. Smith. Academic Press, New York and London.

Solheim, B. (1975). A model of the recognition-reaction between Rhizobium trifolii and Trifolium repens. National Alliance Treaty Organization Conference on Specificity in Plant Diseases, Advanced Study Institute, Sardinia. May 4-16, 1975.

Solheim, B. & Raa, J. (1973). Characterization of the substances causing deformation of root hairs of Trifolium repens when inoculated with Rhizobium trifolii. Journal of General Microbiology 77, 241-247.

Speck, M.L. (1976). Interactions among lactobacilli and man. Journal of Dairy Science 59 (2), 338-343.

Stahmann, M.A. & Demorest, D.M. (1973). Changes in enzymes of host and pathogen with special reference to peroxidase interaction. p. 405-420 in Fungal pathogenicity and the plants response Ed. by R.J.W. Byrde and C.V. Cutting. Academic Press, London and New York.

Tarr, S.A.J. (1972). The principles of plant pathology. The MacMillan Press, London and Basingstoke.

Wardlaw, A.C. (1963). The complement-dependent bacteriolytic activity of norman human serum. Canadian Journal of Microbiology 9, 41-52.

Watson, L., Knox, R.B. & Creasar, E.H. (1974). Con A differentiate among grass pollen binding specifically to wall glycoproteins and carbohydrates. Nature 249, 574-576.

Wiese, L. (1973). Nature of specific glycoprotein agglutinins in Chlamydomonas. Annals of the New York Academy of Sciences 234, 383-395.

DISCUSSION

Chairman: B. Solheim

Bateman: I would like to raise a question regarding binding of pathogens to host cells as it relates particularly to your implications with regard to fungal pathogens. Is it not true that most of the binding which takes place initially, is with the plant cuticle as opposed to the plant wall per se, and that this may be non-specific in terms of plant pathogens? I am just raising this as a question in point, since I think your implication was mainly to the host cell wall proper.

Raa et al. - Cell surface biochemistry

Raa: It is of course true that the cuticle is the first surface the pathogen is exposed to, but I think the ability to adhere firmly to the surface is specific for pathogens.

Bateman: In many instances pathogens will adhere to the surface of non-hosts, but the specificity comes at the penetration. Whereas, non-pathogens will not adhere at all, in most instances that I am aware of. So what I am suggesting is that the adherence mechanism so far as host surface is concerned is initially between the pathogen and the cuticle. Whereas the specificity comes later after penetration of the host surface. Then it either becomes virulent or avirulent.

Raa: From our experience saprophytes or non-host pathogens do not adhere so firmly, even to the cuticle, that they remain there after fixation for electron microscopy. This adhesion must accordingly be very loose. I agree that the process of coalescence and specific adhesion must involve some wall structure or wall compound. It may be that cell wall components are exposed through the cuticle so that these can be receptor sites for pathogens.

Bateman: From the literature I understand that some pathogens can even adhere to glass surfaces. For example, germinating uredospores of _Puccinia graminis_ will adhere to glass surfaces.

Raa: Yes, but I think this process is different from the binding to the surface of a living plant.

Albersheim: I want to make a correction concerning elicitors, at least with regard to our research. Some time ago we speculated that elicitors might be the products of the avirulence genes of the pathogens. However, when we did the research the speculations proved to be false. In a series of papers by Ayres, Abel, Valent and myself that were published in May of this year, which I realize that you have not yet received, we give evidence that elicitors which do activate defence mechanisms in the host, do not have specificity with regard to this activation and we will talk about this in some detail later in this symposium. The other comment I would like to make is that you listed a series of cell-cell or pathogen-host, etc. types of interactions in plants and animals. You said that these were well established interactions based on certain molecules and you ascribed certain functions to certain molecules. I would like to suggest that most of these are still quite hypothetical.

PECTIC ENZYMES: INVOLVEMENT IN PATHOGENESIS AND POSSIBLE RELEVANCE TO
TOLERANCE AND SPECIFICITY

Harry Mussell and Larry L. Strand, Boyce Thompson Institute for Plant
Research, 1086 North Broadway, Yonkers, New York 10701, U.S.A.

Pectic Enzymes and Pathogenesis

The occurrence and probable participation of pectic enzymes in
pathogenesis have been documented for over half a century (Jones, 1909;
Brown, 1915). It is now almost universally accepted that pectic enzymes
are responsible for the tissue maceration commonly referred to as "rot".
Jones and Brown also suggested that the cell death associated with soft
rots might also be attributed to the action of pectic enzymes.

In reviewing the evidence then available on pectic enzymes and cell
killing, Byrde & Archer (1973) concluded that there must be two mechanisms
operative in the killing of plant cells by these enzymes. These authors
postulated that the physiological effects of these enzymes could be
explained on the basis of either a direct effect or an indirect mechanism
of cell killing. Although the ideas upon which some of this thinking was
based have since been demonstrated to be improbable, the concept is still a
useful one and might be used to explain the differences in the function of
pectolytic enzymes in biotrophic and necrotrophic diseases. For the
purposes of this discussion, necrotrophic parasites will be defined as
pathogens which cause extensive host tissue disruption as an early manifestation of infection, i.e. the pathogens which incite soft rot diseases,
lesion diseases, cankers etc. Biotrophic parasites, on the other hand,
will be defined as those pathogens which have the ability to infect and
colonize host tissue without immediate gross manifestations of distress in
the invaded host tissue. This group would include not only the classic
"obligate parasites" but also the fungal vascular wilts, the apple scab
pathogen and other pathogens which have the capacity for establishing

compatible, non-disruptive associations with host tissues during the early phases of infection and colonization. We propose to combine the consideration of pathogen mode of nutrition with the concept of two mechanisms of cell disruption to attempt to present a more clarified view of the potential roles for pectic enzymes in host-parasite physiology.

The terminology for pectic enzymes followed in this paper will be based on that of Bateman & Millar (1966), and we will emphasize the participation of the two depolymerizing enzymes, endopectate lyase (endoPL, E.C. 4.2.2.1) and endopolygalacturonase (endoPG, E.C. 3.2.1.15) in pathogenesis.

Direct Mechanism

There have been three proposals put forward on the possible direct mechanism of cell killing by pectic enzymes. Tribe (1955) proposed that the pectic enzymes might be interacting directly with a substrate in the plant cell membranes. This possibility was investigated by Mount, Bateman & Basham (1970), using a highly purified endoPL. They found that this highly purified pectic enzyme caused the death of cells in treated tissues, and that this death occurred concurrently with tissue maceration. Their investigations also indicated that neither a phosphatidase nor a protease from the culture fluids of a phytopathogenic _Erwinia_ sp. would kill plant cells under conditions identical to those used to demonstrate cell killing by the endoPL. Fushtey (1957) suggested that the enzymatic degradation of the pectic portion of plant cell walls might result in the release of phytotoxic materials, which, in turn, were responsible for the observed cell death. This possibility was somewhat supported by the observations of Lund & Mapson (1970) that partially purified pectic enzymes from culture filtrates of _Erwinia carotovora_ catalyzed the release of several enzymes from the cell walls of cauliflower tissues and that these enzymes were then responsible for the appearance of ethylene and hydrogen peroxide. A third possible mechanism for cell killing was proposed by Wood (1972). He suggested that the enzymatic degradation of the plant cell walls by pectic enzymes renders the walls incapable of mechanically supporting the protoplast.

Byrde & Archer (1973) suggested that the direct cell killing effects might relate to the degradation of a postulated substrate molecule in the plasma membrane, as had been suggested by Mount, Bateman & Basham (1970).

However, the elegant experiments of Tseng & Mount (1974) just about preclude this possibility. Using cell wall-free protoplasts, these authors tested several purified enzymes from culture fluids of a bacterial plant pathogen for the ability to kill the protoplasts. They observed that a purified endoPL, which was capable of causing tissue maceration and cell death when applied to intact plant tissues, did not cause death of the wall-free protoplasts. Their evidence indicates that the sensitivity of living plant cells to the effects of endoPL is in some manner related to the presence of cell walls around the living cells. The only criticism that can be directed at this work is that the authors used very high levels of commercial pectic enzymes to create their protoplasts. It is therefore possible that they were selecting for endoPL-resistant protoplasts during their preparation from the plant tissues. However, this purified endoPL caused the death of all cells when applied to plant tissues, indicating that the presence of cell wall material around the protoplast was the critical factor for cell killing and that, at least in situ, there were no resistant protoplasts in the tissues used.

A report by Ulrich (1975) would seem to support Byrde & Archer's hypothesis about membrane effects and the direct mechanism. Ulrich reported that a PG purified from culture filtrates of Pseudomonas cepacia and labeled with a fluorescent indicator rapidly penetrated the plasma membranes of onion root cells. However, comparison of the specific activity of Ulrich's PG preparation with representative specific activities of pectic enzymes that have been rigorously purified (Table 1), indicates that the specific activity of her preparations were unusually low. Although there was demonstrable PG activity in the preparations used for fluorescent labeling and this label did appear rapidly in cells of treated tissue, proof that the fluorescent label in the cells was attached to the PG awaits more rigorous purification of the PG prior to fluorescent labeling. Also, if Ulrich's conclusions about penetration of PG through cell membranes resulting in cellular damage were correct, Tseng & Mount (1974) should have seen some evidence of this in their protoplast studies. They did not.

Presently the existence of the substrate molecule in the plasma membrane of plant cells postulated by Byrde & Archer (1973) is doubtful. However, the concept of a direct mechanism is still a good one and can still be used to help understand the mechanisms by which pectic enzymes kill plant cells. We suggest that the direct mechanism of cell killing is

Table 1.   Specific activities of endopolygalacturonases purified from various sources.

Source	Specific activity[1]	Authors
Verticillium albo-atrum	3,086,666	Mussell & Strouse (1972)
Verticillium albo-atrum	1,296,750	Wang & Keen (1970)
Aspergillus niger	3,600	Rexova-Benkova & Slezarik (1966)
Aphanomyces euteiches	1,645	Ayers, Papavizas & Lumsden (1969)
Tomato fruit (Lycopersicon sp.)	1,200	Hunter & Elkan (1974)
Pseudomonas cepacia	361	Ulrich (1975)

1) All activities converted to relative viscometric units per mg protein

probably characteristic of the role of pectic enzymes in many necrotrophic diseases. This mechanism can be characterized by the rapidity of the effects of the enzymes on the cells, the substantial efflux of water and salts from treated tissues, the high natural concentrations of pectic enzymes observed at the infection sites and the apparent lack of specificity associated with this effect. We tend to concur with the suggestion of Wood (1972) that the direct mechanism of cell killing is probably associated with a critical alteration of the molecular architecture of the cell walls, rendering the walls incapable of supporting the living protoplast. The reports of Hall & Wood (1973, 1974) and the observations of Cronshaw & Wood (1973) document the rapidity of the direct effect, and these reports, and that of Garibaldi & Bateman (1971), document the magnitude of the water and salt effluxes from treated tissues. The work of Basham & Bateman (1975a, 1975b) has demonstrated that the kinetics of cell wall degradation, maceration, electrolyte leakage and cell death are inseparable. This indicates that all of these effects are probably the result of a single enzymatic act of the highly purified endoPL used in these studies. Although this characterization of the direct effect is somewhat different from the one originally proposed by Byrde & Archer (1973), we choose to retain the designation "direct effect" to emphasize that the cell death observed in these systems is apparently a direct result of the pectolytic degradation of the cell walls; there is apparently no participation of materials released from the host cell walls by the pectic enzymes.

Indirect Mechanism

The indirect mechanism proposed by Byrde & Archer (1973) was character-ized as involving either toxic end products of cell wall degradation, or as resulting from the activation of host wall-bound enzyme systems through the act of solubilization of these enzymes. Available evidence (Fushtey, 1957; Basham & Bateman, 1975a) indicates that there are probably no solubilization products of pectolytic cell wall degradation that are toxic to plant cells; however, the participation of solubilized, activated host wall-bound enzyme systems in the indirect mechanism of cell killing remains a distinct and probable possibility. We would like to suggest that the indirect mechanism can be characterized as occurring more slowly than the direct mechanism, with a distinct lag between the application of the pectic enzymes and the appearance of cell necrosis (symptoms). In addition, there does not seem to be a rapid efflux of water and salts from the affected tissues, and there is some evidence for tissue and host specificity associated with the indirect mechanism.

It is our contention that indirect effects of pectic enzymes produced by vascular parasites are responsible for many of the symptoms observed in the vascular wilt syndrome. Understanding the possible role of pectic enzymes in the vascular diseases is not as easy or as intuitively obvious as in the soft rots, because there is little or no tissue maceration involved in these biotrophic diseases. Although the quantitative data are sometimes contradictory or conflicting, the presence of pectic enzymes during vascular wilt pathogenesis has been conclusively demonstrated (Kamal & Wood, 1956; Deese & Stahmann, 1962; Talboys & Busch, 1970; Mussell & Green, 1970; Wiese, DeVay & Ravenscroft, 1970; Dimond, 1970; Mussell, 1973; Ferraris, Garibaldi & Matta, 1974; Cappellini & Peterson, 1976).

One limitation on attempts to elucidate the indirect mechanism of cell killing by pectic enzymes is the absence of an available, rapid bioassay for these effects, analogous to the neutral red test used to assay for the direct mechanism. Those of us attempting to characterize this indirect mechanism have been limited to using disease symptom generation as our indicator for efficacy of the enzyme preparations being tested. For the most part, therefore, the question of an indirect mechanism of pectolytic enzyme effects on host tissues can be reduced to the question "Do pectic enzymes generate symptoms of any biotrophic diseases, and, if so, do they participate naturally in this symptom generation during pathogenesis?"

Various lines of experimentation have been used to attempt to elucidate the possible participation of pectolytic enzymes in the biotrophic vascular wilt diseases.

I. Correlations of in vitro production of pectic enzymes with pathogenicity. These experiments can be further divided into comparison of pectolytic enzyme production by naturally-occurring species or isolates of the pathogens, and examination of derived mutant strains of an individual pathogen. Although the dangers of attempting to relate observed in vitro production of pectic enzymes to presumed in vivo performance of the pathogen have been repeatedly pointed out (English & Albersheim, 1969; Puhalla & Howell, 1975), many attempts to achieve this correlation have been reported, and the results are as contradictory as would be expected. In vitro production of pectic enzymes by vascular pathogens depends on the isolates used (Talboys & Busch, 1970) and can be influenced by the composition of the culture medium (Mussell, 1968; Wiese, DeVay & Ravenscroft, 1970) and the age of the cultures at the time of harvest (Mussell, 1968). Enzyme production is regulated by a subtle and complex system of induction/ repression effects (Cooper & Wood, 1973, 1975). With respect to Verticillium albo-atrum, the situation is further compounded by the fact that the fungus produces both an endoPG and an exopolygalacturonase in culture, both of which will exhibit pectolytic activity in assays performed at acid pHs (Mussell & Strouse, 1972) and the relative amounts of these two enzymes produced vary with the isolate examined (Mussell, 1973) and the culture medium used (Mussell, unpublished).

It is not surprising that the only consistently confirmed fact in the literature on pectic enzyme production by these pathogens is the fact that all species and isolates of vascular pathogens investigated produce pectic enzymes under the appropriate culture conditions (Dimond, 1970). There is also evidence that the ability to produce pectolytic enzymes can be used to distinguish between pathogenic and saprophytic species of V. albo-atrum (Leal & Villanueva, 1962).

Induced mutants of Fusarium oxysporum f.sp. lycopersici and V. albo-atrum have been investigated for their ability to produce pectic enzymes in vitro and for their relative pathogenicity. McDonnell (1958) and Mann

(1962) investigated the pathogenicity of UV-induced mutants of F. oxysporum f.sp. lycopersici that were deficient for in vitro production of PG. Both of these authors reported that isolates deficient in PG production in culture exhibited reduced pathogenicity in susceptible tomato plants. The authors both emphasized the continued pathogenicity of their mutants, and concluded that production of PG was not necessary for pathogenesis by this fungus. However, the mutants of F. oxysporum f.sp. lycopersici deficient in the production of pectic enzymes were also deficient in their pathogenic abilities, indicating to us that production of pectic enzymes by this pathogen is a prerequisite for competent pathogenesis.

Recently, Puhalla & Howell (1975) undertook the same type of investigation using induced mutants of V. dahliae (= V. albo-atrum). They produced several UV-induced mutants and compared the abilities of the parent and mutant isolates for pectic enzyme production and for their relative pathogenicities. In their original report, these authors indicated that their mutants were unable to produce any pectic enzymes when cultured on several different media. Verticillium normally produces an endoPG (Wang & Keen, 1970; Mussell & Strouse, 1972), and exoPG (Mussell & Strouse, 1972) and an endoPL (Heale & Gupta, 1972; Cooper & Wood, 1973, 1975). Puhalla & Howell also noted that the culture fluids of their mutants contained a heat labile wilt-inducing factor, while other investigators have reported that the wild-inducing factors in culture filtrates of V. albo-atrum are heat stable (Stoddart & Carr, 1966; Keen & Long, 1972). This indicates that the mutant isolates used by these investigators were apparently quite different from the parent cultures. Subsequently it has been reported that these mutants produce low levels of pectolytic activity when culture conditions were altered slightly (Howell & Puhalla, 1975), and symptom development was slower in plants inoculated with the mutant strains than when the parent isolates were used. Therefore, although these authors have an excellent idea in comparing the performance of mutants directly with that of the parent isolates, present information would seem to indicate that reduced ability to produce pectic enzymes also results in reduced pathogenicity of V. albo-atrum.

II. Pectic enzymes in infected plant parts. Several attempts to extract pectic enzymes from host tissues infected with vascular pathogens have been reported, with variable results. Pectic enzyme activity has been detected in extracts of Fusarium-infected tomato plants by Deese & Stahmann

(1962) and Mussell & Green (1970). In both of these reports, the levels of
pectic enzyme activity recovered from susceptible plants were higher than
the levels observed in extracts of resistant plants. Capellini & Peterson
(1976) have also reported detection of pectic enzyme activity in Fusarium-
infected mimosa trees. In this case, the PG activity recovered from infected
trees had a pH optimum similar to the PG activity observed in culture
filtrates of the pathogen. Polygalacturonase activity has been detected in
Verticillium-infected cotton plants by Mussell & Green (1970), Wiese, DeVay
& Ravenscroft (1970) and Keen & Erwin (1971). Mussell & Green reported
observing more PG activity in plants infected with severe isolates of
Verticillium than in plants infected with mild pathotypes, while the latter
two reports indicated either no correlation with pathogenicity or an inverse
correlation. Mussell & Green also reported detection of PG activity in
vascular sap of susceptible and resistant tomato plants inoculated with
Verticillium. The levels of PG observed in susceptible plants were higher
than those observed in resistant tomato varieties. Heale & Gupta (1972)
observed both PG and endoPL activity in Verticillium-infected lucerne shoots,
and the levels of endoPL activity increased rapidly during symptom
development.

All of the above reports stated or implied that the amount of pectolytic enzyme activity recovered was representative of the quantity of
enzyme being produced by the pathogens in infected host tissues. However,
recent work from several laboratories would indicate that two alternative
interpretations must be considered. Plant cell walls have been demonstrated
to contain proteins which specifically inhibit endoPG (Albersheim &
Anderson, 1971; Jones, Anderson & Albersheim, 1972; Fisher, Anderson &
Albersheim, 1973). It is possible that some of the low levels of PG
activity observed in extract of infected plants were a result of inhibition
of endoPG by these inhibitor proteins.

Recent work by Skare, Paus & Raa (1975) indicates that the PG produced
by Cladosporium cucumerinum binds to the cell walls of cucumber tissues and
is very difficult to recover. Should this binding prove to be a general
characteristic of pectic enzymes and host cell walls, the conditions for
extraction of pectic enzymes from infected tissues will have to be very
carefully defined if the resulting extracted activities are to be quantitatively significant. We have investigated the binding of the endoPG from
Verticillium to cotton cell walls. We find that this enzyme binds to cotton

cell walls, and to recover the endoPG from these walls, the walls must be extracted with high concentrations of salt.

Inhibition of pectic enzyme activity by host cell wall materials and binding of pectic enzymes to host cell walls will both have to be more thoroughly investigated before definite conclusions can be drawn about the significance of the levels of pectic enzyme activity observed during vascular pathogenesis. However, the information presently available implicates these enzymes as being present in the appropriate places at the appropriate time to participate in symptom generation in these diseases.

III. Effects of purified pectic enzymes on host tissues. Considering the possible information to be obtained by application of purified pectic enzymes to host plants, it is surprising that this approach has not been tried more often. Keen & Erwin (1971), Mussell (1973) and Cooper (1974) have purified pectic enzymes from V. albo-atrum and applied them to cuttings of the respective hosts, and Mussell & Strand (1974) have applied purified endoPG to intact host plants. Keen & Erwin reported that application of an apparently homogeneous endoPG from cultures of V. albo-atrum did not generate foliar symptoms of Verticillium wilt when applied to cotton cuttings in water. Mussell (1973), using an endoPG purified in a slightly different manner than the enzyme used by Keen & Erwin (Mussell & Strouse, 1972), demonstrated that this enzyme was capable of generating many of the typical foliar symptoms of Verticillium wilt of cotton when applied to cuttings or leaves in a dilute salt solution. The role of the salt solution is not yet understood; however, the salt solution appears necessary for symptom expression (Mussell, 1973). These salts did not affect the pectolytic activity of the endoPG when tested in vitro (Mussell & Strouse, 1972; Mussell, 1973). Apparently the cations in the salt mixture activated some of the proteins released from the host cell walls by endoPG. The reactions catalyzed by these activated proteins apparently lead to the indirect effects of the endoPG on the host tissues.

In an extension of the above work, we have applied purified endoPG to intact cotton plants using vascular transfusion (Mussell & Strand, 1974). In these experiments, we verified the requirement for a dilute salt solution and observed two distinct patterns of symptom expression. Young expanding leaves (which presumably would take up more of the endoPG moving in the translocation stream) developed severe disease symptoms within 48 hours, as had been previously observed using cotton cuttings. Mature foliage did not begin to exhibit symptoms until 7 to 10 days after treatment, and the symptoms in this older foliage were less severe, resembling accelerated and telescoped senescence.

Cooper (1974) purified endoPG and endoPL from cultures of tomato isolates of V. albo-atrum and applied these enzymes to cuttings and leaves of tomato varieties susceptible to Verticillium wilt. Cooper also used a dilute salt solution as their vehicle for application of the enzymes, and also observed many of the typical foliar symptoms of Verticillium wilt. In these experiments, the severity and nature of the symptoms observed corre- lated well with the relative susceptibility of the two tomato varieties tested. All of the above results indicate that both endoPG and endoPL will generate disease symptoms in susceptible host tissues if applied in a dilute salt solution. Symptoms resulting from application of these enzymes are remarkably similar to those observed in Verticillium wilt of tomato and cotton. We conclude from the above results that pectic enzymes produced by vascular wilt pathogens are causally involved in natural pathogenesis by the vascular wilt fungi, and that the indirect effects of these enzymes on host tissues are responsible for some, if not all, of the symptoms observed in these diseases.

Pectic enzymes solubilize some of the proteins bound to plant cell walls (Lund & Mapson, 1970; Stephens & Wood, 1974; Barnett, 1974; Strand & Mussell, 1975; Strand, Rechtoris & Mussell, 1976). We feel that the solubilization (and probable concurrent activation) of these wall-bound enzymes initiates several host metabolic sequences at physiologically inappropriate times and places, disrupting coordinate physiology of the host organism. The detection of peroxides in bathing solutions over tissues treated with endoPG (Mussell, 1973) and the solubilization of peroxidases from tissues and cell walls treated with pectic enzymes (Barnett, 1974; Basham, 1974; Stephens & Wood, 1974; Strand & Mussell, 1975; Strand, Rechtoris & Mussell, 1976) suggest to us that delocalization of peroxidase enzymes within the host cell walls may be responsible for some of the tissue damage observed after treatment with pectic enzymes.

Although the precise mechanisms of cell killing in both the direct and indirect mechanisms remain to be elucidated, the available evidence supports the conclusion that there are two distinct effects of pectic enzymes on plant tissues. It also seems safe to generalize that the direct effects of pectolytic enzymes are associated with the damage resulting from pathogenesis by many necrotrophic pathogens, while the indirect effects of these enzymes are probably responsible for some or all of the foliar symptoms observed in many biotrophic diseases. In the natural disease situations, these two mechanisms can be differentiated by the rapidity of the deleterious effects and the levels of pectolytic activity produced by the microorganism at the site of the infection.

Specificity, Susceptibility and Tolerance

Symptoms are the manifestation of a disease situation. Any phenomenon that tends to reduce symptom severity reduces the impact of disease on crop yield. Resistant varieties tend to reduce symptom severity by limiting the ability of the pathogen to invade host tissues. Alternatively, symptom severity can be lessened by producing varieties of host plants less sensitive to the pathogenic stimuli of the invading microorganism. This approach has been successful in lessening the impact of Victoria Blight of oats, Frogeye Leaf Spot of sugarcane and Southern Leaf Blight of corn. In the above examples, crop performance was improved by genetically removing sensitivity to host-specific toxins from new cultivars. Since pectolytic enzymes play an important role in pathogenesis, any change in host characteristics which reduced the effectiveness of pectolytic enzymes would lessen the ability of the pathogen to cause disease. Reducing the effectiveness of pectolytic enzymes would not necessarily result in resistance; however, a substantial reduction in the ability of an invading organism to initiate symptom expression in the host might be sufficient to tip the balance of the host-pathogen interaction from one of susceptibility to one of tolerance. Possible mechanisms for altering the effectiveness of pectolytic enzymes include:
1. Differential regulation of pectic enzyme synthesis through differences in host cell wall composition.
2. Inactivation of pectolytic enzymes by components of host cell walls.
   a. Differential binding to host cell walls.
   b. Pectolytic enzyme inhibitors in host cell walls.
3. Differential host tissue sensitivity to the deleterious effects of pectolytic enzymes.

Differential Regulation

Differential regulation of pectic enzyme synthesis through the availability of host cell wall carbohydrates to the pathogen has been discussed in detail by Albersheim, Jones & English (1969) and Albersheim & Anderson-Prouty (1975). They suggested that regulation of pathogen synthesis of cell wall enzymes by induction characteristics of host cell wall carbohydrates is not functional in specificity, only in general resistance. This conclusion was based upon the observation that little or no difference exists between the ability of purified cell walls from susceptible and resistant tissues to induce synthesis of cell wall degrading enzymes (Albersheim & Anderson-Prouty, 1975). These authors did not consider the possibility that there are distinct differences in the availability of important cell wall carbohydrates in situ in susceptible and resistant walls, nor did they consider that repression of enzyme synthesis might be an important characteristic of walls of resistant tissues and hosts.

Cooper & Wood (1973, 1975) have presented evidence for specific regulation of the synthesis of endoPG in V. albo-atrum and F. oxysporum f.sp. lycopersici by galacturonid acid. Their results indicate that galacturonic acid induced the synthesis of endoPG when the monomer was present at very low concentrations in culture media and did not accumulate in the cultures. The synthesis of endoPG was also remarkably sensitive to repression by galacturonic acid. In the case of V. albo-atrum, galacturonic acid levels three times the optimum for induction of endoPG completely repressed the synthesis of this enzyme. The appearance of vascular occlusions is a common occurrence during vascular pathogenesis. These occlusions are largely pectic in nature. Degradation of these plugs by pectic enzymes of the pathogen would result in supraoptimal levels of galacturonic acid in the vascular fluids, resulting in repression of pectic enzyme synthesis.

Field observations suggest that this may, in fact, be occurring. Invasion of resistant tomato varieties by vascular pathogens often results in transient symptom expression, followed by recovery of the host. Subsequent to recovery by the resistant host, the pathogen can still be recovered from host vascular tissue. The pathogen is viable but apparently cannot synthesize the cell wall degrading enzymes required to invade the new vascular tissue differentiated by the resistant host.

Keen & Horton (1966) demonstrated that synthesis of endoPG by

Pyrenochaeta terrestris is subject to repression by glucose. Treatments which resulted in reduced glucose levels in the roots of onion seedlings led to increased disease severity after inoculation (Horton & Keen, 1966). On the other hand, treatments which led to increased glucose levels in the seedling roots resulted in reduced synthesis of PG by the pathogen and reduced disease severity. Patil & Dimond (1968) observed that glucose repressed synthesis of endoPG by F. oxysporum f.sp. lycopersici. Supplying glucose to infected tomato plants increased mycelial growth of the pathogen in the tomato tissues but reduced disease symptoms. The results of these two studies indicate that repression of PG synthesis can function in altering the results of a host-parasite interaction and alter the severity of a disease. This effect may be operative in many "low sugar diseases" (Horsfall & Dimond, 1957).

Inactivation

In vivo inactivation of pectic enzymes from pathogens may result from host proteins which inhibit the activity of these enzymes (Albersheim & Anderson, 1971) or through binding of pathogen enzymes to host walls (Skare, Paus & Raa, 1975). The inhibitor proteins from bean cell walls have been extensively investigated, and it has been determined that one of these proteins inhibits the endoPG produced by Colletotrichum lindemuthianum 40-fold more efficiently than it inhibits the endoPG produced by F. oxysporum f.sp. lycopersici (Fisher, Anderson & Albersheim, 1973). This purified inhibitor did not inhibit the endoPG produced by Sclerotium rolfsii. These observations raise the possibility that plant cell walls may contain inhibitors specific for the endoPGs from pathogens of that host. Although the in vivo physiology of the proteins that inhibit endoPGs has not been characterized, these endoPG inhibitors seem capable of functioning to lessen the effectiveness of the pathogen produced endoPG, thus attenuating disease severity.

The studies of Skare, Paus & Raa (1975) have clearly indicated that pathogen endoPG binds tightly to host cell walls. We have been able to demonstrate that pathogen endoPG and non-pathogen endoPG bind to cell walls differently, and that there is differential binding of pathogen endoPG to resistant cell walls (Figure 1). The significance of this differential binding has not been totally defined; however, this may represent a manner in which the pathogen can "perceive" the difference between susceptible and

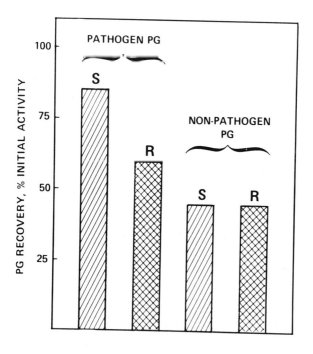

Figure 1. Endopolygalacturonase recovered from purified cotton cell walls
by elution with 1N NaCl. (Pathogen - V. albo-atrum; Non-
pathogen - F. oxysporum f.sp. lycopersici; S - cultivar "M-8",
susceptible to V. albo-atrum; R - cultivar "SBSI", resistant to
V. albo-atrum.)

resistant host tissues. In addition, our inability to recover as much
pathogen endoPG from resistant walls as we could from susceptible walls
implies that this enzyme is bound in an inactive state on the resistant
walls. It seems that the resistant cell walls have a greater capacity to
bind the endoPG, thus rendering it ineffectual in degrading the host cell
walls.

Differential Effects

Little work has been done on the differential ability of pectic
enzymes to affect plant tissues; however, all of the available evidence
suggests that differential susceptibility to maceration exists. Garibaldi
& Bateman (1971) reported that carrot tissue was resistant to one form of
endoPG produced by Erwinia chrysanthemi but was macerated by another form
of endoPG elaborated by the same pathogen. Ishii (1976) investigated the
endoPG and endoPL produced by Aspergillus japonicus and observed that
tissues susceptible to maceration by one of these enzymes were frequently
resistant to maceration by the other. Due to the potential role of
differential susceptibility of plant tissues to pectolytic disruption,
more emphasis should be placed on characterization of the physiologic
differences between multiple forms of pectolytic enzymes produced by
individual pathogens. These differences could be especially important
when the forms of pectic enzymes isolated from infected plants differ from
those elaborated by the same pathogen in culture (Bateman, 1963; Byrde,
Fielding, Archer & Davies, 1973; Hancock, 1976).

We have investigated the physiological effects of endoPGs from
V. albo-atrum and F. oxysporum f.sp. lycopersici in tissues of cotton
susceptible and resistant to Verticillium and immune to F. oxysporum f.sp.
lycopersici. We have found that the physiologic state of the host plant
can determine the appearance and severity of symptoms appearing in cotton
plants inoculated with a severe strain of V. albo-atrum. In comparative
studies, susceptible plants photoinduced to the vegetative state were much
more prone to symptom expression than the same plants photoinduced to the
reproductive state. Resistant plants photoinduced to the vegetative state
sometimes developed transient, mild symptoms of Verticillium wilt, while
resistant plants photoinduced to the reproductive state never developed
symptoms of the disease. This is of interest because, in our studies, the
pathogen was capable of colonizing all plants tested and could be reisolated

for as long as 8 weeks after inoculation, even in reproductive, resistant plants. In contrast, F. oxysporum f.sp. lycopersici did not evoke symptom expression in any cotton variety tested, and we could not recover this fungus from cotton 10 days after inoculation. These experiments seem to indicate that, in the Verticillium-cotton relationship, susceptibility relates more to host capability for symptom expression than it does to an absence of the ability to prevent pathogen colonization.

When purified endoPGs produced by either V. albo-atrum or F. oxysporum f.sp. lycopersici were applied to intact cotton plants using vascular transfusion, results similar to the inoculation studies were obtained. Susceptible, vegetative cotton plants developed symptoms characteristic of Verticillium wilt when treated with the Verticillium endoPG; susceptible reproductive plants developed milder symptoms; resistant, vegetative plants infrequently exhibited very slight foliar symptoms of the disease; and resistant, reproductive plants did not respond to the endoPG. On the other hand, all plants treated with the Fusarium endoPG developed a foliar condition we have labeled "silvering" (Figure 2). This condition appears

Figure 2.   "Silvering" symptoms which appeared on all cotton cultivars after vascular transfusion of 250 relative viscometric units of purified endopolygalacturonase from F. oxysporum f.sp. lycopersici.

to be similar to the silvering symptom described by Naef-Roth, Kern & Toth (1963), which was also attributed to the effects of a pectolytic enzyme. The silvering condition did not seem to adversely affect the development of the plants; they grew and set squares at the same rates as untreated plants.

We interpret these experiments to indicate that the endoPG produced by Verticillium is host specific, causing symptoms only in plants prone to symptom development when inoculated with the pathogen. The effects of the Verticillium endoPG were apparently not simply manifestations of dissolution of the pectic regions of the host cell walls since the Fusarium endoPG, applied at equivalent activity levels, did not elicit these symptoms. Symptom expression in this disease appears to be the result of a combination of a specific enzymatic capability of the pathogen endoPG and the presence of specific sites on host cell walls susceptible to this enzymatic capacity.

Cell wall preparations from susceptible and resistant cotton varieties released proteins when digested with the purified endoPG from either Verticillium or Fusarium (Strand & Mussell, 1975), and these solubilized proteins contained peroxidase activity. The ratio of peroxidase activity to protein concentration in the solubilized fractions varied with the source of the host cell walls and with the source of the endoPG (Strand, Rechtoris & Mussell, 1976), indicating that the phenomenon of protein release from cell walls by endoPG is also qualitatively determined by the specific enzymatic capacity of the endoPG and the nature of the cell walls supplied to the enzyme.

Using gel isoelectric focusing and dual beam scanning, we have compared the pattern of protein release from susceptible and resistant tissues by Verticillium endoPG. The results of this method of analysis (Figure 3) indicate that the endoPG solubilized different proteins from the susceptible than from the resistant cell walls. We deem it significant that the dual beam analyzer recorded the presence of proteins in the susceptible preparation that were not in the resistant fraction, but not vice versa. In contrast, when the proteins released from cell wall preparations from the same varieties by the Fusarium endoPG were analyzed with this method (Figure 4), few discernible differences were noted.

In a similar study, we compared the proteins released from cell walls of cotton tissues susceptible to Verticillium by the Verticillium endoPG and the Fusarium endoPG (Figure 5). This comparison indicated that the

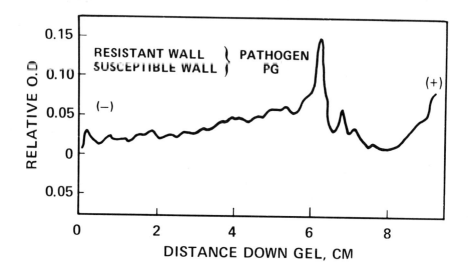

Figure 3. Differential pattern of proteins released from purified cell walls of cotton plants susceptible and resistant to Verti-cillium. Walls were incubated 3 hr with 400 relative viscometric units of purified Verticillium endopolygalacturo-nase, the soluble fraction was electrofocused in acrylamide gels with a pH gradient from 3.5 to 6.0. The trace represents a dual beam scan at 280 nm with the preparation from the resistant tissue in the reference cell and the preparation from the susceptible tissue in the measuring cell.

Mussell & Strand - Pectic enzymes

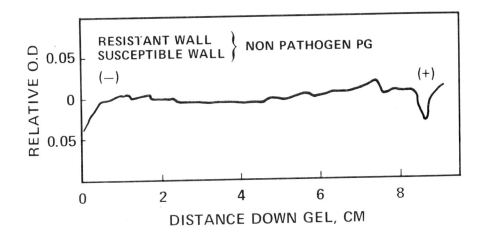

Figure 4. Differential pattern of proteins released from purified cell
walls of cotton plants susceptible and resistant to
Verticillium. Walls were incubated 3 hr with 400 relative
viscometric units of purified Fusarium endopolygalactoronase.
Electrofocusing and analysis as in Figure 5.

- 49 -

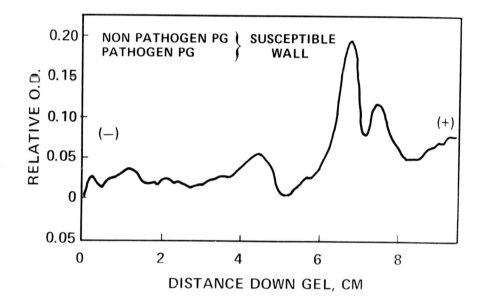

Figure 5.   Differential pattern of proteins released from purified cell
walls from cotton plants susceptible to Verticillium. Walls
were incubated with 400 relative viscometric units of either
purified Verticillium endopolygalacturonase or Fusarium
endopolygalacturonase for 3 hr. Soluble fractions were
electrofocused in acrylamide gels with a pH gradient from
3.5 to 6.0. The trace represents a dual beam scan at 280 nm
with a preparation from the Fusarium enzyme in the reference
cell and the preparation from the Verticillium enzyme in the
measuring cell.

pathogen endoPG solubilized several proteins from susceptible cell walls that were not solubilized by the endoPG from the non-pathogen. Comparison of the differences in the proteins solubilized from susceptible and resistant cell walls by the pathogen endoPG (Figure 3) with the differences in proteins solubilized from susceptible cell walls by pathogen and non-pathogen endoPG (Figure 5) indicates at least two coincident protein peaks. The occurrence of these two peaks only in extracts derived from host-parasite combinations that would have led to symptom expression in vivo indicates to us that the solubilization of these two protein fractions is probably one of the steps leading to symptom expression in this disease.

Because there was peroxidase activity associated with one of the two coincident protein peaks, we have analyzed the isoperoxidases released from susceptible and resistant cotton cell walls by Verticillium endoPG. We have been able to determine that the isoperoxidase pattern released from susceptible walls contains at least one isoperoxidase not released from resistant walls by the endoPG. Although the isoperoxidases solubilized from these cell walls are unstable (Figure 6), the distinctions between the iso-peroxidase patterns released from susceptible and resistant cell walls are readily apparent in both fresh and aged preparations.

Delocalization of peroxidase isozymes within cell walls may play a role in the indirect effects of pectolytic enzymes on host tissues. Two possibilities come to mind. Peroxidases may normally be localized deep within the wall matrix so the products of the peroxidase reaction are separated from the plasmalemma. If this is the case (Figure 7), delocalization may permit peroxidase to diffuse close to the plasmalemma, resulting in membrane damage or disruption by the postulated toxic product(s) of the peroxidase reaction. Alternatively (Figure 8), peroxidase may be localized on the cell walls at the membrane surface to protect the membrane from toxic materials generated at this site. In this case, delocalization of the peroxidases would result in loss of membrane protection, and the resultant membrane disruption would lead to eventual cell death.

Our results indicate that the peroxidase activity in the wall protein fractions represents only a small portion of the total protein released from the cell walls (Strand, Rechtoris & Mussell, 1976). In attempting to characterize other enzymes in the soluble protein fraction, we have observed the presence of indoleacetic acid oxidase activity in the fraction solubi-lized from susceptible cell walls by pathogen endoPG. Indoleacetic acid

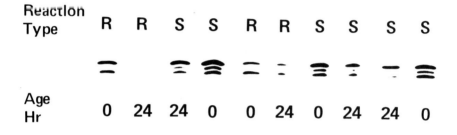

Reaction Type	R	R	S	S	R	R	S	S	S	S
Age Hr	0	24	24	0	0	24	0	24	24	0

Figure 6. Cell wall isoperoxidases solubilized from resistant (R) and susceptible (S) cotton cell walls by Verticillium endopolygalacturonase. Duplicate preparations were applied to the gel immediately after solubilization or after 24 hr storage at 4°.

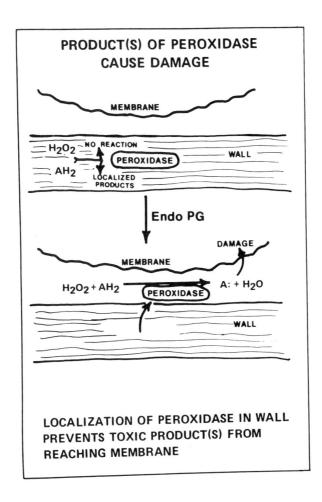

Figure 7. Postulated involvement of peroxidase in endoPG-induced
cellular damage assuming that the products of the
peroxidase reaction are toxic.

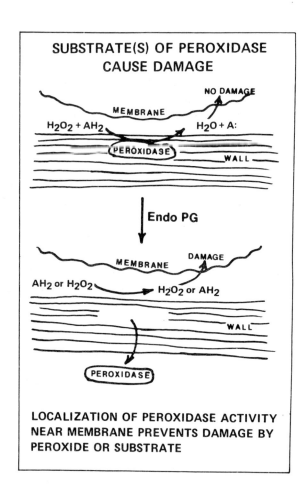

Figure 8. Postulated involvement of peroxidases in endoPG-induced
cellular damage assuming that the peroxidase is localized
at the membrane surface to protect the membrane.

oxidase is not detectable in fractions solubilized from resistant cell walls by Verticillium endoPG, nor is this enzyme detectable in any of the fractions released from cotton cell walls by the Fusarium endoPG. The presence and amount of indoleacetic acid oxidase in the soluble fractions from susceptible cell walls correlate well with the propensity of the host tissues to generate symptoms after inoculation with the pathogen. The amount of indoleacetic acid oxidase in fractions from vegetative material is usually 10- to 20-fold greater than the amount of this enzyme we can observe in fractions prepared from foliage of plants in the reproductive state.

These experiments indicate that one of the characteristics of the cell walls of cotton varieties susceptible to Verticillium is the presence of a wall-bound indoleacetic acid oxidase, and the solubilization of this oxidase is probably one of the early events which leads to symptom expression. We have observed that application of indoleacetic acid to the foliage of infected plants will delay symptom expression for up to 96 hours, while application of naphthaleneacetic acid will delay symptom expression for up to one week.

We infer from the above experiments that one of the events leading to symptom appearance in Verticillium infected cotton plants is the disruption of coordinate host physiology through alterations in the normal levels of indoleacetic acid present in the cotton foliage.

Conclusions

The correlations that we have observed between host predisposition to symptom expression and the specific nature of the proteins released from host cell walls by endoPGs indicate that this solubilization of host wall-bound proteins is probably a critical event in the development of symptoms during vascular pathogenesis.

From these experiments, we conclude that susceptibility, defined as the physiologic ability to generate disease symptoms, can be considered as a distinct set of potential physiological characteristics. By this definition, susceptibility is separate and distinct from the absence of resistance (the ability to restrict or eliminate the pathogen from infected tissues). In the case of Verticillium wilt of cotton, one of the physiological characteristics of susceptible tissues appears to be the qualitative nature of the cell wall-bound proteins susceptible to solubilization by Verticillium endoPG.

In our experience, many cotton varieties that have been classified as resistant do not possess the capability to control colonization by the pathogen; however, these varieties apparently do not possess the appropriate physiological capabilities to develop symptoms in response to pathogen stimulus. These plants are, in fact, tolerant; they can not restrict or eliminate the invading microorganism nor are they capable of responding to the pathogenic stimulus produced by the invading pathogen.

Deletion of susceptibility characters should prove useful for the production of tolerant varieties of many crop species. These tolerant plants would perform up to expectations regardless of the presence or absence of the pathogen. As emphasized by Kenaga (1974), an often over-looked virtue of tolerant plants is the fact that their utilization results in stabilization of pathogen population genetics through reduction in selection pressure for more virulent races. This strategy of pest management (as opposed to pest eradication) should result in the production of crop varieties that are useful for longer periods of time, stabilizing to some degree the annual variations in crop yields attributable to disease losses.

Work supported in part by the U.S. Department of Agriculture, CSRS Grant No. 316-15-42.

References

Albersheim, P. & Anderson, A.J. (1971). Proteins from plant cell walls inhibit polygalacturonases secreted by plant pathogens. Proceedings of the National Academy of Science (USA) 68, 1815-1819.

Albersheim, P. & Anderson-Prouty, A.J. (1975). Carbohydrates, proteins, cell surfaces and the biochemistry of pathogenesis. Annual Review of Plant Physiology 26, 31-52.

Albersheim, P., Jones, T.M. & English, P.D. (1969). Biochemistry of the cell wall in relation to infective processes. Annual Review of Phytopathology 7, 171-194.

Ayers, W.A., Papavizas, G.C. & Lumsden, R.D. (1969). Purification and properties of the endopolygalacturonase of Aphanomyces eutieiches. Phytopathology 59, 925-930.

Barnett, N.M. (1974). Release of peroxidase from soybean hypocotyl cell walls by Sclerotium rolfsii culture filtrates. Canadian Journal of

Botany 52, 265-271.

Basham, H.G. (1974). Injury of plant cells by pectic enzymes. Ph.D. Thesis, Cornell University, Ithaca, N.Y.

Basham, H.G. & Bateman, D.F. (1975a). Killing of plant cells by pectic enzymes: The lack of direct injurious interaction between pectic enzymes or their soluble reaction products and plant cells. Phytopathology 65, 141-153.

Basham, H.G. & Bateman, D.F. (1975b). Relationship of cell death in plant tissues treated with a homogeneous endopectate lyase to cell wall degradation. Physiological Plant Pathology 5, 249-262.

Bateman, D.F. (1963). Pectolytic activities of culture filtrates of Rhizoctonia solani and extracts of Rhizoctonia-infected tissues of beans. Phytopathology 53, 194-204.

Bateman, D.F. & Millar, R.L. (1966). Pectic enzymes in tissue degradation. Annual Review of Phytopathology 4, 119-146.

Brown, W. (1915). Studies in the physiology of parasitism. I. The action of Botrytis cinerea. Annals of Botany 29, 313-348.

Byrde, R.J.W. & Archer, S.A. (1973). Biochemistry of pathogenic cell wall degradation, tissue maceration and cell death. Proceedings of the Second International Congress of Plant Pathology, Abstract No. 0603.

Byrde, R.J.W., Fielding, A.H., Archer, S.A. & Davies, E. (1973). The role of extracellular enzymes in the rotting of fruit tissue by Sclerotinia fructigena. pp. 39-54 in Fungal Pathogenicity and the Plant's Response, ed. by R.J.W. Byrde & C.V. Cutting. Academic Press.

Capellini, R.A. & Peterson, J.L. (1976). Pectic enzymes associated with mimosa wilt. Bulletin of the Torrey Botanical Club, in press.

Cooper, R.M. (1974). Cell wall degrading enzymes of vascular wilt fungi. Ph.D. Thesis, Imperial College, University of London.

Cooper, R.M. & Wood, R.K.S. (1973). Induction of synthesis of extracellular cell-wall degrading enzymes in vascular fungi. Nature 246, 309-311.

Cooper, R.M. & Wood, R.K.S. (1975). Regulation of synthesis of cell wall degrading enzymes by Verticillium albo-atrum and Fusarium oxysporum f.sp. lycopersici. Physiological Plant Pathology 5, 135-156.

Cronshaw, D.K. & Wood, R.K.S. (1973). An analysis of the Mussell and Morré quantitative bioassay for polygalacturonases using pectate transeliminase from Erwinia atroseptica. Annals of Botany 37, 463-471.

Deese, D.C. & Stahmann, M. (1962). Pectic enzymes in Fusarium-infected susceptible and resistant tomato plant. Phytopathology 52, 255-260.

Dimond, A.E. (1970). Biophysics and biochemistry of the vascular wilt syndrome. Annual Review of Phytopathology 8, 302-322.

English, P.D. & Albersheim, P. (1969). Host pathogen interactions: I. A correlation between α-galactosidase production and virulence. Plant Physiology 44, 217-224.

Ferraris, L., Garibaldi, A. & Matta, A. (1974). Polygalacturonase and polygalacturonate trans-eliminase production in vitro and in vivo by Fusarium oxysporum f.sp. lycopersici. Phytopathologische Zeitschrift 81, 1-14.

Fisher, M.L., Anderson, A.J. & Albersheim, P. (1973). Host pathogen interactions. VI. A single protein efficiently inhibits endopolygalacturonase secreted by Colletotrichum lindemuthianum and Aspergillus niger. Plant Physiology 51, 489-491.

Fushtey, S.G. (1957). Studies in the physiology of parasitism. XXIV. Further experiments on the killing of plant cells by fungal and bacterial extracts. Annals of Botany (N.S.) 21, 273-286.

Garibaldi, A. & Bateman, D.F. (1971). Pectic enzymes produced by Erwinia chrysanthemi and their effects on plant tissue. Physiological Plant Pathology 1, 25-40.

Hall, J.A. & Wood, R.K.S. (1973). The killing of plant cells by pectolytic enzymes. pp. 19-31 in Fungal Pathogenicity and the Plant's Response, ed. by R.J.W. Byrde & C.V. Cutting. Academic Press.

Hall, J.A. & Wood, R.K.S. (1974). Permeability changes in tissues and other effects of cell-separating solutions from soft rots caused by Corticum praticola and Erwinia atroseptica. Annals of Botany 38, 129-140.

Hancock, J.G. (1976). Multiple forms of endopectate lyase formed in culture and in infected squash hypocotyls by Hypomyces solani f.sp. cucurbitae. Phytopathology 66, 40-45.

Heale, J.B. & Gupta, D.P. (1972). Mechanism of vascular wilting induced by Verticillium albo-atrum. Transactions of the British Mycological Society 58, 19-28.

Horsfall, J.G. & Dimond, A.E. (1957). Interactions of tissue sugar, growth substances and disease susceptibility. Zeitschrift für Pflanzenkrankheiten (Pflanzenpathologie) und Pflanzenschutz 64, 415-421.

Horton, J.C. & Keen, N.T. (1966). Sugar repression of endopolygalacturonase and cellulase synthesis during pathogenesis by Pyrenochaeta terrestris as a resistance mechanism in onion pink root. Phytopathology 56, 908-916.

Howell, C.R. & Puhalla, J.E. (1975). Significance of an induceable pH 4 optimum endopolygalacturonase to symptom expression in Verticillium wilt of cotton. Proceedings of the American Phytopathological Society 2, 55.

Hunter, W.J. & Elkan, G.H. (1974). Endopolygalacturonase from tomato fruit. Phytochemistry 13, 2725-2727.

Ishii, S. (1976). Enzyme maceration of plant tissues by endo-pectin lyase and endo-polygalacturonase from Aspergillus japonicus. Phytopathology 66, 281-289.

Jones, L.R. (1909). The bacterial soft rots of certain vegetables. II. Pectinase, the cytolytic enzyme produced by Bacillus carotovorus and certain other soft rot organisms. Vermont University Agricultural Station Bulletin 147, 281-360.

Jones, T.M., Anderson, A.J. & Albersheim, P. (1972). Host pathogen interactions. IV. Studies on the polysaccharide degrading enzymes secreted by Fusarium oxysporum f.sp. lycopersici. Physiological Plant Pathology 2, 153-166.

Kamal, M. & Wood, R.K.S. (1956). Pectic enzymes secreted by Verticillium dahliae and their role in the development of the wilt disease of cotton. Annals of Applied Biology 44, 322-340.

Keen, N.T. & Erwin, D.C. (1971). Endopolygalacturonase: Evidence against involvement in Verticillium wilt of cotton. Phytopathology 61, 198-203.

Keen, N.T. & Horton, J.C. (1966). Induction and repression of endopoly-galacturonase synthesis by Pyrenochaeta terrestris. Canadian Journal of Microbiology 12, 443-453.

Keen, N.T. & Long, M. (1972). Isolation of a protein-lipopolysaccharide complex from Verticillium albo-atrum. Physiological Plant Pathology 2, 307-315.

Kenaga, C.B. (1974). Principles of Phytopathology, 2nd ed. pp. 294-295, Balt Publishers.

Leal, J.A. & Villanueva, J.R. (1962). Lack of pectic enzyme production by non-pathogenic species of Verticillium. Nature 195, 1328-1329.

Lund, B.M. & Mapson, L.W. (1970). Stimulation by Erwinia carotovora of synthesis of ethylene in cauliflower tissue. Biochemical Journal 119, 251-263.

Mann, B. (1962). Role of pectic enzymes in the Fusarium wilt syndrome of tomato. Transactions of the British Mycological Society 45, 169-178.

McDonnell, K. (1958). Absence of pectolytic enzymes in a pathogenic strain of Fusarium oxysporum f. lycopersici. Nature 182, 1025-1026.

Mount, M., Bateman, D.F. & Basham, H.G. (1970). Induction of electrolyte loss, tissue maceration and cellular death of potato tissue by an endopolygalacturonate trans eliminase. Phytopathology 60, 924-931.

Mussell, H.W. (1968). Production of polygalacturonase by Verticillium albo-atrum and Fusarium oxysporum f.sp. lycopersici in vitro and in vascular tissue of susceptible and resistant hosts. Ph.D. Thesis, Purdue University, Lafayette, Indiana.

Mussell, H.W. (1973). Endopolygalacturonase: Evidence for involvement in Verticillium wilt of cotton. Phytopathology 63, 62-70.

Mussell, H.W. & Green, R.J. (1970). Host colonization and polygalacturonase production by two tracheomycotic fungi. Phytopathology 60, 192-195.

Mussell, H.W. & Strand, L.L. (1974). Symptoms generated in intact cotton plants by purified endopolygalacturonases. Proceedings of the American Phytopathological Society 1, 78.

Mussell, H.W. & Strouse, B. (1972). Characterization of two polygalacturonases produced by Verticillium albo-atrum. Canadian Journal of Biochemistry 50, 625-632.

Naef-Roth, S., Kern, H. & Toth, A. (1963). Zur pathogenese des parasitogenen und physiologischen silberglanzes am steinobst. Phytopathologische Zeitschrift 48, 222-239.

Patil, S.S. & Dimond, A.E. (1968). Evidence on repression of polygalacturonase synthesis in Fusarium oxysporum f.sp. lycopersici by sugars and its effect on symptom reduction in infected tomato plants. Phytopathology 58, 676-682.

Puhalla, J.E. & Howell, C.R. (1975). Significance of endo-polygalacturonase activity to symptom expression of Verticillium wilt in cotton, assessed by the use of mutants of Verticillium dahliae. Physiological Plant Pathology 7, 147-152.

Rexova-Benkova, L. & Slezarik, A. (1966). Isolation of extracellular pectolytic enzymes produced by Aspergillus niger. Collections of

Czechoslovakian Chemical Communications 31, 122-129.

Skare, N.H., Paus, F. & Raa, J. (1975). Production of pectinase and cellulase by Cladosporium cucumerinum with dissolved carbohydrates and isolated cell walls of cucumber as carbon sources. Physiologia Plantarum 33, 229-233.

Stephens, G.J. & Wood, R.K.S. (1974). Release of enzymes from cell walls by an endopectate-trans-eliminase. Nature 251, 358.

Stoddart, J.L. & Carr, A.J.H. (1966). Properties of wilt-toxins produced by Verticillium albo-atrum Reinke & Berth. Annals of Applied Biology 58, 81-92.

Strand, L.L. & Mussell, H. (1975). Solubilization of peroxidase activity from cotton cell walls by endopolygalacturonases. Phytopathology 65, 830-831.

Strand, L.L., Rechtoris, C. & Mussell, H. (1976). Polygalacturonases release cell wall proteins. Plant Physiology, in press.

Talboys, P.W. & Busch, L.V. (1970). Pectic enzymes produced by Verticillium species. Transactions of the British Mycological Society 55, 367-381.

Tribe, H.T. (1955). Studies in the physiology of parasitism. XIX. On the killing of plant cells by enzymes from Botrytis cinerea and Bacterium aroideae. Annals of Botany (N.S.) 19, 351-368.

Tseng, T.C. & Mount, M.S. (1974). Toxicity of endopolygalacturonate trans-eliminase, phosphatidase and protease to potato and cucumber tissue. Phytopathology 64, 229-236.

Ulrich, J.M. (1975). Pectic enzymes of Pseudomonas cepacia and penetration of polygalacturonases into cells. Physiological Plant Pathology 5, 37-44.

Wang, M.C. & Keen, N.T. (1970). Purification and characterization of endopolygalacturonase from Verticillium albo-atrum. Archives of Biochemistry and Biophysics 141, 749-757.

Wiese, M.V., DeVay, J.E. & Ravenscroft, A.V. (1970). Relationship between polygalacturonase activity and cultural characteristics of Verticillium isolates pathogenic to cotton. Phytopathology 60, 641-646.

Wood, R.K.S. (1972). The killing of plant cells by soft rot parasites. pp. 272-288 in Phytotoxins in Plant Diseases, ed. by R.K.S. Wood, A. Ballio & A. Graniti. Academic Press.

DISCUSSION

Chairman: B. Solheim/C. Ballou

Bateman: You said protoplasts are resistant to pectic enzymes and protoplasts within cell walls are susceptible. I think the distinguishing
feature between resistance and susceptibility relates to the osmotic strength
of a solution that the cells find themselves in.

Mussell: That's a distinct possibility but with the methods we have
available to do this type of experiment, it's very difficult to do a control
on that.

Bateman: You can subject tissue to an osmoticum of 0.25 up to 0.40 molar
strength and in the presence of pectic enzyme the tissue will macerate but
the cells will remain quite alive.

Mussell: Then if you deplasmolize the cells will die.

Bateman: Right, and to maintain protoplasts you have to keep them in a
plasmoticum anyway, when you deplasmolize, they burst. I don't think the
distinction is between the cell wall being present or absent but rather the
presence of an osmoticum.

Mussell: My interpretation is that the distinction is between the surface
of the cell wall being there to do something to the protoplast or it not
being there.

Bateman: I don't think that is the case, at least with potato cells.

Mussell: You think they are simply seeping out through the wall and
bursting because there is no osmoticum to hold them in place?

Bateman: In the presence of pectic enzyme. Apparently the enzyme does
something to the wall, and my interpretation is that it weakens the wall so
that it is no longer able to retain the protoplast which is under osmotic
stress. If you throw the whole system into an osmoticum the cell wall is
degraded, the tissue will macerate and the protoplast will remain alive in
the presence of the wall.

Ballou: My question deals with the experimental problem when you want to compare polygalacturonase production by an isolated organism and then correlate this with what you think happens in the infected tissue. I assume that the enzyme production is induced in the organism by the substrate and the availability of substrate in the plant tissue that is infected would seem to be variable and uncontrollable.

Mussell: There is some question about whether these enzymes are induced in the classic sense. I don't know of any culture medium on which some of these enzymes are not made. There are obviously some media in which more are produced than other media but, especially with respect to Verticillium, regardless of carbon source, there will be some endopolygalacturonases produced. The point I have been trying to make in showing some of the studies that have been done by us and by others is that I don't think that it is safe to try to correlate in vitro production with in vivo behavior of the organism. By manipulating the age of the cultures or anything to do with the culture conditions you can change the amount of enzyme produced and the amount of change that you get will be dependent upon which isolate you are looking at. (R.M. Cooper, this symposium.)

Cooper: In the past, comparison of virulence with enzyme production has usually been done on the wrong medium, this is one problem. You quoted the work of Wiese, DeVay & Ravenscroft (Phytopathology 60, 641-646, 1970), and their medium had about two percent glucose. They couldn't pick up the enzymes. I am not really surprised.

Mussell: It was a mixed substrate too, it also had pectin.

Cooper: It is only recently that this has been appreciated. I think there is still a possibility for this type of study, and I would suggest that possibly the best medium would be extracted host cell walls and to do a very careful time course on it.

Mussell: Now there are two definitions that we need: what is best and what is extracted? Once again, if you'd like I can set you up with cultures that can show correlations with virulence and if you'd rather not, I can give you minor changes in wall extraction and you can show inverse correlations. I just don't think the in vitro stuff is worth pursuing that much more.

Cooper: No! I will still question whether it has been done carefully enough or not, at this stage. With all due respect, I think there is still a possibility. Can I site one example I've already talked to you about and we had a look at. In our country we have aggressive and non-aggressive strains of Ceratocystis ulmi, the Dutch Elm fungus. We attempted a typical correlation with virulence production of various polysaccharidases, and the one that came through, interestingly enough, was xylanase. The six isolates of the aggressive strain produced high loads of xylanase where six isolates of the non-aggressive strain produced much lower levels, so maybe it can work if you use the correct media.

Pegg: You mentioned some work on the use of ultra violet induced autotrophs which has been widely cited as evidence against the involvement of pecto-lytic enzymes. Now there is an innate weakness in this work in that most of the ultra violet mutants which had been used, have been singly deficient for individual pectic enzymes. I think only two of these have been linked such that if one has a deficiency for endopectate lyase one still retains the activity for endo-PG and vice versa. I would like to know how specific the symptoms which you showed are for endo-PG and whether they could be duplicated by endo-pectate lyase.

Mussell: I can't tell you, I have done virtually nothing with purified Verticillium endopectate lyase.

Pegg: I think Heale & Gupta would claim that endopectate lyase is produced first and does have a role in this.

Mussell: Well, it's easier to detect in infected tissue, that's for sure.

Pegg: You've implied that there is a specificity between endo-PG and the symptoms you've shown and there is indeed a very nice correlation between the purified enzyme preparations and the symptoms you've illustrated, compared to the natural occurring ones. Endo-PG will induce ethylene and ethylene predisposes the cells to pectolytic enzyme action and polysaccha-ride action to an order of magnitude greater than the non-ethylene treated tissue. I think if you look at your endo-PG in the absence of polyvalent cations, in ethylene predisposed tissue, you'll find that you will have symptom induction and at levels of endo-PG which you probably only just detect by the method which you are using.

Mussell: I could go along with that because we are still wrestling with what these cations are doing and it's distinctly possible that one of the sets of host cell wall proteins that they are interacting with is an ethylene generating system similar to that which Lund and Mapson (Biochem. J. 119, 251-263, 1970) described for the cauliflower and Erwinia carotovora.

Pegg: There's a further point which I think goes a little bit away from your paper in that the toxin dialogue with respect to Verticillium has tended to polarize between polysaccharide and the pectolytic enzymes. However, the polysaccharide in the absence of ethylene is non toxic, but in ethylene predisposed tissue you can induce all the symptoms which you have shown for the endo-PG. There is clearly a lot going on in the system which is involving ethylene.

Mussell: There are quite a few symptoms we don't stress because we can't cause them with PG, but these are all later symptoms. We can not, for example, defoliate with PG. We can defoliate if we put in PG and then cellulase and this I do not understand. We can not get stem blackening, which is typical of very late symptoms of Verticillium wilt, with PG. We can get stem blackening, once again, if we put in the PG and later come in with cellulase. The cellulase alone in whole plants is totally ineffectual so there is definitely a sequence here and I think it probably relates to the sequence of enzyme secretion or secretion of all kinds of products by the pathogen.

Albersheim: There is certainly information from in vitro studies which I think pertains and which suggests again that the polygalacturonase or pectic degrading enzymes are important. One of these studies now shown by a number of laboratories is that when you grow pathogens on isolated cell walls you get a sequence. The first enzyme being the pectic degrading enzyme, at least in the dicot walls that have been studied, and this is true in a number of pathogens, a number of fungal pathogens and a number of walls. This probably results from the fact that the substrate for the pectic enzyme is the only one that is readily degradeable, and probably hasn't got anything to do with some inherent sequence of control in the fungus. I'm sure we will hear from Cooper about the fact that it is the product of these enzymes which induces, but only low levels of the product, and that's why you have to grow them on cell walls. I would say that studying the growth on cell walls still has a lot to add if people want to work in that area. In any case the fact is that these enzymes are the first ones that come up. I think that adds to the evidence that they are important.

Mussell: As I recall from some of your work also, if you start adding enzymes back to degrade wall, until you get some degradation by the pectic enzymes, the other enzymes are not capable of degrading their respective polymers in the native wall.

Albersheim: That is correct, at least with the dicot. I would like to go back to the argument you had with Bateman. I think you are both right in a sense, but there is really a major point here and that was that Bateman was saying you need a high osmotic concentration to protect the membranes from the enzymes and you were saying that this is because the membrane moves away from the cell wall and you don't have it in conjunction with the products or the enzymes that are degrading the walls and therefore affecting the membrane. I don't think that question can be answered with the evidence that is available. I don't think that Bateman can say that it is high osmotic concentration which prevents the membrane from breaking in the cell nor that you can say that isn't the case. There is no evidence to decide between the two questions really.

Mussell: No. I wasn't trying to make the final decision on that. The point that I was attempting to make is that in this direct rapid cell killing effect, the whole thing can be attributed to whatever the enzyme is doing in the wall per se and it is not that a product from the wall is killing the cell and it is not that the enzyme is attacking the membrane of the plant cell.

Albersheim: I don't think we know the answer to that.

Mussell: We don't know the answer but we know an awful lot of non answers. We have eliminated a lot of possibilities.

Bateman: I would like to comment with respect to Ballou's question. I think it's a very valid question, with regard for appropriate controls for the types of diseases that Dr. Mussell is working with. I also wanted to point out that when we talk about pectic enzymes in plant diseases we perhaps should divide our thinking into two areas. One related to the diseases known as soft rots and in those cases we can have appropriate controls, namely, the analysis of cell walls during the infection process and this has been done several times. But, when it comes to the wilt diseases and diseases where we do not have massive tissue breakdown, as is the case with the Verticillium wilt, then I think there is a valid question as to what type of control is appropriate and how does one go about incorporating a control for these types of studies.

Mussell: Yes, this is something we wrestle with all the time. The best I can give you is that the exo-PG of Verticillium and the charge isomer mixture of the Fusarium endo-PG applied to the same plants grown under the same conditions do not elicit any of the symptoms that we can get with the Verticillium endo-PG. Autoclaved Verticillium endo-PG doesn't do it. We run out of controls rapidly after that. We have run proteins through trying to match up isoelectric points with the Verticillium endo-PG, and histones with molecular weights and isoelectric points around the same range as the Verticillium endo-PG do nothing, so we don't think this is a charge or intercalation effect. Beyond that I am at a loss for more controls.

Bateman: One other comment in relation to Albersheim's comment regarding cell wall breakdown, in relation to action of pectic enzymes. I think there is emerging at least some evidence, and I will report later in this symposium on the xylanase system which does in fact release a considerable amount of its substrate from isolated cell walls without the action of a pectic enzyme. I agree with Albersheim on the sequence, this has been repeated in our laboratory as well as his, and the sequence does seem to be definite.

Albersheim: Yes, but there is no xylan in the primary cell walls so you must be working on a defense mechanism.

Bateman: Yes, and with corn.

Cooper: On the same theme we found every enzyme of Verticillium albo-atrum which we have induced specifically to study this process is able to degrade cell walls, and that includes arabinases, β-galactosidases and so on. That's without the prior action of pectic enzymes.

Unestam: I would like to switch over to another point of Dr. Mussell's paper and that was the tolerance problem, the hypothesis that you brought up was very interesting. I wonder if anyone has any idea of how common microorganisms are inside trees or plants as non-pathogenic organisms just sitting and growing there. I don't think anyone knows. In the animal world at least, I can off-hand mention a few, let's say the malaria protozoan in mosquito does not cause any harm to the mosquito as far as I know. In lobster they have found that in the blood stream there are bacteria running around without doing any harm. Athlete's foot of man for instance is probably this kind of organism. It is found on almost anyone, but it causes harm only to a few due to allergic reaction. That is, you have a disease only when you have a reaction on the part of the host. It has been reported that there are plenty of bacteria, maybe even fungi, inside the woody parts of some trees. We don't know how these micro-organisms do get in there. They might cause diseases sometimes, but often they don't so it's a good question as to whether it is true or not.

Mussell: Yes. I think one thing we often lose sight of is the fact that susceptibility is a very unusual case and in most of nature you don't have susceptibility. I am not going to say that you have resistance but you have the absence of disease. With reference to when there are organisms growing in plants without symptoms; until fairly recently most evaluation of so-called resistance for vascular parasites has been the presence or absence of symptoms in the field. We have now, I think, in tomato three separate genes that have been identified as resistance genes that I don't think are. The plants do get colonized but symptoms don't show up. There is one situation in which this occurs in Fusarium wilt of tomato, and in just about everything that Steve Wilhelm has looked at in Verticillium wilt of cotton there is colonization. You can go out into a field that he maintains as a Verticillium nursery and you see apparently very healthy, robust plants and start isolating and find that the fungus is present, from the tip of the roots to the tip of the growing point and out in all the branches and in the leaves, but the plants are not getting sick. I think that the genetic potential for manipulating susceptibility has already been demonstrated, it is just that we haven't recognized it. In vascular diseases there is no superficial growth of the pathogen, so the presence of symptoms is the only distinction that the field people make between susceptible and other. There is recognized tolerance in rust; but in rust you see a little spot and there are spores there. In many other diseases, you can't tell if the pathogen is there unless you go through some laborious isolation.

Knee: I'd like to make two observations firstly about the question of osmotic stress and cell death. It is fairly familiar to me, as a worker on ripening fruit, that in a number of ripe fruits where autolytic degradation of the cell walls occurred it is easy to show by a variety of methods that the cells are very vulnerable to osmotic shock and that they will die if placed in solutions of low osmotic strengths but will survive much longer in a solution which is isotonic with the cell contents. The other point I wanted to make was to second what Dr. Cooper said that cell wall degrading enzymes other than polygalacturonase have a capacity to degrade plant cell walls. I have found this with an arabinosidase which Dr. Bryde, of Long Ashton, supplied me with, and also with galatanase.

Mussell: Could I make a point on cell walls? There apparently is no standard definition of what a plant cell wall is. It is whatever you make up in your lab and you call plant cell walls. In some cases these are produced using only aqueous techniques and in some cases organic solvents are used. In some cases there is massive tissue disruption before preparation of the walls and in others not. These various methods result in very, very, different particulate fractions, unfortunately, all of which get called plant cell walls when they get into the literature. I would just like to point out that some of these "plant cell walls" have some very viable living material in them after they have been prepared, and others have been totally stripped of just about all enzymatic capacity that is normally associated with the wall through the action of organic solvents and other materials. The question is, how do you make a plant cell wall? What is it?

Lippincott: In that connection you mentioned the use of binding your enzymes to purified plant cell walls. What cell walls were you using?

Mussell: We've modified and doubled the technique used by Barnett (Can. J. Bot. 52, 265-271, 1974) involving an extraction in an aqueous buffer with detergents followed with an aqueous buffer with high salts followed by massive rinse with deionized water followed by a French pressure cell at 12,000 PSI followed by repeating all the above steps. After producing this, we do have a wall that has protein on it that has some enzymatic capacity that we have been able to measure. By examining and using several of the marker enzymes for cytoplasm and using the electron microscope we can detect no cytoplasmic or membrane or nuclear contamination in these preparations. I am not saying that it is all native state wall but I think it is a fair approximation. Now, how you can sterilize that, and grow something on it I am not too sure. Everything we do to sterilize it does mess it up again.

Hiruki: I would like to add one comment on the importance of cell walls against plasmalema. It is well known that if you isolate protoplast from a plant which carries hypersensitive genes such as Nicotiana glutinosa, this isolated protoplast does not produce any necrosis. In other words, after removing the cell wall the protoplast supports the virus reproduction without showing any necrosis. This seems to me to indicate the importance of a cell wall. However, if you also make rather careful observations using hypersensitive tissue, the necrosis seems to occur when the plasmalema starts to deteriorate. In other words, also the intactness or the condition of plasmalema seems to be very important to maintain normal cellular processes. So! I would say that both cell walls and plasmalema are important.

DEGRADATION OF WOOD CELL WALLS BY THE ROT FUNGUS <u>SPOROTRICHUM PULVERULENTUM</u>
- ENZYME MECHANISMS

Karl-Erik Eriksson, Swedish Forest Products Research Laboratory, Box 5604,
S-114 86 Stockholm, Sweden

It has long been known that the main degradation of wood is caused by
fungi. The strong wood-degrading effect that fungi have, in part, depends
upon the organisation of their hyphae, which gives the organisms a
penetrating capacity. Different types of fungi give rise to different
types of wood rot. You normally distinguish between soft-rot, brown-rot
and white-rot fungi. The blue-staining fungi are also associated with wood
damage. They do not, however, cause wood degradation.

The term soft-rot emanates from the fact that there is a softening of
the surface layer when wood is attacked by this group of fungi. In the
secondary wall of the attacked wood, there appear cylindrical cavities with
conical ends. The term soft-rot is now used whenever this characteristic
cavity pattern appears even if no softening of the surface layer has taken
place. Soft-rot is more common in hardwood than in softwood. It has been
suggested that the reason for this is the qualitative differences in the
lignin of hard- and softwood. The methoxyl content of hardwood lignin is
usually higher, about 21%, than in softwood lignin, where the methoxyl
content is about 14%. Soft-rot is caused by fungi belonging to Ascomycetes
and Fungi Imperfecti. Like all other wood-degrading fungi, soft-rot
demethylates lignin. This seems to be the main mechanism of attack on
lignin of both soft- and brown-rot fungi.

Brown-rot and white-rot attack on wood is mainly caused by fungi
belonging to Basidiomycetes. The hyphae of the fungi are normally localized
in the cell lumen and these hyphae penetrate from one cell to another
through openings or by producing boreholes themselves. Brown-rot fungi
degrade cellulose and the hemicelluloses in wood. In an early stage of

degradation, they depolymerise cellulose faster than the degradation products are utilized. Brown rotters begin the degradation process from the hyphae within the cell lumen and first degrade the S2 and then, later, the S1 wall.

The white-rot group of fungi is a rather heterogeneous group of organisms. These organisms have in common that they can degrade lignin as well as all the other wood components. They also have in common the ability to produce enzymes which oxidize phenolic compounds related to lignin. This is the reason that phenolic compounds have been utilized for the identification of white-rot fungi. The relative amounts of lignin and cellulose degraded and utilized by these fungi vary and so does the order of preferential attack. The normal method of wood degradation by white-rot fungi is that the cellulose and the lignin are attacked simultaneously. In Figure 1, however, is given an example where wood is degraded by the rot

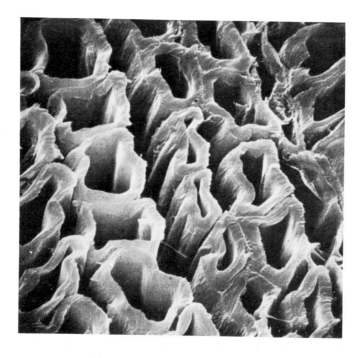

Figure 1.   Specific delignification of wood by the rot fungus <u>Phellinus</u> <u>isabellinus</u>.

fungus <u>Phellinus isabellinus</u>. The lignin is here seen to be attacked
specifically in the middle lamella. It is, however, questionable if an
absolutely specific attack on the lignin can be undertaken. It seems
from our work, as well as from work of other researchers (Ander & Eriksson,
1975; Ander & Eriksson, 1976; Hiroi & Eriksson, 1976; Kirk & Moore, 1972;
Kirk, Connors & Zeikus, to be published), that one of the polysaccharides
in wood must be degraded simultaneously with lignin. This is most likely
so because it takes so much energy to degrade lignin that an additional,
easily accessible energy source is also necessary. This is demonstrated
in Figure 2, where the loss of wood components after 10 weeks rotting with

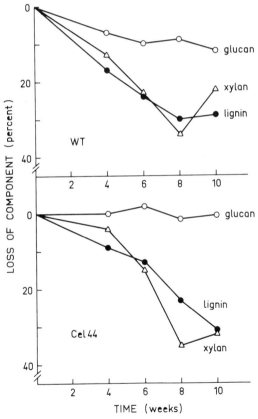

Figure 2. Degradation of wood components in birch by WT and Cel 44.
(o) Glucan. (•) Lignin. (Δ) Xylan.

a wildtype and a cellulase-less mutant strain, Cel 44, of the rot fungus
Sporotrichum pulverulentum is demonstrated. The rotting with the wildtype
fungus involves a decrease in all three wood components cellulose, hemicellulose and lignin, while a degradation with the cellulase-less mutant
causes a degradation of only xylan and lignin. The xylanase-producing
power is very small in the cellulase-less mutant Cel 44 compared to the
wildtype. In spite of this, a considerable loss of xylan is achieved
simultaneously with the loss of lignin. In Figure 3 the carbon dioxide
evolution is given when wood is rotted with wildtype and Cel 44 of
Sporotrichum pulverulentum. A considerably higher carbon dioxide evolution
is obtained when the wildtype fungus is used.

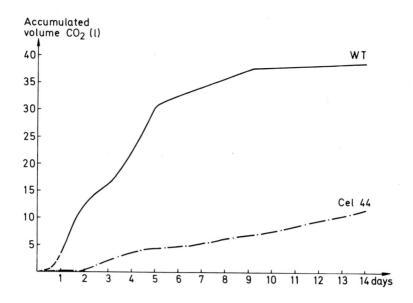

Figure 3. Carbon dioxide evolution during rotting of birch wood chips
with wildtype and a cellulase-less mutant of _Sporotrichum
pulverulentum_.

Degradation of cellulose and lignin by the rot fungus <u>Sporotrichum pulverulentum</u> has been studied on the molecular level for a long time. The fungus degrades cellulose and lignin more or less simultaneously. We have also found an enzyme, cellobiose: quinone oxidoreductase, which utilizes degradation products from both lignin and cellulose simultaneously as substrates (Westermark & Eriksson, 1974a, 1974b).

The total picture we now have of the enzyme mechanisms for the degradation of cellulose and lignin by the rot fungus <u>Sporotrichum pulverulentum</u> and the factors involved in extracellular regulation of these enzymes is given in Scheme 1.

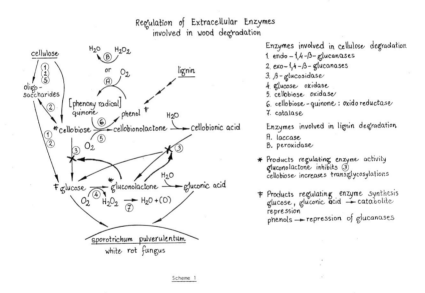

Scheme 1

Crystalline cellulose is degraded to water-soluble monomeric and oligomeric sugars and acids by the cooperation of the enzymes ①, ② and ⑤*. The endo-1,4-β-glucanases ①, of which <u>S. pulverulentum</u> produces five different ones, attack β-1,4-glucosidic bonds randomly over the cellulose chains. A synergistic action exists between the endo-glucanases

---

*Figures and letters within circles refer to Scheme 1.

and a simultaneously produced exo-glucanase. The exo-glucanase splits off glucose or cellobiose from the non-reducing end with an inversion of configuration in the products from β to α (Streamer, Eriksson & Pettersson, 1975).

We have also demonstrated that in vitro degradation of cotton cellulose with a culture solution from S. pulverulentum goes twice as fast in O_2-compared to N_2-atmosphere. A new oxidizing enzyme, at present called cellobiose oxidase ⑤, has subsequently been discovered (Eriksson, Pettersson & Westermark, 1975). This enzyme is now under purification. We postulate, before this enzyme has been purified, that it also oxidizes the reducing end group formed when a β-glucosidic bond is split through the action of the endo-glucanases. By doing so, the oxidative enzyme, in cooperation with the exo-glucanase, effectively prevents a broken β-1,4-glucosidic bond from reforming in the crystalline part of the cellulose.

Sporotrichum pulverulentum has two metabolic routes to oxidize cellobiose to cellobionic acid via cellobiono-δ-lactone. Route 1 is dependent upon lignin. Cellobiose is then oxidized, concomitant with the reduction of quinones and/or phenoxy radicals, by the enzyme cellobiose: quinone oxidoreductase ⑥ (Westermark & Eriksson, 1974a, 1974b). Phenoxy radicals and quinones have been formed from lignin phenols by their oxidation through mechanism Ⓐ or Ⓑ. It has been demonstrated that in the absence of phenol oxidases lignin is not degraded and that phenols present in wood cause a repression of the production of cellulose-degrading enzymes (Ander & Eriksson, 1976). Route 2 for information of cellobionic acid is air oxidation by the recently discovered enzyme cellobiose oxidase ⑤ (Ayers & Eriksson, to be published).

Cellobiose and cellobionic acid are split by a β-1,4-glucosidase ③, most probably the same enzyme, to glucose in the first case and a combination of glucose and glucono lactone in equilibrium with gluconic acid in the second case (Ayers & Eriksson, to be published). The glucono lactone has been demonstrated to be an efficient inhibitor of this β-glucosidase.

High concentrations of glucose simultaneously cause metabolite repression of the hydrolytic endo- and exo-glucanases (Eriksson, Hamp & Szajer, to be published) and induction of the enzymes glucose oxidase ④ and catalase ⑦. The result is that glucono lactone is formed by oxidation of glucose. The lactone inhibits the β-glucosidase and this causes an increase in the cellobiose concentration (Ayers & Eriksson, to be published).

At sufficiently high concentrations of cellobiose, the transglycosylation mediated by the hydrolytic enzymes will predominate (Streamer, Eriksson & Pettersson, 1975), i.e. higher oligosaccharides will be formed from cellobiose. Water-soluble cellodextrins may even precipitate (> DP 7). The hydrogen peroxide formed by glucose oxidation by glucose oxidase can be utilized a) by the enzyme peroxidase ⓑ for phenol oxidation, or b) by the enzyme catalase ⑦ to release oxygen from hydrogen peroxide. The use of every oxygen atom present in the normal environment of the fungus, i.e. the inside of a dead trunk, must be of the utmost importance.

In order to obtain a more complete knowledge of the extracellular regulation of the enzymes presented in Scheme 1, it is presently being studied to what degree the different monosugars cause catabolite repression of the endo-1,4-β-glucanases. The results we have reached hitherto (Eriksson, Hamp & Szajer, to be published), show that the repression level for glucose and mannose is 50-100 mg/l, for xylose approximately 1000 mg/l and for galactose > 1000 mg/l. Cellobiose, on the other hand, turns out to be an inducer also at so relatively high concentrations as approximately 1000 mg/l.

We have recently found that the same enzymes which we have considered to be completely extracellular, also are bound to the fungal cell wall to a higher degree than we had expected. By the aid of a 1 M solution of NaI the enzymes can be released in their active form from the cell walls. After dialysis of NaI, the enzymes can again be fixed on fragments of cell walls in their active form. By isoelectric separation on thin-layer chromatography in combination with an earlier developed zymogram technique, we now investigate if different properties of the enzymes, for instance glycoproteins, versus non-glycoproteins, give varying affinities for the cell walls.

References

Ander, P. & Eriksson, K.-E. (1975). Influence of carbohydrates on lignin degradation by the white-rot fungus Sporotrichum pulverulentum. Svensk Papperstidning 78, 643.

Ander, P. & Eriksson, K.-E. (1976). The importance of phenol oxidase activity in lignin degradation by the white-rot fungus Sporotrichum pulverulentum. Archives of Microbiology 109, 1.

Ayers, A. & Eriksson, K.-E. To be published.

Eriksson, K.-E., Pettersson, B. & Westermark, U. (1975). Oxidation: an important enzyme reaction in fungal degradation of cellulose. FEBS LETTERS 49, 282.

Eriksson, K.-E., Hamp, S. & Szajer, C. To be published.

Hiroi, T. & Eriksson, K.-E. (1976). Microbiological degradation of lignin. Part 1. Influence of cellulose on the degradation of lignins by the white-rot fungus Pleurotus ostreatus. Svensk Papperstidning 79, 157.

Kirk, T.K. & Moore, W.E. (1972). Removing lignin in wood with white-rot fungi and digestability of resulting wood. Wood and Fiber 4:2, 72.

Kirk, T.K., Connors, W.J. & Zeikus, J.G. Requirement for a growth substrate during lignin decomposition by two wood-rotting fungi. Applied Environmental Microbiology. To be published.

Streamer, M., Eriksson, K.-E. & Pettersson, B. (1975). Extracellular enzyme system utilized by the fungus Sporotrichum pulverulentum (Chrysosporium lignorum) for the breakdown of cellulose. 4. Functional characterization of five endo- and one exo-1,4-β-glucanases. European Journal of Biochemistry 59, 607.

Westermark, U. & Eriksson, K.-E. (1974a). Carbohydrate-dependent enzymic quinone reduction during lignin degradation. Acta Chemica Scandinavica B 28, 204.

Westermark, U. & Eriksson, K.-E. (1974b). Cellobiose: quinone oxidoreductase, a new wood-degrading enzyme from white-rot fungi. Acta Chemica Scandinavica B 28, 209.

DISCUSSION

Chairman: D. Delmer

Cooper: I think you mentioned that three enzymes, xylanase, cellulase and mannanase, are regulated by one regulatory gene. Could you elaborate on this because you said you would come back to it and didn't?

Eriksson: These results are published by myself and Goddell (Can. J.

Microbiol. 20, 371-378, 1974). We found that pleiotropic mutants are
produced by UV irradiation of spores of different white rot fungi. We have
cellulase-less mutants from several different white rot fungi. The aim of
this work was mainly to use these mutants in the technical project I
mentioned: specific deletion of lignin from wood chips. We screened for
the mutants by plating out the irradiated spores on cellulose agar plates.
Here we of course had to obtain colonial growth by the addition of different
colonial growth forming compounds, different for different fungi. We also
had to add a little glucose to allow growth of the cellulase-less mutants.
Non-mutants create a clearance zone in the cellulose around the colony
within 10 days; colonies that do not create clearance zones are transferred
to test tubes with a similar medium and kept there for a month. If they
don't cause a clearance zone in the cellulose within this period we call
them cellulase-less mutants. These mutants we then transferred to other
test tubes where we used xylan or mannan as carbon sources. The cellulase-
less mutants were found not to produce xylanase or mannanase either. Only
a very slight amount of the original xylanase (probably 1%) was produced by
some mutants. As you could see in some of my slides here some xylan was
degraded by the mutants simultaneously with the lignin. I think it is a
necessity of energy reason for the fungi to degrade one of the polysaccharides simultaneously with the lignin. We can stand to lose some hemicellulosic material, it doesn't matter. If we take away 2% of the lignin
we would probably degrade a similar amount of xylan. It is not of any great
economical importance. Does that answer your question?

Cooper: The final question was - you were talking about catabolite
repression and synthesis. Do you think this normally occurs while the
degradation of an insoluble polymer is going on? Do you think sugar levels
do build up to high enough levels to have that effect? I wondered if you
had ever looked at monomer levels in culture for example, whether they have
already built up to sufficient levels.

Eriksson: I don't think that monosugar levels are built up that cause
catabolite repression. This is, however, under investigation. It is very
important for us to find out how the K_m for growth of S.p. on monosugars
correlates with the level of catabolite repression and the level of monosugars obtained when we grow the fungus on different carbon sources. This
is, of course, particularly important for us to know when in the fermentation

process, we ferment a mixture of monosugars, water-soluble oligomeric sugars and solid lignin-cellulosic particles or fibres.

Brown: I'm a little confused about what happened to cellulose oxidase. Is what you are calling now cellobiose oxidase and cellulose oxidase the same?

Eriksson: We have demonstrated that the oxidizing enzyme oxidizes cellobiose to cellobionic acid. It also oxidizes lactose. We have not been able to free this enzyme as yet from hydrolytic enzyme. So therefore, it doesn't make much sense to use higher oligomeric substances as substrates for the enzyme since hydrolytic enzymes would always produce cellobiose anyway.

Brown: So in your scheme you showed on the slide you didn't have cellulose oxidase anywhere in that scheme?

Eriksson: That is right. We postulate (and maybe I did not point this out strongly enough - that this is purely a postulation) that the enzyme demonstrated to be a cellobiose oxidase may also participate in the initial attack of the cellulose by oxidizing the reducing end group formed when a β-1,4-glucosidic bond is cleaved by an endoglucanase.

Brown: I remember talking with you about this approximately a year ago and you were thinking that glucuronic acid was produced. Do you still think so?

Eriksson: The enzyme that Art Ayers is now purifying does form uronic acid. There might be an additional enzyme that causes this reaction but we doubt that right now.

Ballou: Is there any evidence that the cellulose molecule gets oxidized at any site other than possibly at the reducing end?

Eriksson: When we first discovered the importance of an oxidizing enzyme, we thought that this enzyme introduced uronic acid moieties into the crystalline cellulose, thereby causing swelling of the cellulose. We published a preliminary paper on this in FEBS Letters where we suggested that this might be the case. We had some evidence that uronic acid was formed. The method we used for the assay of uronic acid was, however, not very sensitive. Consequently we were very careful to interpret what the enzyme did. We said: it might be that uronic acid moieties are introduced in the cellulose. But for the moment, we think that the mechanism I

Eriksson - Degradation of wood cell walls

suggested here, oxidation of the reducing endgroups, is the more likely.

Brown: If you don't have oxidation and subsequent swelling, we're still back to the elusive C_1 enzyme. Do you have any comments?

Eriksson: I certainly do not think that we have an enzyme that you can characterize as a "hydrogen bondase".

Selvendran: I would like to ask three closely related questions. Does your purified cellulase preparation have any xylanase activity associated with it? If it does, could you comment on it? Wood contains about 20% xylan - why is it that the enzymic degradation of xylan is not studied so comprehensively? Is there any evidence from your work on lignin-degrading enzymes that xylan is covalently linked to lignin?

Eriksson: There is no trace of xylanase activity in our purified enzymes - either in the endo-1,4-β-glucanases or in the exoglucanases. In addition to testing for purity by different techniques like isoelectric focusing, SDS gels, etc. we also test for other enzyme activities. There is no trace of xylanase activity nor of any other polysaccharide-degrading activity in enzymes that we claim are pure. Your second question was if we had an evidence for an association between lignin and xylan. We don't have it, but I know that U. Eriksson and B. Lindgren at our Institute have evidence for this.

Selvendran: The third question was on why so little work is done on the degradation of xylan.

Eriksson: We have done quite a bit of work on xylanases also but it did not fit in very well with this lecture. We haven't particularly investigated the enzyme mechanisms involved in xylan degradation, I think, however, that they are very similar to what is going on in cellulose degradation. There certainly are endoxylanases. We even have some slight evidence for the existence of a xylobiose oxido-reductase. I wouldn't say that it does exist but it is a possibility. I don't think that the organism would have any use for an oxidative mechanism for degradation of xylan. You know that one of the differences between cellulose and xylan is that cellulose is crystalline in certain parts whereas xylan is not. Therefore, I wouldn't think it is as necessary to have a synergistic action of a number of enzymes for xylan degradation as it is for cellulose degradation.

Selvendran: The reason I asked the first question was that P.M. Wood published some work on the cellulases of Trichoderma viridi, and he made no mention of xylanase activity. And I actually asked him whether he found any xylanase activity, and he said it was always associated with the cellulase and he couldn't separate it. He said he just wondered whether the cellulase would in fact break down xylan in Trichoderma.

Eriksson: It certainly has been true that it has been hard sometimes to separate xylanase and cellulase activity. In this particular case, the Sporotrichum fungus, it wasn't. When we, however, purified the cellulase from Stereum sanguinulentum, we thought for a long time that the enzyme in question had both activities. Finally we could, however, separate the two activities. Both the cellulase and the xylanase are carefully described as to amino acid composition, molecular weight, isoelectric points etc. (Eriksson, K.-E. & Pettersson, B., Arch. Biochem. Biophys. 124, 142-148, 1968; Eriksson, K.-E. & Pettersson, B., Int. Biodeter. Bull. 7, 115-119, 1971). The cellulase has a molecular weight of 23200, the xylanase, 21600. The amino acid composition is also similar.

Delmer: It is my impression from talking to Dr. Eriksson that he has established through the years very good criteria for how to assess purity of polysaccharide degrading enzymes. Karl-Erik, I wish you would comment for just a moment on what you consider the best criteria for purity when you are working with these types of enzymes.

Eriksson: We use of course the usual biochemical methods for testing purity as SDS gels, isoelectric focusing, etc. In addition to the use of these methods we always also test for additional enzyme activities that can be expected to exist. If we have purified a cellulase, we incubate it for a long time (24 hours) with xylan, mannan, starch, etc. In the unpurified culture solution, we try to find out the number of enzyme activities that we have. These enzymes are again tested for when the enzyme we are interested in is considered pure. The number of protein bands that we can obtain with isoelectric focusing is also a very useful technique. When we grow the fungus on crystalline cellulose as the only carbon source, we obtain approximately 20 protein bands. We can now identify 16 of these by

Eriksson - Degradation of wood cell walls

the aid of zymogram technique. These methods will be extremely useful when
we now start to work more intensively on lignin-degrading enzymes to find
out how many additional bands we obtain when the fungus is grown on a
lignocellulosic substrate compared to a pure cellulose substrate.

Unestam: Talking about the purity of the enzymes - do you think that there
is any other unrelated enzyme activity associated with any of these purified
enzymes? Let's say, protease activity by a cellulase. Is that possible?

Eriksson: Our pure cellulases are not contaminated by any proteolytic
activity.

Bateman: There has been described enzymatic activity which gives rise to
short fibres. Would your preparations also give rise to short fibres?

Eriksson: Let me say that if we perform _in vitro_ digestion of cotton
fibres, with concentrated culture solution from S.p., we obtain what you
call short fibres in oxygen atmosphere, but not in a nitrogen atmosphere.

Bateman: One more question. It relates to the oxidation mechanism in
terms of where it may occur within the cellulose chain. You indicate only
at the reducing end. If this be the case, then how does one really explain
the fragmentation of the crystalline fibre by having it act only at this
location?

Eriksson: We now think that it acts both on cellobiose transferring it to
cellobionic acid, but also to oxidize the reducing endgroup formed when the
endo-glucanases split a 1,4-β-glucosidic bond. To the short-fibre formation
I have no comments.

THE ORIGIN AND SIGNIFICANCE OF PLANT GLYCOPROTEINS

Deborah P. Delmer and Derek T.A. Lamport, MSU/ERDA Plant Research
Laboratory, Michigan State University, East Lansing, Michigan 48824, USA

General Introduction

One of the most dramatic biochemical advances over the past ten to
fifteen years must surely be our enhanced understanding of glycoproteins.
Previously shunned both by protein and carbohydrate chemists too intent on
purity (Miller, Mellman, Lamport & Miller, 1974), glycoproteins have
evolved from a chemical "no-man's land" into a frontier of significant
research. And, although the glycoproteins of animal cells have been
intensively studied in recent years and we now know a great deal about their
structures, how they are synthesized, and what physiological functions they
serve, in contrast to this, we know far less about the glycoproteins of the
plant kingdom. For many years it was widely believed that amino sugars,
one of the common constituents of glycoproteins, were only rarely, if ever,
found in plant cells. However, within the past ten years, our knowledge
has expanded considerably, and we are beginning to recognize that glycoproteins of a variety of types, including those containing amino sugars,
are commonly found in essentially all types of plant cells. With the
exception of sialic acid, all of the sugars found in animal glycoproteins
have now been shown to exist in various glycoproteins of plant origin
(Sharon, 1973).

Since there are only a few plant glycoproteins which have been purified
and characterized extensively, it is still perhaps premature to generalize
too much about differences or similarities of plant glycoproteins to those
of the animal kingdom. However, a few facts are apparent. As for animal
glycoproteins, it is clear that the glycoproteins of plants can have a
variety of structures and functions. Table 1 shows a list of some of the
general classes of glycoproteins isolated from plant sources. Glycoproteins

serving such diverse roles as enzymatic, structural, reserve polymer, and possible primer molecules have been demonstrated. For many others, most notably the lectins and the soluble peptide-linked arabinogalactans, we still have no clear idea about their function, but their widespread occurrence certainly suggests that they serve some special roles, and we shall speculate about this later.

Table 1. Glycoproteins from the plant kingdom

General type	Examples	Type of linkage	Sugars present
STRUCTURAL	Extensin	O-glycosidic to hyp, ser	ara, gal
	Clamydomonas cell wall protein	O-glycosidic to hyp	ara, gal, glu
ENZYMES	Bromelain, Ficin	N-glycosidic to AsN	glcNAc, man, xyl, fuc (gal)
	Peroxidase	N-glycosidic to AsN	glcNAc, man, (fuc, xyl)
STORAGE PROTEINS	Vicilin, Legumin 7S Soybean protein	N-glycosidic to AsN	glcNAc, man
LECTINS	Soybean agglutinin	N-glycosidic to AsN	glNAc, man
PEPTIDE-ARABINO-GALACTANS	Sycamore cell extra-cellular glycoprotein	O-glycosidic to hyp	ara, gal
	Maple sap glycoprotein	O-glycosidic to hyp	ara, gal
	Potato lectin and "all β-lectins"	uncertain (peptides contain hyp)	ara, gal (glcNAc)
PRIMER PROTEIN	Glycoproteic starch primer	uncertain	glu

A survey of the types of linkages involved between the carbohydrate and protein of plant glycoproteins (Table 1) shows that the two most commonly identified in plants are an N-glycosidic linkage between the amide nitrogen of asparagine and the C-1 of N-acetyl glucosamine (to which mannose residues are usually attached) and the O-glycosidic linkage between hydroxyproline and a sugar, usually arabinose. The former type of structure is quite commonly found as the core structure of animal glycoproteins; the latter, however, seems to be unique to the plant kingdom, and Lamport & Miller (1971) have shown that arabinosides attached to hydroxyproline are widespread throughout the plant kingdom.

In the presentation to follow, we shall be discussing some of our thoughts on the origin and significance of plant glycoproteins - first, by reviewing the current status of our knowledge about the biosynthesis of glycoproteins in plant cells, and, second, to discuss in a broader context, our thoughts on the evolutionary origin of the glycoproteins, thoughts which also lead us to speculate on the functional role of glycoproteins in plants.

Pathways of Biosynthesis

Glycoproteins Containing N-Acetylglucosamine and Mannose (N-Glycosidic Linkage)

Several studies have shown that radioactivity from ^{14}C-glucosamine, when fed to a variety of living tissues, can be incorporated into high molecular weight materials, presumably glycoprotein (Roberts, Cetorelli, Kirby & Ericson, 1972; Basha & Beevers, 1976; Roberts, Connor & Cetorelli, 1972; Roberts & Pollard, 1975). From these studies it appears that most (if not all) plant tissues have the capacity to synthesize glucosamine-containing proteins; such proteins were found to be localized in all cellular fractions, including the cell wall.

The mechanism of biosynthesis of animal glycoproteins containing a core region of N-acetylglucosamine and mannose residues has been shown to involve the participation of lipids as intermediate sugar carriers (Lennarz, 1975). The basic mechanism for this type of glycosylation sequence is diagrammed in Figure 1. The types of lipids which have been shown to participate in these reactions in mammalian systems are dolichols - long chain polyisoprenoids containing anywhere from 18 to 22 isoprenoid units. Although polyprenols are known to occur in plants (Wellburn & Hemming, 1966) it is only recently that evidence has been obtained which strongly supports a role for these lipids in the glycosylation of plant proteins. Kauss (1969) has demonstrated the synthesis of a compound with properties resembling a mannosyl-phosphoryl-polyprenol using GDP-mannose as substrate with enzyme preparations from mung bean seedlings; however, synthesis of glycoprotein from this compound was not demonstrated in this system.

1. UDP-GlcNAc + Lipid-P → Lipid-PP-GlcNAc + UMP
2. UDP-GlcNAc + Lipid-PP-GlNAc → Lipid-PP-(GlcNAc)$_2$ + UMP
3. GDP-Mannose + Lipid-P → Lipid-P-mannose + GDP
4. n(Lipid-P-mannose) + Lipid-PP-(GlcNAc)$_2$ → Lipid-PP-oligosaccharide + n(Lipid-P)
5. Lipid-PP-oligosaccharide + Protein Acceptor → Glycoprotein + Lipid-PP

Where structure of Lipid-P is:

$$H - [CH_2 - \overset{\overset{\textstyle CH_3}{|}}{C} = CH - CH_3]_x - O - \overset{\overset{\textstyle O}{|}}{\underset{\underset{\textstyle O^-}{|}}{P}} - OH$$

where x = 18 to 22 isoprenoid units

Figure 1. Lipid intermediate involvement in the biosynthesis of glycoproteins containing an N-glycosidic linkage.

Forsee & Elbein (1973) have also shown that membrane preparations from
cotton fibres can synthesize a similar mannolipid from GDP-mannose. They
have recently gone on to show that, in addition to this compound, the
cotton fibre preparations also catalyze the synthesis of a lipid-oligosaccharide as well as a low percentage of mannosylated protein (Forsee &
Elbein, 1975). The kinetics of the reactions are consistent with the
sequence of glycosylations outlined in Figure 1 for glycoprotein synthesis
in mammalian systems.

Dr. Mary Ericson and one of us (D.P.D.) have been studying glycoprotein
synthesis in developing cotyledons of Phaseolus vulgaris harvested during
the time of active synthesis of vicilin (Ericson & Delmer, 1976). Vicilin
is the major storage protein of these seeds and is a glycoprotein containing
N-acetylglucosamine and mannose. The results of these studies will be
presented in more detail elsewhere in this symposium, and they can be
summarized by stating that it appears that glycoprotein synthesis in this
plant system also follows the same pattern as that outlined in Figure 1.
Figure 2 shows an example of the type of kinetics we obtain when we follow
the incorporation of GDP-^{14}C-mannose into mannolipid and glycoprotein.
Incorporation of radioactivity into mannolipid occurs very rapidly and
reached a steady state within a few minutes, whereas incorporation into

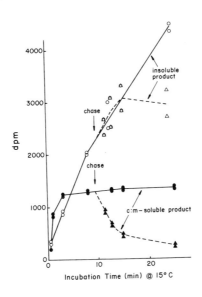

Figure 2. Kinetics of incorporation of radioactivity from GDP-^{14}C-mannose
into mannolipid and insoluble products (including glycoprotein).
Incubations were performed using crude extracts of Phaseolus
vulgaris cotyledons harvested during the active synthesis of
vicilin. Solid line = pulse; dashed line = chase (addition of
100-fold excess of unlabeled GDP-mannose).

insoluble products (in our system, a mixture of lipid oligosaccharide,
glycoprotein, and other material, probably mannan) occurs essentially linearly with time. Upon addition of unlabeled GDP-mannose, radioactivity is
chased out off the mannolipid and into insoluble products in a fashion
consistent with a precursor-product relationship between the mannolipid
and insoluble products. The properties of the mannolipid are consistent
with it being a mannosyl-phosphoryl-polyprenol. Using appropriate solvents,
a lipid-oligosaccharide (also chaseable with unlabeled GDP-mannose) can be
extracted from the insoluble residue. Up to 40% of the insoluble product
has been shown to be glycoprotein, and the radioactivity is not β-
eliminateable, suggesting, but not proving, that the carbohydrate is
attached via an N-glycosidic linkage. Similar data have also been obtained
by Dr. Ericson using UDP-^{14}C-N-acetylglucosamine as substrate.

Thus, it would appear that the pathway for the biosynthesis of this
type of glycoprotein has been conserved in the evolution of both plants

and animals. At present, neither we, nor Elbein's group, are sure just what proteins we are glycosylating in these systems, but the glycoprotein products remain tightly membrane-bound suggesting that they may either be nascent chains of incompleted proteins or completed membrane proteins. The current interest in lectins, many of which are glycoproteins containing mannose and N-acetylglucosamine, suggests that a study of the biosynthesis of these molecules would be both feasible and worthwhile.

Biosynthesis of Hydroxyproline-containing Glycoproteins

Our knowledge about the biosynthesis of extensin, the hydroxyproline-rich glycoprotein found in cell walls, was recently reviewed by Chrispeels (1976). The peptide moieties are synthesized on polysomes, with hydroxylation of proline occurring as a post-translational modification. In another biochemical analogy between plants and animals, it is interesting that the proline hydroxylase of plants has properties which quite closely resemble that of its animal counterpart, and we shall have more to say about this later. Pulse/chase experiments using ^{14}C-proline suggest that extensin passes through some type of smooth membrane vesicles while on its way to the cell wall. Karr (1972) has demonstrated the synthesis of hydroxyproline arabinosides from UDP-arabinose using membrane preparations isolated from cultured sycamore cells, but the precise localization of this activity has not been determined. The attachment of galactose to the serine residues has not been studied in vitro. Hydroxyproline-containing proteins with considerably larger amounts of carbohydrate than that found in extensin have also been demonstrated as cytoplasmic constituents of cultured cells (Lamport, 1970); it is possible that these represent precursors, not to cell wall protein, but to the soluble extracellular glycoprotein secreted by these cells in culture.

The Origin and Evolution of Plant Glycoproteins

First and foremost glycoproteins typify the eukaryotes. In prokaryotes glycoproteins are a highly unusual exception rather than the rule (Mescher & Strominger, 1976). This profound divergence reflects the main lines of evolution: chemical rather than morphological in the prokaryotes, and morphological rather than chemical in the eukaryotes. This implies that glycoproteins play an important role in morphogenesis. That is an under-

statement. The extracellular matrix of all eukaryotes is based on glycoproteins, and recognition phenomena and cell-cell interactions, such as
adhesion and mating of the simplest and the most complex eukaryotes, also
involve glycoproteins. One can therefore speculate that the origin of
glycoproteins was a key factor in the transition from prokaryotic to
eukaryote organization.

There are many many glycoproteins, but metazoan organization is built
around one major matrix glycoprotein only, namely collagen. Virtually all
green plants have extensin as a matrix (or cell wall) glycoprotein. Because
these major matrix glycoproteins of both plants and animals are hydroxyproline rich, one can also speculate, as has Aaronson (1970), that these
glycoproteins have a common origin among primitive photosynthetic organisms,
these being the first to produce (and hence utilize) the molecular oxygen
required for the biosynthesis of hydroxyproline. This implies that plants
and animals also have a common origin among photosynthetic organisms, a
view not in vogue at the moment, but encouraged by the intuitive notion
that the origin of collagen predicates the origin of the metazoa. (That is
to say, if we know the origin of collagen, we know the origin of the
metazoa.)

The involvement of matrix glycoproteins in morphogenesis is self-
evident. How this is so in plants is not so clear. That is why one of us
(D.T.A.L.) has for some years studied extensin. If the foregoing preamble
shows you how extensin catalyses wild ideas, be reassured of the accompanying substance. We shall proceed by asking some innocent questions and
attempt to answer them.

Here are the main questions:

1. How did glycoproteins originate?
2. What reasons are there for believing that hydroxyproline-rich
    glycoproteins appeared relatively late on the scene?
3. In which groups of photosynthetic eukaryotes do hydroxyproline-
    rich glycoproteins occur?
4. What is the structure of these hydroxyproline-rich glycoproteins?
5. Do the structures of plant glycoproteins throw any light on
    function?

1. How did glycoproteins originate?

Answer: through the formation of glycopeptide linkages. But when did they occur and which glycopeptide linkage was first? in which organism, and what was the selection pressure?

Basically there are only two types of glycopeptide linkages, namely N-glycosidic and O-glycosidic, and these involve a total of only five amino acids, namely asparagine (N-glycosidically linked) and serine, threonine, hydroxylysine, and hydroxyproline - all O-glycosidically linked. Two arguments favour selection of asparagine as the primitive glycopeptide link:

First, N-glycosidic linkages appeared on the evolutionary scene in nucleic acids. Second, Sinohara and Maruyama's analysis (1973) comparing the frequency of certain tripeptides in prokaryotes and eukaryotes shows that the recognition code Asn-X Thr/Ser occurs much less frequently in eukaryotes than in prokaryotes implying that Asn-X-Thr/Ser occurrence is restricted in eukaryotes because of its special function. This should convince anyone that glycoproteins do not and cannot appear in prokaryotes, reminiscent perhaps of those immortal lines:

> But scientists who ought to know
> Tell us that it must be so ....
> ... Oh let us never never doubt
> What nobody is sure about!

Quite obviously Mescher & Strominger (1976) were not aware of the above because their recent paper gives detailed evidence for a cell surface glycoprotein in the prokaryotic Halobacterium. This is an exciting discovery for several reasons. It is the first prokaryote glycoprotein to be convincingly characterized. It contains the primitive glycopeptide linkage, i.e. asparagine-N-acetylglucosamine. It is an extracellular product and replaces the peptidoglycan component of the cell wall found in virtually all bacteria. Halobacterium could easily represent a transition organism which lost its peptidoglycan because its saline environment posed no osmotic problem, but gained a much less rigid replacement which could go on to serve other functions.

The loss of a rigid cell wall can lead to a variety of consequences. First, these cells could now develop the capacity for phagocytosis. Armed with this possibility, the way opens up for endosymbiont capture and the establishment of organelles. But now the cells may have run up against a problem of a different type. As pointed out by Winterburn and Phelps (1972), the interior of organelles is really akin to an extracellular environment. How does the cell ensure that the proper cytoplasmically-synthesized proteins find their way to the interior of these organelles? The answer could be, by the synthesis of proteins specifically tagged with carbohydrate as a means of coding for the proper topological location of these proteins. (This is really the "raison d'etre" for glycoproteins originally proposed by Winterburn & Phelps, 1972). Thus, by this reasoning, the evolution of the eukaryotic cell and the evolution of glycoproteins go hand in hand. Concomitant with this event, the excretion of glycoproteins to the cell surface would open up the vast array of possibilities for cell surface interactions with which we presently associate many present-day glycoproteins. It is important to note here that the mechanism of polyprenol-mediated sugar transfers already exists in the membrane of procaryots where it functions for the synthesis of cell wall constituents including peptidoglycan (see Lennarz, 1975).

2. What reasons are there for believing that hydroxyproline-rich glycoproteins appeared relatively later on the scene?

The acquisition of glycoproteins represents a quantum jump in evolution resulting from the vast increase of interactive possibilities created by the simple addition of sugar units to a polypeptide backbone. However, this by itself was not sufficient for mainline eukaryotic evolution. The appearance of green plants and the metazoa apparently required a new glycoprotein type (i.e. collagen or extensin) involving the utilization of molecular oxygen in its biosynthesis. Thus proline hydroxylase of both plants and animals (and one would suppose photosynthetic protists too) involves fixation of molecular oxygen. The earliest organisms able to evolve this remarkable enzyme were the blue-green bacteria (Cyanophyta), and there are indeed reports of small amounts of peptide bound hydroxyproline in these organisms (Gottelli & Cleland, 1968). There are no reports of eukaryotic glycoproteins in the Cyanophyta, and there are no confirmed reports of hydroxyproline-rich proteins in the protozoa. It could be argued that very early on there existed an alternative pathway for hydroxyproline biosynthesis which did not involve molecular oxygen. If the alternative pathway existed, one would expect to find (but one does not find)

      a) the pathway in some relic microorganisms

      b) some prokaryotes besides Cyanophyta with hydroxyproline

      c) a codon assignment for hydroxyproline

If the endosymbiont capture hypothesis is correct, and there is overwhelming

MOLECULAR EVOLUTION OF EUKARYOTES,
HYPOTHETICAL SCHEME.

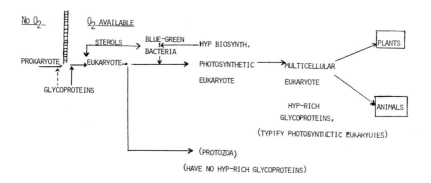

Figure 3.   Molecular evolution of eukaryotes.   Hypothetical scheme.

evidence in support (Margolis, 1970), an originally non-photosynthetic eukaryote acquired the entire genome of a cyanophyte.   According to our
hypothesis (outlined in Figure 3) this would introduce at least one (probably two) new hydroxyamino acids into the new organism which presumably
already had the capability of glycosylating hydroxyamino acids such as
serine and threonine.   Thus, the enzymic apparatus to glycosylate the new
amino acids was already at hand needing at most only minor changes in
specificity.   Thus, on the basis of this hypothesis, the hydroxyproline-
rich glycoproteins originated soon after the acquisition of the chloroplast.
This implies that all hydroxyproline-rich proteins (including those as
widely separated as collagen and extensin) are members of the same super-
family.   Following our discussion of the structure of extensin and related
glycoproteins, we shall discuss this idea more fully.

3.   In which groups of photosynthetic eukaryotes do hydroxyproline-rich
glycoproteins occur?

Still the best survey here is the work of Gotelli & Cleland (1968) who
reported the virtual absence of peptide-bound hydroxyproline in the red
algae (Rhodophyta) and only extremely low levels in brown algae (Phaeophyta).
However, most other photosynthetic protists do contain peptide-bound hydro-
proxyline: Haptophyceae, Chlorophyceae, Chrysophyceae, and diatoms, etc.

One interesting exception is the family <u>Characeae</u> which lacks hydroxyproline, but does contain hydroxylysine (Morikawa, 1974). We would like to know very much, in view of their primitive character, whether the dinoflagellates contain peptide-bound hydroxyproline, but we have to wait awhile for sufficient material. For all other higher forms of photosynthetic life, hydroxyproline is ubiquitous (Lamport & Miller, 1971).

Now we can turn to question that has occupied one of us for some years:

4. What is the structure of these hydroxyproline-rich glycoproteins?

First let us confine ourselves to the higher plants where two or three main classes of hydroxyproline-rich glycoproteins occur, viz:

    (i)   Cell wall

    (ii) Cytosol

    (iii) Soluble macromolecules secreted into the medium (or
          xylem sap)

The cell wall glycoprotein (extensin) is characterized by its extreme resistance to removal by non-degradative techniques. For example, anhydrous hydrogen fluoride, one of the best protein solvents known, dissolves 90% of the isolated tomato cell wall and leaves the protein as an insoluble residue, but with a morphology still reminiscent of a cell wall (Mort & Lamport, 1976). Generally most of the hydroxyproline residues of extensin are glycosylated by a tri- or tetrasaccharide of arabinose, the two inner residues being 1-2 linked (probably α according to Knee, 1976), the non-reducing terminus being 1-3 linked (Lamport, 1974). Most of the serine residues are galactosylated (Lamport, Katona & Roerig, 1973), and recent evidence from chlorite oxidation of cell walls suggests that this represents the attachment region for an arabinogalactan which in turn is linked to other wall polysaccharides (Mort & Lamport, 1975).

The amino acid sequence of extensin is still incomplete. Removal of arabinose residues by treatment of tomato cell walls at pH 1 rendered the protein tryptic labile. All the tryptic peptides contained at least one unit of the pentapeptide Ser-Hyp-Hyp-Hyp-Hyp (Lamport, 1973; see Figure 4). We also demonstrated the presence of this pentapeptide in tryptic peptides obtained from cell walls of sycamore maple (Lamport, unpublished) and therefore conclude that Ser-Hyp-Hyp-Hyp-Hyp is intrinsic to extensin. Here we should note that the frequent occurrence of hydroxyproline will lead to

A. Ser-Hyp-Hyp-Hyp-Hyp-Ser-Hyp-Ser-Hyp-Hyp-Hyp-Hyp-("Tyr"-Tyr)-Lys
   Exists with 3, 2, or 1 galactose residues.

B. Ser-Hyp-Hyp-Hyp-Hyp-Ser-Hyp-Lys
   Exists with 2, 1, or 0 galactose residues.

C. Ser-Hyp-Hyp-Hyp-Hyp-Thr-Hyp-Val-Tyr-Lys
   Exists with 1 or 0 galactose residues.

D. Ser-Hyp-Hyp-Hyp-Hyp-Lys
   Exists with 1 or 0 galactose residues.

E. Ser-Hyp-Hyp-Hyp-Hyp-Val-"Tyr"-Lys-Lys
   Exists with 1 or 0 galactose residues.

Fig. 4. Peptides isolated from acid-stripped tomato cell walls by treatment with trypsin.

a helix typical of polyproline II (Isemura, Okabayashi & Sakakivara, 1968)
(3 residues per turn, with a pitch of 9.4 A), and the molecule will tend to
be rod-like. If, as we suspect, the hydroxyproline-arabinose residues are
α-linked, molecular models of extensin peptides show a distinct tendency
for the tetra-arabinoside to fold back upon the helical peptide backbone,
hydrogen bond with it, and thereby contribute to the overall stability of a
rod-like macromolecule.

   Now the question arises - are the other, i.e. <u>soluble</u> hydroxyproline-
rich glycoproteins related to extensin - closely, distantly, or not at all?
We find these arabinogalactin-alanine-rich glycoproteins not only in the
growth medium of sycamore maple and tomato cultures (Pope & Lamport, 1976),
but also in the xylem sap of sugar maple in a fraction previously described
as arabinogalactan (Adams & Bishop, 1960). Some of these glycoproteins
show remarkable differences from extensin: first an arabinogalactan polysaccharide is directly attached to hydroxyproline (i.e. we have galactosyl
hydroxyproline), and, secondly, the alanine content is about equal to
hydroxyproline.

   As the simplest working hypothesis, we suggest that they <u>are</u> related
to extensin, since they do, in spite of the differences, share a number of
similarities. If this is correct we suspect that these glycoproteins are
amphipathic molecular "tadpoles" having a hydrophilic head consisting of
extensin-like sequences (SerHyp$_4$ pentapeptides?) and hydrophobic alanine-

rich tail. Such an arrangement of hydrophobic/hydrophilic domains typifies membrane proteins in general (Weathers, Jost & Lamport, 1976). Now we can discuss our final question:

Does the Structure of Plant Glycoproteins throw any Light on Function?

We have already arrived above at a working hypothesis for the general structure and possible function of some of the soluble hydroxyproline-rich glycoproteins with attached arabinogalactan. Can we be any more specific about function? Although similar structure implies similar function, it is too early to do more than list a few possibilities for these soluble glycoprotein.

Certainly their occurrence as secretory products in the growth medium of tissue cultures suggests more than a trivial role, because plant tissue cultures are notoriously poor producers of secondary plant products. Even more persuasive of a non-trivial role is our demonstration that these glycoproteins occur in the xylem sap of sugar maple, and presumably other plants when we take time to look for it. The occurrence of these glycoproteins in xylem sap also hints strongly at the possibility of a message-like macromolecule produced by the roots perhaps as part of a control system involved in regulating the balance between root and shoot growth. On this basis these glycoproteins could be involved in cell extension (plasticizers? growth limiting protein?) or they might show mitogenic or other activities. The work of Brown & Kimmins (1973, 1975) suggests a possible role for these types of glycoproteins as a defense mechanism against viral infection. The "all β-lectins" described by Jermyn & Yeow (1975) appear to fall in this structural class of glycoproteins, and their ubiquitous nature and conservation of properties in a variety of plants certainly implies some rather significant functional role for these molecules. Since these molecules recognize β-linked glucose residues, one could speculate that they may be recognition molecules for the β-glucan elicitors (Ayers, Valent, Ebel & Albersheim, 1976) involved in pathogenic interactions between fungi and plants. Clearly there is great scope here for experimental inquiry.

Turning back now to extensin and similar hydroxyproline-rich glycoproteins of the primitive green algae, it seems reasonable to conclude that these macromolecules play an important structural role in the extracellular matrix. In Chlamydomonas, for example, the cell wall consists solely of a crystalline glycoprotein lattice (Roberts, 1974) while in <u>Volvox</u> the cells are embedded in the same sort of glycoprotein (Lamport, 1974). In higher plants, current evidence based on chemical degradation of cell walls favours the view that extensin crosslinks cell wall polysaccharides.

Now we can return to our earlier suggestion that extensin and collagen share a common ancestry as well as a similar function. Figure 5 outlines the remarkable similarities between these two proteins when examined at a fundamental level. In fact, collagen and extensin share most features in common except amino acid sequence. This seems to be yet another example of the observation that tertiary structure is more highly conserved than amino acid sequence during evolution.

1. Position: Extracellular matrix
2. Function: Structural
3. Chemically: Hydroxyproline-rich
4. Structure: Glycoproteins
   Helix type: Polyproline II (9.4 A pitch, 3 residues/turn)
5. Biosynthesis: <u>No</u> hydroxyproline codon (implies "late" evolution)
   Post translational modification
   Involves very similar proline hydroxylases
   Utilising: Molecular O_2
   Cofactors: Fe^{++} ascorbate   α-ketoglutarate
6. Occurrence: Extensin-like proteins occur in most photosynthetic protists
   Collagen does <u>not</u> occur in protozoans
   (i.e. hyp occurrence correlates with photosynthetic ability)
7. Evolutionary Branch Point: Volvox-like organism?   (Origin of the blastula?)

Figure 5.   Do collagen and extensin share a common origin?   (I.e. are they in the same protein "super family"?)

What about the functions of the N-glycosidically-linked glycoproteins of plants? The recent suggestions of a role for the lectins of this type in the establishment of the symbiotic relationship between <u>Rhizobium</u> and legumes is certainly intriguing. Although there is some evidence which

suggests that the carbohydrate portion of these lectins is not essential for
the specificity of sugar binding (Sharon, 1973), this moiety may certainly
function in the general role proposed by Winterburn & Phelphs (1972) - as a
code for ultimate localization of the molecules within the plant. In this
regard, it is also interesting to note that the storage proteins of legumes,
similarly tagged, are stored in membrane-bound organelles called protein
bodies. Is the carbohydrate attached in this case to ensure the proper
localization of these proteins during seed development?

In conclusion, the advances made so far in our knowledge of plant
glycoproteins can only serve to whet our appetite for future research on
these ubiquitous and highly interesting macromolecules.

References

Aaronson, A. (1970). Molecular evidence for evolution in the algae: a
possible affinity between plant cell walls and animal skeletons.
Annals of the New York Academy of Science 175, 531-540.

Adams, G.A. & Bishop, C.T. (1960). Constitution of an arabinogalactan
from maple sap. Canadian Journal of Chemistry 38, 2380-2386.

Ayers, A.R., Valent, B., Ebel, J. & Albersheim, P. (1976). Host-pathogen
interactions. XI. Composition and structure of wall-released elicitor
fractions. Plant Physiology 57, 766-774.

Basha, S.M.M. & Beevers, L. (1976). Glycoprotein metabolism in the
cotyledons of Pisum sativum during development and germination.
Plant Physiology 57, 93-97.

Brown, R.G. & Kimmins, W.C. (1973). Hypersensitive resistance. Isolation
and characterization of glycoproteins from plants with localized
infections. Canadian Journal of Botany 51, 1917-1922.

Brown, R.G., Kimmins, W.C. & Lindberg, B. (1975). Structural studies of
glycoproteins from Phaseolus vulgaris. Acta Chemica Scandinavica
B29, 843-852.

Chrispeels, M.J. (1976). Biosynthesis, intracellular transport, and
secretion of extracellular macromolecules. Annual Review of Plant
Physiology 27, 19-38.

Ericson, M.C. & Delmer, D.P. (1976). Glycoprotein synthesis in plants. I.

The role of lipid intermediates. Plant Physiology, in press.

Forsee, T. & Elbein, A.D. (1973). Biosynthesis of mannosyl- and glucosyl-phosphoryl-polyprenols in cotton fibres. Journal of Biological Chemistry 248, 2858-2867.

Forsee, T. & Elbein, A.D. (1975). Glycoprotein biosynthesis in plants. Demonstration of lipid-linked oligosaccharides of mannose and N-acetyl-glucosamine. Journal of Biological Chemistry 250, 9283-9293.

Gotelli, I.B. & Cleland, R. (1968). Differences in the occurrence and distribution of hydroxyproline-proteins among the algae. American Journal of Botany 55, 907-914.

Isemura, T., Okabayashi, H. & Sakakivara, S. (1968). Steric structures of L-proline oligopeptides. I. Infrared absorption spectra of the oligopeptides and poly-L-proline. Biopolymers 6, 307-321.

Jermyn, M.A. & Yeow, Y.M. (1975). A class of lectins present in the tissues of seed plants. Australian Journal of Plant Physiology 2, 501-531.

Karr, A.L. (1972). Isolation of an enzyme system which will catalyze the glycosylation of extensin. Plant Physiology 50, 275-282.

Kauss, H. (1969). A plant mannosyl-lipid acting in reversible transfer of mannose. FEBS Letters 5, 81-84.

Knee, M. (1976). Soluble and wall-bound glycoproteins in apple fruit tissue. Phytochemistry, in press.

Lamport, D.T.A. (1970). Cell wall metabolism. Annual Review of Plant Physiology 21, 235-270.

Lamport, D.T.A. (1973). The glycopeptide linkages of extensin: O-D-galactosyl serine and O-L-arabinosyl hydroxyproline. pp. 149-164 in Biogenesis of Plant Cell Wall Polysaccharides by F. Loewus (ed.). Academic Press, New York.

Lamport, D.T.A. (1974). The role of hydroxyproline-rich proteins in the extracellular matrix of plants. pp. 113-130 in Macromolecules regulating growth and development. 30th Symposium of the Society of Developmental Biology.

Lamport, D.T.A., Katona, L., & Roerig, S. (1973). Galactosylserine in extensin. Biochemical Journal 133, 125-131.

Lamport, D.T.A. & Miller, D.H. (1971). Hydroxyproline arabinosides in the plant kingdom. Plant Physiology 48, 454-456.

Lennarz, W.J. (1975). Lipid-linked sugars in glycoprotein synthesis. Science 188, 986-991.

Margolis, L. (1970). Origin of the Eukaryotic Cell. Yale University Press, New Haven.

Mescher, M.F. & Strominger, J.L. (1976). Purification and characterization of a procaryotic glycoprotein from the cell envelope of Halobacterium salinarium. Journal of Biological Chemistry 251, 2005-2014.

Miller, D.H., Mellman, E.S., Lamport, D.T.A. & Miller, M. (1974). The chemical composition of the cell wall of Chlamydomonas gymnogama and the concept of a plant cell wall protein. Journal of Cell Biology 63, 420-429. (see p. 426 footnote).

Morikawa, H. (1974). Studies on the ultrastructure of the plant cell wall in Nitella. Ph.D. Thesis, Kyoto University, Japan.

Mort, A.M. & Lamport, D.T.A. (1975). Evidence for polysaccharide attachment to extensin in cell walls obtained from tomato cell suspension cultures. Plant Physiology 56, S-80.

Mort, A.M. & Lamport, D.T.A. (1976). Specific deglycosylation via anhydrous hydrogen fluoride. Plant Physiology 57, S-297.

Pope, D.G. & Lamport, D.T.A. (1974). Hydroxyproline-rich material secreted by cultured Acer pseudoplatanus cells: evidence for polysaccharide attached directly to hydroxyproline. Plant Physiology 54, S-82.

Roberts, K. (1974). Crystalline glycoprotein cell walls of algae: their structure, composition and assembly. Philosophical Transactions of the Royal Society of London B 268, 129-146.

Roberts, R.M., Connor, A.B. & Cetorelli, J.J. (1972). The formation of glycoproteins in higher plants. Specific labeling using D-1-^{14}C-glucosamine. Biochemical Journal 125, 999-1008.

Roberts, R.M., Cetorelli, J.J., Kirby, E.G. & Ericson, M. (1972). Location of glycoproteins that contain glucosamine in plant tissues. Plant Physiology 50, 531-535.

Roberts, R.M. & Pollard, W.E. (1975). The incorporation of D-glucosamine into glycolipids and glycoproteins of membrane preparations from Phaseolus aureus hypocotyls. Plant Physiology 55, 531-536.

Sharon, N. (1973). Plant Glycoproteins. pp. 235-252 in Plant Carbohydrate Chemistry, ed. by J.B. Pridham. Academic Press, New York.

Sinohara, H. & Maruyama, T. (1973). Evolution of glycoproteins as judged by the frequency of occurrence of the tripeptides Asn-X-Ser and Asn-X-Thr in proteins. Journal of Molecular Evolution 2, 117-122.

Weathers, P.J., Jost, M. & Lamport, D.T.A. (1976). The gas vacuole membrane of <u>Microcystis aeurginosa</u>. A partial amino acid sequence and proposed functional model. Archives of Biochemistry and Biophysics, in press.

Wellburn, A.R. & Hemming, F.W. (1966). Gas-liquid chromatography of derivatives of naturally-occurring mixtures of long-chain polyisoprenoid alcohols. Journal of Chromatography 23, 51-66.

Winterburn, P.J. & Phelps, C.F. (1972). The significance of glycosylated proteins. Nature 236, 147-154.

DISCUSSION

<u>Chairman</u>: C. Ballou

<u>Brown</u>: I don't know what this does to your theory, but we've recently been working with the capsular material from <u>Ruminococcus albus</u>, a rumin anaerobic bacterium that binds the bacterium specifically to cellulose fibres for digestion. Mind you, it is an extracellular product - it is capsular - but it is a glycoprotein containing both N and O glycosidic bonds.

<u>Delmer</u>: That's very interesting. I think, certainly, it's going to be true that glycoproteins are not common in the procaryotes, but now that people are looking better they are finding them.

<u>Kimmins</u>: In relation to your question of the basic significance of glycoproteins, you mention the "tagging function". I think if you say it's a tagging function, you're excluding a lot of significant functions of glycoproteins. One has to go down one level and look at the bonding capabilities. If you look at the proteins that have magnesium as a prosthetic group, the reason is that none of the available amino acid R-groups is a strong electrophilic agent. If you look then at what advantages a protein would gain by having attached to it a carbohydrate residue, it simply increases dipole-dipole interactions. None of this excludes tagging, but it does explain why storage proteins are glycosylated and it rationalizes the role of glycoproteins as a fish antifreeze.

<u>Delmer</u>: Yes. As a matter of fact, Dr. Mary Ericson, who is working in my laboratory on glycoprotein synthesis, has the idea that storage proteins are

glycosylated to prevent crystallization within the seed. It is well known that glycoproteins are difficult to crystallize, even those with low carbohydrate content. I argue with her about this because, in fact, you can demonstrate that these proteins crystallize a bit in the seed. But maybe they would crystallize better if they weren't glycosylated. We're looking for a way to inhibit glycosylation to try to answer some of these questions. But I didn't mean to imply this is the only function. What we were fancifully speculating was that the tagging function might have been one of the things that sped up the evolution of glycoproteins during eucaryote development, and that it still plays a major role though not the only one.

Albersheim: On table 1, where you summarized all the plant glycoproteins, I think you left off an important group. Mushrooms are plants, and in yeasts there are mannoproteins including the cell surface immunochemical determinants and the sexual mating type factors (Crandall, M.A. & Brock, T.D., Bacteriol. Rev. 32, 139-163, 1968). The reason I brought this up is that there have been experiments with yeast mutants in which the extracellular enzyme secreted by the yeast lack all or large parts of the glycoportion (Smith, W.L. & Ballou, C.E., Biochem. Biophys. Res. Commun. 59, 314-321, 1974). I hope Clint Ballou will bear this out. In any case, the selected enzymes lack a good portion of their carbohydrate. Regardless, they still end up on the outside of the cell, and if these were keys to tagging, as you suggest, then this modification didn't destroy that tag.

Delmer: First of all, I'd like to clear it with Clint Ballou whether they are completely deglycosylated.

Ballou: I can comment on that point. I think one function for glycosylation of proteins, illustrated in yeast, is to make the protein very large and thus immobilize it outside the plasma membrane or in the cell wall. Plants must have the same problem - to immobilize some enzymes between the plasma membrane and the cell wall. Now, in our yeast mutants that still secrete glycoprotein, the mutations only affect the outer part of the carbohydrate chain and do not alter the core structure. I believe the core of these glycoproteins probably is very critical for their translocation or processing, whatever you want to call it.

Delmer: Some of you were at the ACS meeting in New York this Spring when Dr. Robert Schimke, who works on ovalbumin synthesis, presented some evidence that supports Eylar's original hypothesis (J. Theoret. Biol. 10, 89-113, 1965) that perhaps all proteins which are secreted are glycosylated. One of the classic examples is serum albumin, which is not glycosylated but is found in abundance as a secreted protein. Dr. Schimke has evidence that serum albumin is glycosylated when it is synthesized inside the cell, and that the glyco-portion is clipped off after secretion (Schimke, R.T., Kiely, M.L., McKnight, G.S. & Clark, S., Abstracts ACS Meeting, New York, April 5-9, 1976).

THE USE OF HYDRAZINOLYSIS FOR STRUCTURAL STUDIES OF PLANT GLYCOPROTEINS

Robert G. Brown and Jeffrey L.C. Wright, Department of Biology, Dalhousie University and Atlantic Regional Laboratory, National Research Council of Canada, Halifax, N.S., Canada, B3H 4J1.

Hydrazine is a poor base but a good nucleophile. Use of this reagent with hydrazine sulfate to act as a catalyst (Yosizawa, Sato & Schmid, 1966) has been used extensively in studies of glycoproteins to selectively depolymerize the polypeptide chain while preserving carbohydrate constituents. Application of this technique to the study of hydroxyproline deficient glycoproteins from Phaseolus vulgaris resulted in a number of experimental discrepancies. For instance, gel filtration indicated a higher molecular weight for the carbohydrate containing hydrazinolysis products that methylation analysis (Brown, Kimmins & Lindberg, 1975). It has been noted that two adjacent hydroxyproline residues in a polypeptide yield a cyclic diketopiperazine which retains carbohydrate side chains (Heath & Northcote, 1973). Formation of this diketopiperazine would contribute to an explanation of the molecular weight discrepancy referred to above; however, the cyclic diketopiperazine isolated by Heath & Northcote was not present in the hydrazinolysate. In our more recent studies amino acid analysis of the analogous high molecular weight products following hydrazinolysis of P. vulgaris glycoproteins revealed substantial enrichment of both hydroxyproline and serine. The ratio of these two amino acids (1:1) indicates more enrichment of hydroxyproline than serine in comparison to the original glycoprotein. The quantity of serine, which was very much greater than any other amino acid present except hydroxyproline, implicates this amino acid in protein - carbohydrate linkages. Using labelled serine hydrazide (^{14}C) and hydroxyproline hydrazide (^{3}H), formation of diketopiperazines (isolated by gel chromatography using Sephadex G-10) under the same conditions used for hydrazinolysis of glycoproteins was studied. The yield of purified serine diketopiperazine and hydroxyproline diketopiperazine was 53 and 32% respectively. The structure of these diketopiperazines was confirmed by ^{13}C-NMR spectrometry and gas liquid chromatography-mass spectrometry (g.l.c.-m.s) of trimethylsilyl derivatives. This result is hardly surprising since diketopiperazine formation is known to occur fairly readily. Thus, having adjacent residues of hydroxyproline in a polypeptide is not an absolute requirement for hydroxyproline diketopiperazine formation during hydrazinolysis although may facilitate it. An equimolar mixture of the methyl esters of serine and hydroxyproline were subjected to hydrazinolysis and the resultant mixture of dimeric products was examined by ^{13}C NMR spectrometry and g.l.c. of trimethylsilyl derivatives. The mixture consisted mainly of serine diketopiperazine.

Enrichment of serine in the high molecular weight fraction of the hydrazinolysate of hydroxyproline deficient glycoproteins implies retention by serine of carbohydrate side chains during hydrazinolysis. Lamport (1974) has reported destruction of substituted serine residues on treatment with hydrazine. The hypothesis we propose is that if diketopiperazines are formed before elimination of carbohydrate substituents, subsequent elimination of these substituents during hydrazinolysis might be unfavourable. Formation of diketopiperazines would probably be more favourable during hydrazinolysis of glycoproteins than amino acid hydrazides (see figure 3, Brown, Kimmins & Lindberg, 1975), but even the hydrazide gave a yield of 53% in the case of serine. Currently, the stability of the di-o-acetyl derivative of serine diketopiperazine to hydrazinolysis is being studied.

References

Brown, R.G., Kimmins, W.C. & Lindberg, B. (1975). Structural studies of glycoproteins from Phaseolus vulgaris. Acta Chemica Scandinavica B 29, 843-852.

Heath, M.F. & Northcote, D.H. (1973). A hydroxyproline-containing glycopeptide released from the walls of Sycamore tissue - culture cells by hydrazinolysis. Biochemistry Journal 135, 327-329.

Lamport, D.T.A. (1974). The glycopeptide linkages of extensin: o-D-galactosyl serine and o-L-arabinosyl hydroxyproline. p. 160 in Biogenesis of plant cell polysaccharides, ed. by F. Loewus. Academic

Press, New York.

Yosizawa, X., Sato, T. & Schmid, K. (1966). Hydrazinolysis of α-acid glycoprotein. Biochemica et Biophysica Acta 121, 417-418.

DISCUSSION

Chairman: D. Delmer

Delmer: I should be more familiar - I know that Derek Lamport has some explanation of his own for the discrepancy between his results and Northcote's, and I believe that the conditions are really somewhat different for their hydrazinolyses.

Brown: They're quite different because Derek didn't include any hydrazine sulfate, he just used straight hydrazine.

STRUCTURAL STUDIES OF GLYCOPROTEINS FROM PHASEOLUS VULGARIS

Robert G. Brown, Dept. of Biology, Dalhousie University, Halifax, Nova
Scotia, Canada, B3H 4J1.

The production and structure of hydroxyproline-poor glycoproteins
isolated from primary leaves of Phaseolus vulgaris following sham or virus
inoculation have been studied (Brown & Kimmins, 1973; Kimmins & Brown, 1973;
Brown, Kimmins & Lindberg, 1975; Kimmins & Brown, 1975). Recent investigations sought to further elucidate the structural features of glycoproteins
isolated following sham inoculation. Hydroxyproline has been implicated in
protein-carbohydrate linkage but definitive proof is still lacking. To
obtain glycoproteins in an early stage of biosynthesis the light regime of
sham inoculated plants was changed from 16 h. light - 8 h. dark to 12-12.
Amino acid analysis of the high molecular weight portion of a hydrazinolysate
of the glycoproteins demonstrated the presence of both hydroxyproline and
serine (approximately 1:1) which together accounted for 63% of the amino
acids present. Thus, both hydroxyproline and serine appear to be involved
in protein - carbohydrate linkage. The presence of serine was unexpected as
Lamport has demonstrated destruction of serine residues which have carbohydrate side chains during hydrazinolysis. (A proposal to explain the origin
of serine in the high molecular weight portion of the hydrazinolysate was
discussed in the previous paper in this symposium.) Knowledge that both
hydroxyproline and serine seem to be involved in protein-polysaccharide
linkage posed two questions. Firstly, how much of the carbohydrate attached
to serine is lost from the glycoproteins during extraction (1 M NaOH, $2^{o}C$,
approximately 30 min.)? Secondly, which carbohydrate moieties are attached
to serine and which are attached to hydroxyproline? The first question was
investigated using sodium sulfite (^{35}S) in alkali and extracting at two
temperatures ($2^{o}C$ and $50^{o}C$). The effect of temperature on the extent of ^{35}S
labelling of purified glycoproteins and amino acid compositions was compared.

Glycoproteins extracted at 50° were more extensively labelled with ^{35}S than those extracted at 2^OC, but ^{35}S was present in the latter glycoproteins indicating that some carbohydrate was lost even at 2^OC. The proposed approach to the second question was to treat purified glycoproteins with base and isolate the released carbohydrate side chains. Before treating with base, the glycoproteins were rechromatographed on Sepharose CL-6B to ensure an homogeneous starting preparation (which had been prepared in 3 batches) and provide greater ease of isolation of the released carbohydrate. A substantial portion of the glycoprotein preparation was bound strongly to the top of the Sepharose column. This is the first demonstration of binding of hydroxyproline-poor glycoproteins to any substrate. The carbohydrate and amino acid composition of glycoproteins which were not bound by Sepharose and those which were, have been compared. Glycoproteins which were bound by Sepharose had more glucose than those which were not bound (90% vs. 15%). Current studies include treatment of Sepharose binding and non-binding glycoproteins with base in the presence of sodium sulfite (^{35}S) to convert the olefin derivative of serine to cysteic acid. Carbohydrate released by alkali is also being studied to determine which carbohydrate moieties are attached to serine.

References

Brown, R.G. & Kimmins, W.C. (1973). Hypersensitive resistance. Isolation and characterization of glycoproteins from plants with localized infections. Canadian Journal of Botany 51, 1917-1922.

Brown, R.G., Kimmins, W.C. & Lindberg, B. (1975). Structural studies of glycoproteins from Phaseolus vulgaris. Acta Chemica Scandinavica B 29, 843-852.

Kimmins, W.C. & Brown, R.G. (1973). Hypersensitive resistance. The role of cell wall glycoproteins in virus localization. Canadian Journal of Botany 51, 1923-1926.

Kimmins, W.C. & Brown, R.G. (1975). Effect of a non-localized infection by Southern Bean Mosaic Virus on a cell wall glycoprotein from Bean leaves. Phytopathology 65, 1350-1351.

Lamport, D.T.A. (1974). The glycopeptide linkages of extensin: o-D-galactosyl serine and o-L-arabinosyl hydroxyproline. p. 160 in Biogenesis of plant cell polysaccharides, ed. by F. Loewus. Academic Press, New York.

GLYCOPROTEIN SYNTHESIS IN PLANTS: A ROLE FOR LIPID INTERMEDIATES

Mary C. Ericson and Deborah P. Delmer, MSU/ERDA Plant Research Laboratory,
Michigan State University, East Lansing, Michigan 48824, USA

Glycoprotein synthesis has been intensively studied in animal systems, and recent evidence has shown that the carbohydrate moieties are often transferred to the protein via lipid intermediates (Lennarz, 1975). Glycoproteins also exist in plants, but except for one recent paper by Forsee & Elbein (1975), very little is known of their mode of biosynthesis. Vicilin, one of the major storage proteins in the seeds of legumes, has recently been characterized by Ericson & Chrispeels (1973,1976) and shown to be a glycoprotein with linkage and sugar composition analogous to that of many animal glycoproteins [Asn-(GlcNAc)$_x$(Man)$_x$]. Thus, we have been studying the enzymatic processes involved in glycoprotein synthesis using extracts obtained from developing cotyledons of Phaseolus vulgaris harvested at the time of active deposition of vicilin, the only major storage protein in this species.

Using crude extracts, radioactivity from GDP-^{14}C-mannose can be incorporated into a single chloroform: methanol-soluble product as well as into insoluble products. The kinetics of these reactions, as determined by pulse and pulse/chase experiments, are consistent with the radioactive glycolipid serving as an intermediate in the synthesis of the insoluble produc(s). (See Figure 2 in the article by Delmer & Lamport, this symposium).

The characteristics of the mannolipid synthesized in this system are summarized in Table 1. All of these characteristics are consistent with this product being a mannosyl-phosphoryl-polyprenol. The mannolipid migrates as a single peak in three separate solvent systems during thin-layer chromatography, with mobilities consistent with published values for glycosyl-phosphoryl-polyprenols. Also consistent with this type of compound

Table 1. Properties of C:M soluble product.

1. <u>TLC.</u> C:M soluble product formed a single peak in three solvent systems:

Solvent system	R_f
neutral	0.44
basic	0.19
acidic	0.90

2. The accumulation or synthesis of the mannolipid was found to be reversed by the addition of GDP but not GMP.

3. Radioactivity was not released from the labeled mannolipid by alkali.

Radioactivity in lipid product before alkali treatment:	Radioactivity in lipid product after alkali treatment:
20,500	21,000

4. Radioactivity was completely released from the labeled mannolipid by mild acid (0.01 N HCl at 100°C for 10 min.).

Radioactivity in lipid product before acid treatment:	Radioactivity in lipid product after acid treatment:
17,140	0

5. Radioactivity released by acid hydrolysis was identified as <u>mannose.</u>

are its resistance to alkaline saponification and its extreme acid lability. The fact that synthesis is reversible by the addition of excess GDP, but not GMP, indicates that the mannose is attached to the lipid via a monophosphate linkage.

Of the radioactivity in the insoluble product(s), about 20% is pronase-digestible during a "pulse experiment", and about 40% is pronase-digestible after a chase with unlabeled GDP-mannose. The other 60% is as yet uncharacterized but could be a mannan. The difference in susceptibility to pronase digestion in product obtained from either a pulse or a chase is best explained by the presence of a radioactive lipid-oligosaccharide present in the insoluble product fraction during the pulse; this radioactivity is then transferred to protein during a chase, thus increasing the percent of insoluble product that is glycoprotein after a chase. This conclusion is supported by the fact that a radioactive product, soluble only in a mixture of chloroform: methanol: H_2O (1:1:0.3), may be extracted from the insoluble residue obtained during a pulse, but this radioactivity decreases during a chase. This product stays at the origin during thin-layer chromatography in lipid solvent. Thus we suggest that this product may be a lipid oligosaccharide, the final intermediate in glycoprotein synthesis.

Preliminary data indicate that radioactivity from UDP-N-acetyl-^{14}C-glucosamine can be incorporated into both chloroform: methanol-soluble and insoluble products. Results with both GDP-^{14}C-mannose and UDP-N-^{14}C-glucosamine clearly support a role for lipid intermediates in the glycosylation of protein in this higher plant system and indicate that the mechanisms of glycoprotein synthesis in higher plants are quite similar to those reported for mammalian systems. It seems quite probable that those plant lectins containing mannose and N-acetyl-glycosamine, glycoproteins of special interest to this symposium, would also be synthesized by a similar mechanism. (Supported by ERDA under Contract No. E(11-1)-1338)

References

Ericson, M.C. & Chrispeels, M. (1973). Isolation and characterization of glucosamine-containing storage glycoproteins from the cotyledons of Phaseolus aureus. Plant Physiology 52, 98-104.

Ericson, M.C. & Chrispeels, M. (1976). The carbohydrate moiety of mung bean vicilin. Australian Journal of Plant Physiology, in press.

Forsee, T. & Elbein, A.D. (1975). Glycoprotein biosynthesis in plants. Demonstration of lipid-linked oligosaccharides of mannose and N-acetyl-glucosamine. Journal of Biological Chemistry 250, 9283-9293.

Lennarz, W.J. (1975). Lipid-linked sugars in glycoprotein synthesis. Science 188, 986-991.

DISCUSSION

Chairman: K.-E. Eriksson

Kauss: I would only take the opportunity to advertise a paper which is forthcoming soon in Archives of Biochemistry and Biophysics by Lehler, Padocic, Tanner and me where we find lipid diphosphate-bound N-acetyl-glucosamine, bound glucosamine dimer, and the trimer containing the mannose in addition. The only point which I would add in addition to your data is that there is some evidence in our system, which is mung bean, that the lipid part is a dolichol type. It's not a ficaprenol, but a dolichol type. So that's another similarity to animal systems. And that's done mainly in comparison to yeast. We have used yeast lipid as a standard.

Ballou: I was interested in one of the last experiments you described in which you concluded that the carbohydrate in the insoluble product is not released by conditions that cause β-elimination. And yet that insoluble product presumably also contains the oligosaccharide lipid pyrophosphate which I would think might be degraded and release some water-soluble label.

Delmer: I should have explained that β-elimination was performed on chased product where all of the radioactivity had been chased through lipid oligo-saccharide into glycoprotein. So it contained essentially no radioactive lipid oligosaccharide at the point that experiment was done.

ANALYSIS OF CELL WALL MATERIAL FROM LIGNIFIED AND PARENCHYMA TISSUES OF
PHASEOLUS COCCINEUS

R.R. Selvendran, S.G. Ring & J.F. March, Agricultural Research Council,
Food Research Institute, Colney Lane, Norwich NR4 7UA, England

Cell wall material (CWM) from lignified tissues

During maturation runner beans develop lignified tissues, namely
"parchment layers" and "strings". These tissues were separated from the
mature bean pods and their CWM was prepared by sequentially extracting the
wet ball-milled tissues with 1% Na-deoxycholate and phenol/acetic acid/
water (2:1:1, w/v/v) as described by Selvendrar. (1975a). The wall material
was then delignified with sodium chlorite-HOAc for 4 h at $70^{o}C$. The
holocellulose was fractionated into the various groups of polysaccharides
by sequential extraction with 0.5% ammonium oxalate for 3 h at $95^{o}C$
(Fraction 1), dimethyl sulphoxide for 18 h at $20^{o}C$ (Fraction 2) and 1 and
4N KOH for 2 h each at $20^{o}C$ (Fractions 3 and 4 respectively) leaving the
α-cellulose residue. The polymers were isolated from the various fractions
(including the chlorite-HOAc soluble fraction) after acidification and
extensive dialysis by precipitation with alcohol. The 1N KOH soluble
polymer was further purified by formation of the copper complex (Adams,
1965).

The sugars obtained on acid hydrolysis of the polymers were analysed
by ion-exchange chromatography (Davies, Robinson & Couchman, 1974) and after
aldehyde reduction as their alditol acetates by GLC (Jones & Albersheim,
1972). These studies showed that the main polymer in Fractions 1-4 and the
chlorite-HOAc soluble fraction was a xylan. IR studies showed that alkali
treatment caused deacetylation of the xylans. Cellulase and xylanase present
in extracts of Trichoderma viridi hydrolysed the polymers (including α-
cellulose) mainly to the monosaccharides, whereas the corresponding enzymes
from Trichosporon pullulans gave approximately equal amounts of monosaccha-

rides and disaccharides.

The xylans were further characterized by methylation studies. They were methylated with methyl sulphinyl anion-methyl iodide in DMSO and the methy- lated xylans hydrolysed. The partially methylated sugars formed were analysed after aldehyde reduction, as their alditol acetates by GLC-mass spectrometry (Lindberg, 1972).

These studies showed that the bean xylans are mainly linear unbranched molecules composed of β (1→4) linked chains of xylose residues. The average chain length of the xylans in Fractions 1-3 varied from 70-85 and that in Fraction 4 was about 200. The properties and composition of the xylans from the "parchment layers" were comparable to those from the "strings".

Cell wall material from parenchyma

Parenchyma was obtained by scraping the pods (split in half lengthwise) up to the parchment layer with a spoon and the wall material prepared as before. This was fractionated by two methods (Selvendran, 1975b). In Method 1 fractionation was by sequential treatments with 0.5% oxalate for 3 h at 95^{0}C, chlorite-HOAc for 2 h at 70^{0}C and 1 and 4 N KOH for 2 h each at 20^{0}C, leaving the α-cellulose residue. In Method 2 the chlorite-HOAc treatment was omitted. This gave α'-cellulose which was rich in protein containing hydroxyproline (HP) and associated arabinangalactan, most (80%) of which could be liberated by treatment with chlorite-HOAc for 2 h at 70^{0}C. The polysaccharides and glycoproteins were hydrolysed and the resulting sugars were analysed as before. The amino acids from the hydrolysates of the glycoproteins were estimated by ion-exchange chromatography and by GLC as their heptafluorobutyric-n-propyl derivatives (March, 1975).

At least two main types of wall glycoprotein appear to have been present. One glycoprotein was rich in HP and was associated with α'- cellulose; the other was poor in HP and this was associated with the alkali soluble polymers. Cellulase from T. viride unlike that from T. pullulans hydrolysed α-cellulose and α'-cellulose mainly to glucose. The HP-rich wall glycoprotein, associated with α'-cellulose did not appear to be hydrolysed and could be isolated from the incubation mixture. These studies have thrown light on the composition of polysaccharides from different tissues of beans and have also provided a "non degradative" procedure for the isolation of HP-rich wall glycoprotein from parenchyma of mature beans.

Selvendran et al. - Plant cell wall material

References

Adams, G.A. (1965). Xylans. pp. 170-175 in Methods in carbohydrate
    chemistry Vol. 5, ed. by R.L. Whistler. Academic Press, New York,
    London.
Davies, A.M.C., Robinson, D.S. & Couchman, R. (1974). Accelerated ion-
    exchange chromatography of reducing sugars in neutral borate buffers.
    Journal of Chromatography 101, 307-314.
Jones, T.M. & Albersheim, P. (1972). A gas chromatographic method for the
    determination of aldose and uronic acid constituents of plant cell
    wall polysaccharides. Plant Physiology 49, 926-936.
Lindberg, B. (1972). Methylation analysis of polysaccharides. pp. 178-
    195 in Methods of Enzymology Vol. 28, ed. by V. Ginsburg. Academic
    Press, New York, London.
March, J.F. (1975). A modified technique for the quantitative analysis of
    amino acids by gas chromatography using heptafluorobutyric n-propyl
    derivatives. Analytical Biochemistry 69, 420-442.
Selvendran, R.R. (1975a). Analysis of cell wall material from plant
    tissues: extraction and purification. Phytochemistry 14, 1011-1017.
Selvendran, R.R. (1975b). Cell wall glycoproteins and polysaccharides of
    parenchyma of Phaseolus coccineus. Phytochemistry 14, 2175-2180.

DISCUSSION

Chairman: C. Ballou

Jarvis: It is very nice to see somebody looking at cell walls prepared
from different cell types. This is something we should all have been doing
a long time ago. Let me ask you, was there any lignin or phenylpropenyl
material in the parenchyma preparation, because it obviously affects how one
interprets the release of protein by what accompanies the delignification.

Selvendran: No, we don't find any phenylpropenyl material, but Roy
Hartley does find it in the grasses by fluorescence microscopy.

STRUCTURAL STUDIES ON THE EXTRACELLULAR ACID POLYSACCHARIDES FROM RHIZOBIUM
BACTERIA

Randi Sømme, Department of Chemistry, Agricultural University of Norway,
N-1432 Ås-NLH, Norway

Bacteria from the genus Rhizobium infect the roots of leguminous plants
and form nodules. The nitrogen fixed in these nodules is then transferred
to the plant. Different species of Rhizobium show high specificity with
respect to the host plant species. The specificity was found to be determined by the capsular polysaccharide produced by the bacteria as it was
postulated that the polysaccharide itself was able to induce the formation
of a polygalacturonase in the plant roots thus facilitating the entrance of
bacteria into the host plant. This has not been possible to verify.
Recently it has been reported that the bacteria are bound to the host plant
by a specific interaction between plant lectins and the bacterial polysaccharides (Dazzo & Hubbell, 1975).

The polysaccharides of the different strains contain glucose, galactose
and glucuronic acid bound with β-D-interglycosidic linkages, ketal bound
pyruvic acid and O-acetyl groups. The polysaccharides have (1-4), and (1-3)-
linked glucose residues, (1-3)-linked galactose, and (1-4)-linked glucoronic
acid residues. One exception is the polysaccharide from Rh. meliloti which
lacks glucuronic acid, and has a (1-6)-linked glucose residue in addition.
Six strains have two β- (1-4)-linked D-glucuronic acid residues, β-linked
to D-glucose at position 4 (Sømme, 1975).

The periodate-oxidation values obtained, indicate that the polysaccharides of the non-cooperative strain Coryn, and the non-infective strain Bart
A have a linkage pattern different from the other strains. They indicate a
higher content of (1-3)-linkages or that one of the vicinal hydroxyl groups
in (1-4)- or (1-6)-linked components is acetylated. This might be very
important indications.

The polysaccharides are not amenable to gel filtration because of the high viscosity of their solution even after deionization.

By ultra centrifugation experiments it was impossible to determine the molecular weight. It was found to be in the order of at least 20-40 10^6. Even a partial degraded product obtained by autohydrolysis was excluded with the void volume on a Biogel P-300. column.

References

Dazzo, F.B. & Hubbell, D.H. (1975). Crossreactive antigens and lectin as determinants of symbiotic specificity in the Rhizobium-clover association. Applied Microbiology 30, 1017-1033.

Sømme, R. (1975). Fragmentation analysis of extracellular acid polysaccharides from seven Rhizobium strains. Carbohydrate Research 43, 145-149.

DISCUSSION

Chairman: C. Ballou

Dazzo: I'd like to say that, in our experience analyzing acid hydrolysates of these Rhizobium trifolii polysaccharides, we found no differences in the gas chromatographic patterns from the infective and the noninfective strains. Indeed, one of the noninfective strains we examined, the same Bart A strain you used, we found no differences in the monomer sugar compositions when the polysaccharides were acid-hydrolyzed. We reported that in our paper (Dazzo, F.B. & Hubbell, D.H., App. Microbiol. 30, 1017-1033, 1975).

Albersheim: Your finding of a high molecular weight acidic polysaccharide or exopolysaccharide is not in complete agreement with the literature, which suggests that many of them are of low molecular weight. We have found, in working with these, that they associate with each other and one gets very large apparent molecular weights; but if we put them in salt solution, NaCl, they all have a very low molecular weight and are homogenous in size. So, I think you can dissociate your large polymers into small ones, which may facilitate your work.

Sømme - Rhizobium polysaccharides

Brown: One way of examining polysaccharides that contain uronic acid is by Conrad reduction (Whistler, R.L., Conrad, H.E. & Hough, L., J. Am. Chem. Soc. 76, 1668-1670, 1954). Have you considered its use?

Sømme: Yes. I have tried it, but I have not been able to get complete reduction.

Brown: It wasn't clear in your periodate uptake studies what the two curves indicate.

Sømme: If one does a periodate oxidation, after some time the free aldehyde groups in the polysaccharide will form hemiacetals, and one has to reduce it and start over again. That's why we get a higher value for the periodate consumed.

SURFACE POLYSACCHARIDES OF RHIZOBIUM LEGUMINOSARUM

K. Planqué, J.W. Kijne & A.A.N. van Brussel, Botanisch Laboratorium,
Rijksuniversiteit, eiden, Nederland

In the host cells of pea root nodules the rod-shaped Rhizobium
leguminosarum A 171 changes into a swollen, branched, X and Y shaped form,
the so-called bacteroid. Van Brussel (1973) observed a greater sensitivity
of bacteroids to lysozyme and trypsin as measured by osmotic lysis after
incubation and shock. Fractionation of bacterial (b) and bacteroidal (boid)
cell walls by the phenol extraction method of Westphal et al. (1952) revealed
a strongly reduced amount of polysaccharide in boid cell walls. Fucose,
heptose and an unknown neutral sugar in b- and boid-LPS, estimated from
gas-liquid-chromatography (glc) analysis, were in the same ratio, and no
serological difference could be found by ouchterlony immuno-diffusion tests
between b- and boid-LPS against heat-killed-bacterial rabbit antiserum.
After repeated high speed centrifugation and mild hydrolysis b-LPS appeared
to be heterogeneous as shown by fractionation on Sephadex G50. Glc of
neutral sugars indicated the presence of a glucan.

References

Van Brussel, A.A.N. (1973). The cell wall of bacteroids of Rhizobium
    leguminosarum Frank. Ph.D. Thesis, Rijksuniversiteit, Leiden.
Westphal, O., Lüderitz, O. & Bister, F. (1952). Über die Extraktion von
    Bakterien mit Phenol/Wasser. Zeitschrift für Naturforschung 7b,
    148-155.

Planqué et al. - Rhizobium polysaccharides

DISCUSSION

Chairman: C. Ballou

Graham: Lawrence Rothfield has shown that inhibition of protein synthesis in E. coli results in the excretion of lipopolysaccharide (Rothfield, L. & Pearlman-Kothenz, M., J. Mol. Biol. 44, 477-492, 1969). I wonder if something like that might not happen in the bacteroids.

Kijne: That, of course, is one of the questions - how far are these changes an effect of an existing structure or how far are they the result of altered biosynthesis. I don't think that protein synthesis is affected in bacteroid formation, because there is a lot going on. There is at least 5 times more DNA present in a bacteroid and there should be a large amount of protein synthesis as well. It has to do a lot for nitrogen fixation. But, of course, biosynthesis can also have something to do with the changes.

Graham: Yes, LPS is excreted from the cell very easily, and many things can affect the excretion.

Kijne: Indeed, there is an excretion of LPS in normal free-living cultures. When you spin the cells down, you will find LPS in the culture filtrate.

Lippincott: I was wondering, first of all, why you prefer the idea that the plant is removing the LPS, rather than that synthesis is just repressed.

Kijne: We haven't chosen between the two at this moment.

Lippincott: You seem to think that the very first polysaccharide coming off is an exopolysaccharide, mainly a glucan, after it is hydrolyzed.

Kijne: I don't know if it's an exopolysaccharide. It is indeed a part of the LPS, for in high purification steps it moves with LPS. Exopolysaccharides are mainly excreted in the culture medium and are easy to wash off.

Lippincott: Do you know anything about the nature of the mucilage in the infection thread?

Kijne: No, and I don't think anything is known about this.

Lippincott: It appears, in fact, to be kind of digested away from around the bacterium.

Kijne: It's not a digestion. I think this is merely a question of shrinking during the microscopic technique, because in freeze etching you won't see it. This has been worked out for elder nodules, too.

Albersheim: I agree with your main results, but you just made the comment that exopolysaccharides wash off easily from the cells, and we don't find that. We find that it binds very tightly and is hard to eliminate. We also find similar peaks as you do on separation according to size, but you can get rid of them again in high salt, which means that there is binding in relatively low salt. We have examined the same species of LPS, and three others, for their sugar compositions, and agree with your results. We have decided that we don't need a rough strain to find out what is in the 0-antigen portion, because the sugars which vary from strain to strain should be the species-specific sugars. We do find that every species we've looked at has at least one unique sugar that is not present in the other three. They do contain dideoxyhexoses. One that you left off from your chart, which is present in all of these, is a hexosamine; and they all have glucosamine in them.

Kijne: We only looked for amines present but not specified.

Albersheim: We also find roughly the same proportions of sugars, if I remember correctly, and the same ones you have reported; so our results agree with yours. We have not looked at the bacteroids.

Goodman: We've worked with Rhizobium japonicum, looking ultrastructurally at the disappearance of the apparent wall of the bacterium. Over a time course of investigation, that is from 10 to 27 days following the disappearance of the wall surrounding the infection thread and eventually even that surrounding the individual bacteria, I've seen what appears to be a degradation of the wall material that doesn't really conform to a shrinkage per se. Indeed, it appears to reflect the dissolution of wall material which then drifts away from its original position in vesicles and is eventually destroyed.

Kijne: I can confirm this. I did some electron microscopy on soybean root nodules too, and the situation at the tip of the infection thread is completely different from the situation in pea root nodules. There you see a very clean-cut disappearance of the cell wall material, while in the soybean root nodule there is a very fibrillar substance present.

Goodman: This material is fibrillar and also slightly granular, but it seems to be a dissolution rather than a shrinkage as such. Would you show us the electron micrograph where you had loss of the wall material. It was an enlargement and there was an amorphous material you didn't mention that was very interesting to me.

Kijne: You are interested in this material here, and you want to know what it is. This should be the mucilagenous content of the infection thread, and during the infection process this material looks the same after glutaraldehyde and osmium tetroxide fixation. It could be composed in part from cell wall degradation products, but that's only a guess.

Goodman: It seems quite different in texture from the material that surrounds the infection thread, which is more or less indistinguishable from plant wall material per se. This is reminiscent of some things that I have seen with Rhizobium japonicum. The cell wall begins to degrade or be dissolved by hydrolytic enzymes of some kind, and in a matter of 10 to 16 or 17 days the entire envelope of the bacterium is gone and all you have is that empty shell of host plasmalemma around the bacterium, and the bacterium inside with a space in between.

Kijne: This need not be only degradation, because this infection thread is growing into the cell, and it could merely be that the growth of the infection thread exceeds the ability of the plant to produce cell wall material against it. But, indeed, in soybean nodules there is a lot of fibrillar material that looks like degraded cellulose and things like that. However, that is certainly different from this grayish material in our pictures.

Hubbell: I have one short point pertinent to that slide. You will see more slides very similar to this on Thursday, but as long as the subject is up, this looks quite familiar to me. I'm not an electron microscopist, but Dr. Napoli, the senior author on the paper I'm presenting, is, and she sees essentially the same things in the Rhizobium trifolii clover association. Quite frequently, she sees the infection thread with bacteria directly surrounding the cell, and she attributes this to material being unstable in the electron beam in very thin sections. How thin were your sections?

Kijne: 500 Å.

Hubbell: O.K. She finds in sections ranging from 100 to about 1500 Å that one can get a clear area around the cell, and she sees essentially the same thing that you have there.

AN INTRODUCTION TO THE RESEARCH ON ELICITORS OF PHYTOALEXIN ACCUMULATION

Peter Albersheim, Department of Chemistry, University of Colorado, Boulder,
Colo. 80302, USA

Plants are resistant to the vast majority of the microorganisms with
which they come in contact. In many cases, the critical event which
determines whether the plant will be resistant or susceptible occurs after
penetration by the invading microorganism. Many plants have been shown to
produce phytoalexins, compounds which inhibit growth of microorganisms, in
response to invasion by a fungus, a bacterium, or a virus; phytoalexins are
produced whether or not the invading organism is a pathogen. The phytoalexins have been shown to accumulate in the plant in concentrations more
than sufficient to inhibit growth of microorganisms in vitro. Many plants
produce two or more different phytoalexins. A microorganism which stimulates
phytoalexin accumulation in a plant tissue appears to stimulate accumulation
of all of the phytoalexins known to be produced by that plant tissue, and
each phytoalexin is a compound which inhibits the growth of a wide spectrum
of microorganisms. It can be concluded that the ability to produce phytoalexins is a widespread mechanism by which plants attempt to defend themselves against microorganisms.
    The first clues to the molecular nature of compounds of microorganisms
which trigger the accumulation of phytoalexins came from studies of
pathogens grown in defined media. Cell-free filtrates from cultures of a
number of pathogens were demonstrated to elicit phytoalexin accumulation in
the hosts of the pathogens. The pathogen-synthesized molecules responsible
for stimulation of phytoalexins in their hosts have been termed "elicitors".
Several early attempts were made to purify and identify the molecules
responsible for triggering the phytoalexin reaction, but these efforts were
unsuccessful.
    One elicitor that has been purified was obtained from the mycelia of

Monilinia fructicola (Cruickshank & Perrin, 1968), a fungal pathogen of stone
fruit trees. The M. fructicola elicitor, called monilicolin A, is reported
to be a small peptide and is active on a tissue of the green bean, a non-host
of M. fructicola. It is doubtful that monilicolin A has a physiological role
in disease resistance for this elicitor has only been demonstrated to be
active on a single non-host of M. fructicola. Other non-host species which
are resistant to M. fructicola do not accumulate phytoalexin in response to
this peptide, and no host of this pathogen has been shown to be responsive
to the elicitor.

Phytoalexin accumulation in response to monilicolin A may actually
reflect a relatively non-specific disruption of normal plant metabolism. A
broad spectrum of reagents capable of inducing phytoalexin accumulation in
plants by such non-specific means include the salts of heavy metals,
inhibitors of RNA and protein synthesis, and compounds which intercalate
in DNA.

Purification of elicitors which stimulate phytoalexin accumulation in
the host plant of the pathogen from which the elicitor was obtained has
now been achieved. A glucan isolated from the extracellular culture filtrate and from the mycelial walls of the fungus Colletotrichum lindemuthianum stimulates cells from the host of the fungus, Phaseolus vulgaris, to
accumulate phytoalexins (Anderson-Prouty & Albersheim, 1975). A better
characterized system is the Phytophthora megasperma var. sojae (Pms)-soybean
system. The breakthrough which allowed the Pms elicitor to be characterized
was the development of three quantitative assays for the elicitor activity
(Ayers, Ebel, Finelli, Berger & Albersheim, 1976; Ebel, Ayers & Albersheim,
1976). These assays were used to purify the extracellular elicitor from
Pms cultures. The comparison of the composition and structure of this
elicitor to the polysaccharide components of the mycelial wall of Pms
provided the evidence which suggested that pathogen wall components are
elicitors. Subsequently, elicitor was released from purified mycelial walls
of Pms using a techniqie originally designed to remove the surface antigens
from another fungus, the yeast Saccharomyces cerevisiae.

The development of reliable assays for the Pms elicitor resulted in
several significant findings concerning the recognition of an elicitor by
plant cells. First, soybean seedlings respond to very small amounts
(hormone levels) of the Pms elicitor. Second, the reaction of soybean cells
of elicitor is quite discriminating with regard to the structure of the

elicitor. Third, the cotyledons, hypocotyls, and cell suspension cultures of soybeans give equivalent responses to Pms elicitor. Fourth, the rate of glyceollin accumulation in each of the three biological assays which utilize three different tissues of soybean seedlings is identical. Fifth, the stimulation of glyceollin accumulation is not linked to wounding of the soybean tissues, since soybean cell suspension cultures exhibit the same response as the tissues of soybean seedlings. Thus, it can be concluded that the ability of the Pms elicitor to stimulate the accumulation of glyceollin is a general characteristic of soybean cells.

The three races of Pms are distinguished by differing abilities to infect various soybean cultivars. A Pms-soybean combination which supports the growth of a specific race of Pms is termed a compatible interaction and, conversely, an interaction is incompatible if the growth of the pathogen is restricted. Keen has suggested (1975) that glyceollin accumulates more rapidly in an incompatible Pms-soybean interaction than in a compatible interaction, and that this accounts for the resistance of soybeans to incompatible races of Pms. According to Keen's theory, elicitors from incompatible races of Pms would be expected to stimulate glyceollin accu- mulation, whereas corresponding molecules isolated from compatible races of Pms should not stimulate glyceollin accumulation.

The experiments of our laboratory have shown that the elicitors of phytoalexin accumulation are not the specificity determining factors in the Pms-soybean system. The elicitor obtained from each of the three races of Pms purified in exactly the same manner, and the major structural features of the elicitors from the three races are identical (Ayers, Valent, Ebel & Albersheim, 1976). The activities of the elicitors purified from the three Pms races were carefully examined using the three separate assays, and all three assays gave the same result: the activities of the elicitors from different Pms races are identical. These findings demonstrate that the three races of Pms are equally effective at stimulating phytoalexin accumulation in resistant and susceptible host tissues.

The results of another type of experiment support our conclusion that elicitors are not responsible for race-specific resistance in the Pms- soybean system. The onset and the rate of glyceollin accumulation in seedlings inoculated with compatible mycelia is indistinguishable from the onset and rate of glyceollin accumulation in seedlings inoculated with either incompatible mycelia or purified elicitor (Ayers, Ebel, Valent &

Albersheim, 1976). The finding that the onset and rate of glyceollin accumulation in response to elicitor are indistinguishable from the onset and rate of glyceollin accumulation in response to live mycelia from both compatible and incompatible races of Pms indicates that the elicitor purified in this study is the molecule in the mycelia responsible for glyceollin accumulation. These results demonstrate that difference in rates of glyceollin accumulation in response to different races of Pms do not account for the resistance or susceptibility of various soybean cultivars to the Pms races.

Evidence has also been obtained that the elicitor isolated from the mycelial walls of C. lindemuthianum is, like the elicitor from Pms, not race specific (Anderson-Prouty & Albersheim, 1975). The C. lindemuthianum elicitors are structurally indistinguishable regardless of the races from which they were isolated. The elicitors from several C. lindemuthianum races have the same ability to stimulate phaseollin accumulation in both susceptible and resistant cultivars of their host, true beans.

The available evidence does indicate that elicitors have a critical role in plant disease resistance even though elicitors are not a determinant of race specificity. Part of this evidence is the observation that the elicitor isolated from Pms is capable of protecting soybean hypocotyls from infection by a compatible race of Pms if the elicitor is applied to the hypocotyls 9 hr prior to inoculation with the live pathogen (Ayers, Ebel, Valent & Albersheim, 1976; and unpublished results of this laboratory). The ability of the purified Pms elicitor to protect soybean tissue from a pathogen of that tissue is consistent with results of cross-protection experiments described by many laboratories. Inoculation of plant tissues with nonpathogens, heat-killed or attenuated pathogens, or viruses, 12 to 48 hours before inoculation of the same plant tissue with a compatible pathogen of that tissue protects the plant tissue from the pathogen. All of these protecting organisms have in common the ability to elicit phytoalexin accumulation in the host tissues. In conclusion, many results indicate that the ability of plant tissues to accumulate phytoalexins in response to microorganisms is a general mechanism by which plants resist microorganisms.

Pathogens grow successfully on plants in spite of a plant's ability to produce phytoalexins. Thus, pathogens must have evolved a mechanism of avoiding the potential toxic effects of their host's phytoalexins. There are several plausible mechanisms for such avoidance by successful pathogens. One such mechanism might be simply the ability of a compatible strain of a pathogen to grow away from the area in which the plant accumulates toxic levels of phytoalexin. This possibility seems likely as an explanation for the avoidance of the effects of glyceollin in soybean plants by compatible races of Pms (Ayers, Ebel, Valent & Albersheim, 1976). What this means is that incompatible races of Pms must somehow trigger a response from a plant which slows down the growth of the incompatible race. This slowing down of the growth of the incompatible race must result from a positive interaction between the products of a resistance gene of the host and an avirulence gene of the pathogen.

There are other mechanisms by which a pathogen might prevent a plant from stopping the growth of the pathogen by accumulation of phytoalexin. For example, a pathogen might kill the plant cells in the region of the pathogen before those cells are capable of synthesizing the enzymes necessary for synthesis of the phytoalexin. Still another mechanism by which a pathogen might prevent a plant from accumulating sufficient phytoalexins might simply be to repress synthesis of enzymes involved in phytoalexin synthesis or else to inhibit the enzymes once they are synthesized. A demonstrated mechanism by which some pathogens overcome phytoalexin inhibition is metabolizing the phytoalexins to less toxic or unstable compounds. An additional mechanism might be the production by the pathogen of proteins or other molecules which specifically inhibit the enzymes of the host which solubilize elicitors from the mycelial walls of the pathogen. Evidence suggestive of this type of mechanism has also been obtained.

One might propose still another mechanism by which pathogens could avoid phytoalexins: that the pathogen mutate its cell wall polysaccharides to a form which no longer elicits phytoalexin production; but if this could occur, then pathogens would have routinely been selected which lack elicitors. Plants are likely to have evolved pathogen detection systems by evolving receptors for molecules of pathogens which could not be structurally modified. Therefore, elicitors probably have structures which cannot be significantly altered without such a deleterious effect to the pathogen that the pathogen could not survive. The fact the the elicitors of C. lindemuthianum, Pms, and yeast are structural components of mycelial walls and are polymers present in a wide variety of fungi supports this hypothesis.

An important objective of the current research will be to determine

just how widespread is the response of plant tissues to elicitors in general
and to a single elicitor in particular. In our laboratory, elicitors have
been purified from two fungal pathogens of plants, P. megasperma var. sojae
and C. lindemuthianum, and from brewer's yeast (S. cerevisiae), which is not
a pathogen of plants. The elicitors from these three fungi are glucans,
with structures similar to the major, insoluble, structural component of the
mycelial walls of the fungus from which the elicitor was obtained. Closely
related fungi have the same major structural components in their mycelial
walls. In fact, a correlation has been established between chemical composition of the cell wall and major taxonomic groupings of fungi elaborated
on morphological criteria. Since the mycelial walls of related fungi are
so similar, and since plants are unlikely to possess sufficient genetic
information to code for a unique receptor for every pathogen or non-pathogen
with which it comes in contact, related fungi probably possess the same
elicitors.

Bacterial elicitors have not been characterized, although plants do
respond to bacteria by producing phytoalexins. Since almost all bacterial
pathogens of plants are gram negative, our work will focus on an attempt to
isolate elicitors from gram negative bacteria. The common structural components of the bacterial cell wall are being examined for elicitor activity.
Michael Hahn in our laboratory has obtained preliminary evidence that the
lipopolysaccharide component of the outer membrane of E. coli is an elicitor
of soybean phytoalexin.

The range of plant species responding to a particular elicitor molecule
is also being examined. The Pms elicitor stimulates phenylalanine ammonia-
lyase activity in cell-suspension cultures of such phylogenetically diverse
plants as soybean, parsley and sycamore (Ebel, Ayers & Albersheim, 1976).
The question of whether the elicitor-stimulated increase of phenylalanine
ammonia-lyase activity in parsley and sycamore cells is related to the
synthesis of phytoalexins has not yet been answered. Efforts are now under
way to determine whether the Pms and yeast elicitors stimulate phytoalexin
accumulation in potato, pea, pepper and true bean plants, all of which
accumulate well-characterized phytoalexins.

Another major question is the determination of the detailed molecular
structure of elicitors. Methylation analysis, optical rotation measurements, and degradation by a purified exo-β-1,3-glucanase have demonstrated
that the Pms elicitor is predominantly a β-1,3-glucan with branches at

carbon 6 of some of the glucosyl residues (Ayers, Valent, Ebel & Albersheim, 1976; and unpublished results). An enzyme-produced, highly branched fragment of the Pms glucan retains all of the activity present in the undigested elicitor.

A long-range goal of our studies of elicitors is the development of a new type of biological pesticide. The elicitors which we have studied are carbohydrate in nature and are likely to represent a class of ecologically relatively safe pesticides. The stimulation of phytoalexin accumulation in plants by elicitors could protect the plants against many potential pathogens and not just the pathogen from which the elicitor was obtained. In addition, since the response of plants to elicitors appears to be a general phenomenon, an elicitor pesticide developed for one plant could be equally effective on a wide range of plants.

Many problems will have to be confronted before elicitors are useful as pesticides. The elicitor will have to get into the plant, and this may require chemical modification of an active elicitor fragment or the synthesis of an analogue or derivative of a naturally occurring elicitor with the requisite transport characteristics. The effect of large-scale phytoalexin accumulation on the productivity and health of the plant will have to be investigated and the presence of elicitor or phytoalexin residues in potential food sources will have to be determined, and, if they are present, their harmful effects, if any, on man will have to be ascertained. However, the possibility of obtaining a class of ecologically safe, highly effective biochemical pesticides makes the effort extremely worthwhile.

This research was supported by the Energy Research & Development Administration E(11-1)-1426, the Herman Frasch Foundation, New York City, the National Science Foundation (BMS73-02208).

References

Anderson-Prouty, A. & Albersheim, P. (1975). Host-pathogen Interactions VIII. Isolation of a pathogen-synthesized fraction rich in glucan that elicits a defense response in the pathogen's host. Plant Physiology 56, 286-291.

Ayers, A., Ebel, J., Finelli, F., Berger, N. & Albersheim, P. (1976). Host-pathogen Interactions IX. Quantitative assays of elicitor

activity and characterization of the elicitor present in the extracellular medium of cultures of <u>Phytophthora megasperma</u> var. <u>sojae</u>.
Plant Physiology 57, 751-759.

Ayers, A., Ebel, J., Valent, B. & Albersheim, P. (1976). Host-pathogen
Interactions X. Fractionation and biological activity of an elicitor
isolated from the mycelial walls of <u>Phytophthora megasperma</u> var.
<u>sojae</u>. Plant Physiology 57, 760-765.

Ayers, A., Valent, B., Ebel, J. & Albersheim, P. (1976). Host-pathogen
Interactions XI. Composition and structure of wall-released elicitor
fractions. Plant Physiology 57, 766-774.

Cruickshank, I. & Perrin, D. (1968). The isolation and partial characterization of monilicolin A, a polypeptide with phaseollin-inducing
activity from <u>Monilinia fructicola</u>. Life Sciences 7, 449-458.

Ebel, J., Ayers, A. & Albersheim, P. (1976). Host-pathogen Interactions
XII. Response of suspension-cultured soybean cells to the elicitor
isolated from <u>Phytophthora megasperma</u> var. <u>sojae</u>, a fungal pathogen
of soybeans. Plant Physiology 57, 775-779.

Keen, N. (1975). Specific elicitors of plant phytoalexin production:
determinants of race specificity in pathogens? Science 187, 74-75.

STIMULATION OF HOST PLANT DEFENSES BY CELL WALL COMPONENTS OF A FUNGAL PATHOGEN

Arthur A. Ayers, Chemistry Department, Swedish Forest Products Research Laboratory, Box 5604, S-114 86 Stockholm, Sweden

Cultures of phytopathogenic fungi have been shown in many instances to contain molecules which stimulate the production of phytoalexins in host plant tissues. These stimulatory molecules, called elicitors, have been detected in cultures of Phytophthora megasperma var. sojae (Pms) by their ability to stimulate glyceollin accumulation in soybean tissues. The stimulation of glyceollin accumulation is the basis for three biological assays which were developed and characterized in this study of the Pms elicitors. Two of these assays utilize organs of soybean seedlings, the cotyledons and the hypocotyls, and the third employs soybean cell suspension cultures. Each of these plant materials responds similarly to Pms elicitors, indicating that the response is a general feature of soybean cells.

Elicitors are present in the extracellular medium and the mycelial walls of Pms cultures. The elicitors from Pms cultures were purified by ion-exchange, molecular seiving and affinity chromatography, and were characterized by structural analysis of the carbohydrate components. Elicitor activity was stable to extremes of pH, heat and pronase, but was sensitive to periodate and could be blocked by α-methyl mannoside. The Pms elicitors, which differ from each other in chemical properties, each possess a common carbohydrate constituent responsible for elicitor activity.

The elicitors present in the mycelial walls of Pms were extensively studied. The mycelial wall elicitors were isolated from three races of Pms which differ in their ability to infect various cultivars of soybean. The corresponding elicitors from each race were identical in their activity in each of the three elicitor assays. Structural analysis of the carbohydrate components of the elicitors from the three races also indicates that they are identical. Similarly, the extracellular elicitors, which are structurally related to the mycelial wall elicitors, are the same in the three races of Pms tested. There is, therefore, no indication that any of the elicitors of Pms are responsible for race-specific resistance. Moreover, elicitors are now believed to be common components of the cell walls of many microorganisms.

The role of elicitors in resistance to infection was investigated by comparing the rate of accumulation of glyceollin in soybean hypocotyls treated with elicitor or inoculated with either a compatible or an incompatible race of Pms. The rate of accumulation of glyceollin was identical in each case for the first 15 hr of incubation. These experiments indicate that elicitors are present and result in equivalent glyceollin accumulation during the initial stages of both compatible and incompatible infections. The large difference in glyceollin accumulation which is observed at later stages may be attributed to tissue damage subsequent to the interactions which determine the course of the infection.

The role of elicitors and glyceollin was further studied by simultaneous and sequential applications of elicitor and a compatible race of Pms to soybean hypocotyls. Simultaneous application resulted in a compatible infection. Elicitor applied 10 hr or more prior to inoculation, however, resulted in an incompatible interaction. The presence of elicitor is, therefore, sufficient to prevent the spread of a compatible race of Pms, but only if glyceollin reaches a toxic concentration before the Pms becomes established in the tissue.

It is proposed that race-specific resistance of soybeans to Pms is dependent upon a "specificity factor" which is detected in the incompatible but not in the compatible interaction. The result of this detection is believed to be a hypersensitive response which decreases the rate of growth of the incompatible race of Pms, limiting the invading fungus to the infection site where the accumulation of glyceollin subsequently kills it. In the compatible interaction only the elicitor is recognized and the invading fungus can penetrate new tissue before glyceollin accumulation in response to the elicitor becomes inhibitory.

DISCUSSION

Chairman: J. Paxton

Raa: You assume that the elicitor reaches the host cell membrane, but the elicitor may not reach it. Then you would have a compatible interaction.

Ayers: The reason why I have omitted this possibility is that the data demonstrate that we get glyceollin production with the same kinetics with elicitors or Pms. The plant is responding to the elicitors which are present. If the elicitors weren't detected by the hosts in the compatible case, then you wouldn't expect glyceollin to accumulate.

Raa: But this is in a semi-in-vitro system where you apply the elicitors to a cut surface.

Ayers: This is true, but we still get the same symptoms. The compatible and incompatible fungus, whole, live mycelia were applied.

Raa: But, onto injured tissue.

Ayers: Right, onto injured tissue. But we observe the same symptoms. In the compatible case the seedlings collapse and show all the symptoms. In the incompatible case the plant is essentially unaffected. It shows a little bit of stunting of growth; there may have been a certain amount of damage just in phytoalexin production.

Raa: Have you checked this with the electron microscope? Do the host cell walls become thicker in the compatible than incompatible interaction?

Ayers: We haven't done the studies on the ultrastructure which are required to demonstrate where the pathogen is and this, of course, is important if we are going to make a strong point.

Raa: We have seen in compatible interactions that the host cells respond by producing thicker cell walls than in the incompatible interaction. This might also be one of the factors which could slow down the growth of the pathogen.

Ayers: There may be different responses in different host-pathogen systems.

Ayers - Fungal elicitors

Paxton: There have been some electron microscopic studies of this interaction between Pms and the soybean cell and to my knowledge they didn't
show any particular differences in the cell walls of the resistant or
susceptible combination.

Ayers: This is a very general model and we are just saying that the
elicitors, by themselves, do not account for specificity. We need some
other factor. We think the hypersensitive interaction is a very good
candidate and the results of this interaction may give us some good grounds
for this idea.

Delmer: How can you answer the argument that in this case phytoalexins
don't play any role at all?

Ayers: This is difficult. I know that they can be effective in preincu-
bation experiments. This is a complex interaction.

THE CHEMISTRY AND BIOLOGICAL SPECIFICITY OF ELICITORS

Barbara Valent, Department of Chemistry, University of Colorado, Boulder, Colo. 80302, USA

Soybeans (Glycine max) respond to the presence of the fungal pathogen, Phytophthora megasperma var. sojae (Pms), the causal agent of root and stem rot in soybeans, by producing the antifungal phytoalexin, glyceollin. The molecules of pathogen origin, termed elicitors, which trigger glyceollin accumulation by soybean tissues, are polysaccharides found in the mycelial walls of Pms (Ayers, Ebel, Finelli, Berger & Albersheim, 1976; Ayers, Ebel, Valent & Albersheim, 1976; Ayers, Valent, Ebel & Albersheim, 1976; Ebel, Ayers & Albersheim, 1976). Soybean tissues also accumulate glyceollin in response to polysaccharides from two nonpathogens of soybeans, brewer's yeast, Saccharomyces cerevisiae, and another yeast, Hansenula wingei. Characterization of the elicitors from Pms and brewer's yeast indicates that the elicitors from these two unrelated microorganisms are similar, if not identical.

The elicitors from both Pms and brewer's yeast are glucans. The Pms elicitor is released from purified cell walls of the pathogen using a technique originally designed to remove the surface antigens from yeast (Raschke & Ballou, 1972). The yeast elicitor, on the other hand, has been purified from Bacto-Yeast Extract, a commercially available autolysate of brewer's yeast. The yeast elicitor is precipitated from a solution of yeast extract in 80% ethanol, but is soluble in 60% ethanol. Further purification of the yeast elicitor is accomplished by the same techniques used to purify the Pms elicitor. Mannans, which comprise 92% of the carbohydrate in the 80% ethanol precipitate of the yeast extract, do not possess elicitor activity. Further characterization of the yeast elicitor is in progress.

Convincing evidence has been obtained to demonstrate that the Pms elicitor is a polysaccharide (Ayers, Ebel, Finelli, Berger & Albersheim, 1976; Ayers, Ebel, Valent & Albersheim, 1976; Ayers, Valent, Ebel & Albersheim, 1976). Glycosyl linkage and sugar compositions, optical rotation measurements, and degradation by a purified exo-β-1,3-glucanase have demonstrated that the Pms elicitor is predominantly a β-1,3-glucan. The exo-β-1,3-glucanase used in our studies was purified from Euglena glacilis var. bacillaris by the procedure of Barras & Stone (1969). This enzyme, acting from the nonreducing end of the glucan chain, cleaves 75% of the glycosidic bonds in the elicitor and releases 90% of the sugar residues as monosaccharides, disaccharides and trisaccharides. The exoglucanase-treated elicitor fragment is highly branched. Over half of the glycosyl residues in the backbone of the elicitor fragment are substituted with glycosyl side-chains as compared to only one glycosyl residue out of every 8 backbone residues having a sidechain in the larger undigested elicitor. The highly branched glucan fragments remaining after exoglucanase treatment possess as much activity as the undigested elicitor. Periodate oxidation of the Pms elicitor followed by mild acid hydrolysis indicates that terminal sugar residues are involved in the active site of the elicitor.

Our efforts are now focused on finding or producing enzymatically the smallest elicitor-active molecule possible. The exoglucanase-treated Pms elicitor consists of around 50 glycosyl residues (molecular weight about 10,000). An even smaller elicitor (molecular weight about 5,000) has been obtained from the extracellular fluid of old Pms cultures. The molecular weight of the yeast elicitor varies from about 2,000 to 10,000. If enzymes could be found to further reduce the size of the elicitor, the determination of the detailed structure of the active site of the elicitor would be facilitated.

The finding that two unrelated fungi possess similar, if not identical, elicitors supports the hypothesis that plants respond to microorganisms by recognizing a few common elicitors. Experiments are under way to determine if other plants, in particular, potato, pea, pepper and true bean, respond to the Pms and yeast elicitors by accumulating their characteristic phyto-alexins.

References

Ayers, A., Ebel, J., Finelli, F., Berger, N. & Albersheim, P. (1976). Host-pathogen Interactions IX. Quantitative assays of elicitor activity and characterization of the elicitor present in the extracellular medium of cultures of Phytophthora megasperma var. sojae. Plant Physiology 57, 751-759.

Ayers, A., Ebel, J., Valent, B. & Albersheim, P. (1976). Host-pathogen Interactions X. Fractionation and biological activity of an elicitor isolated from the mycelial walls of Phytophthora megasperma var. sojae. Plant Physiology 57, 760-765.

Ayers, A., Valent, B., Ebel, J. & Albersheim, P. (1976). Host-pathogen Interactions XI. Composition and structure of wall-released elicitor fractions. Plant Physiology 57, 766-774.

Barras, D. & Stone, B.A. (1969). β-1,3-glucan hydrolases from Euglena gracilis. II: Purification and properties of the β-1,3-glucan-exohydrolase. Biochimica Biophysica Acta 191, 342-353.

Ebel, J., Ayers, A. & Albersheim, P. (1976). Host-pathogen Interactions XII. Response of suspension-cultured soybean cells to the elicitor isolated from Phytophthora megasperma var. sojae, a fungal pathogen of soybeans. Plant Physiology 57, 775-779.

Raschke, W.C. & Ballou, C.E. (1972). Characterization of a yeast mannan containing N-acetyl-D-glucosamine as an immunological determinant. Biochemistry 11, 3807-3816.

DISCUSSION

Chairman: J. Paxton

Mussell: You indicated that the elicitors from Pms are active against sycamore and parsley cells, and your measurement of this was increased PAL activity. Do you have any kind of a generalization about this increase in PAL?

Valent: The only experiments we have are the three examples I mentioned, soybeans, sycamore and parsley. They all three respond.

Hiruki: Do you have a control where you injure a plant mechanically and find out to what extent this glucan is produced, and also the amount of glucan induced by injury? Have you determined the structure of this glucan induced by injury in the plant?

Valent: This glucan is a structural component of the mycelial walls of pathogens.

Hiruki: You said that elicitor production is general, not species or even genus specific.

Valent: Yes.

Paxton: Glucans like this are produced by the plant upon injury. They don't appear to be involved in phytoalexin stimulation because injury does not provoke the plant to produce phytoalexins.

Valent: There is no glyceollin in the injured plants.

Hiruki: Can you isolate a glucan from an injured plant and then elicit phytoalexin production with it?

Valent: We tried several β-glucans from other sources, one of these, laminarin, has about 1/1000 the activity of our Pms elicitor. It's not very branched, and is predominately a β-1,3-linked linear molecule.

Albersheim: There is no evidence that callose is an elicitor. Sycamore cells and cells in culture do not have any β-1,3 glucans that we have been able to detect when we have looked at their walls. I don't know if there are other glucans in there, but they do not have phytoalexins so if they do have β-1,3 glucans, they could not be elicitors unless they did not get to the right site. They are not injured by the elicitors, physically, so there is no injury response and they do respond and accumulate phytoalexin.

Selvendran: You said that on removing the terminal 3-linked-glycosyl residue of the elicitor by periodate oxidation, borohydrate reduction and acid hydrolysis, the activity decreases by 50%. This is a very considerable decrease; however, the table showed that the elicitor has about 2% 4-linked glycosyl residues. I assume these are 1,4-linked. Would that account for the marked decrease in activity as well?

Valent: It's possible, but it's also possible that you need a certain size chain on the glucan backbone to have activity. I've gotten a certain amount of over-oxidation in these experiments. That could account for the reason I didn't recover all of my activity. It does vary somewhat from time to time when I do it, but it's always been around 50%.

Albersheim: One of the major points is that you stimulate glyceollin accumulation in the cells in culture, where you have no injury whatsoever. You are adding microgram amounts of the elicitor to the cells growing in 2% sucrose under sterile conditions and they respond.

Pegg: There is evidence for instantaneous 1-3 linked glucan accumulation in plants in the pollen-stigma incompatible reaction where callose is induced within ten minutes. How do you envision the release of these glucans from the pathogens into the plant?

Valent: The elicitors in the culture filtrate are indeed the same as the elicitors we get from the wall. We have found that in suspension culture soybean cells we could remove enzymes from the cell walls. These enzymes are able to release elicitor-active molecules from purified cell walls of the fungus. This is our main evidence right now. We have not characterized the enzymes.

Eriksson: Is it known whether phytoalexins are fungistatic or fungicidal? It would be very important to know the fungistatic or fungicidal concentrations of the phytoalexins both toward the pathogens and toward other microorganisms as well. It would be very important to know the concentration at which phytoalexins stop plant cell synthesis.

Albersheim: It's known that most of these are fungistatic or bacteriostatic and not fungicidal or bacteriocidal. The amount of these materials necessary to stop growth in vitro has been examined and in many systems it works at a very low concentration.

PHYTOPHTHORA INDUCERS OF SOYBEAN PHYTOALEXIN PRODUCTION

Jack D. Paxton, Department of Plant Pathology, University of Illinois, Urbana, Illinois 61801, USA

Frank & Paxton (1971) discovered an inducer of phytoalexin production in soybeans that is produced in culture filtrates of Phytophthora megasperma var. sojae. This lead to several attempts to identify this inducer (or elicitor) reported here and elsewhere by Albersheim, Ayres, and Valent, and by Keen (1975).

We produce the elicitor by growing Phytophthora megasperma var. sojae on filter sterilized plant juice. This represents, as closely as we can simulate, the conditions under which the fungus grows in the plant. Using these conditions, races 1, 2, 3, 4, 5, and 6 of P. megasperma var. sojae produce compounds which induce phytoalexin production in both normally resistant or susceptible soybean cotyledons. Similar inducer activity is obtained whether the fungus is grown on juice from the shoots or roots of 7-day-old plants. This age plant shows the maximum differential in reaction to the fungus.

Since no differential is shown in the reaction of soybean plants to these culture filtrates it is possible the reaction is caused by phytotoxic mycolaminarans (Keen et al., 1975); not unlike the response caused by heavy metal salts. However, the toxicity level of mycolaminarans on soybeans reported by Keen (1-2 mg/ml) is considerably less than the phytoalexin inducing activity reported by Ayers et al. (1976) of 0.01 µg/ml. We have confirmed the activity of Ayres inducer I on soybeans, using a sample supplied to us by Ayres, and the inactivity of laminarin and other β-1,3 linked glucans including paramylum at concentrations as high as 1 mg/ml.

Optical and electron microscopy studies on cotyledons treated with Phytophthora megasperma var. sojae culture filtrates or Ayres purified inducer I indicate that these solutions are apparently non-phytotoxic for

- 147 -

24 hours after treatment and during this time of virtually maximum phytoalexin production.

Inducer production during short time periods up to 12 hours after
inoculation of plant juice (the period of distinction between an incompatible or compatible reaction between the fungus and plant) is more pronounced on juice from a resistant plant. However, these studies have been
plagued by variable responses.

References

Ayres, A.F., Ebel, J., Finelli, F., Berger, N. & Albersheim, P. (1976).
    Host-pathogen Interactions IX. Quantitative assays of elicitor
    activity and characterization of the elicitor present in the extra-
    cellular medium of cultures of Phytophthora megasperma var. sojae.
    Plant Physiology 57, 751-759.
Frank, J.A. & Paxton, J.D. (1971). An inducer of soybean phytoalexin
    production and its role in the resistance of soybeans to Phytophthora
    rot. Phytopathology 61, 954-958.
Keen, N.T. (1975). Specific elicitors of plant phytoalexin production:
    determinants of race specificity in pathogens? Sciende 187, 74-75.
Keen, N.T., Wang, M.C., Bartnicki-Garcia, S. & Zentmyer, G.A. (1975).
    Phytotoxicity of mycolaminarans β-1,3 glucans from Phytophthora spp.
    Physiological Plant Pathology 7, 91-97.

DISCUSSION

Chairman: P. Albersheim

Kijne: In your pictures of treated cells, the contents of the endoplasmic
reticulum are black, which is certainly not normal. Are they black in
nontreated cells?

Paxton: There are black contents in nontreated cells. We believe phenolic
compounds accumulate and react with osmium tetraoxide leading to the deposition of these black materials. They appear to be compartmentalized in
the cytoplasm. As this point we do not know what they are or what their role
is.

Pegg: Dr. Albersheim suggested in his opening talk that there were phyto-
alexinases and that there was an enzyme basis for the breakdown of phytoalexins. Is there any evidence for this?

Paxton: In terms of the soybean system, no - in terms of other phytoalexin
systems, yes. We have some preliminary evidence that glyceollin can be degraded by Phytophthora megasperma var. sojae. In a recent article in Physiological Plant Pathology (8, 189-194, 1976), Bailey et al. indicate that you
have to be careful with in vitro assays because the fungus may initially not
grow, but ultimately start growing as soon as it can produce the enzymes that
are necessary to degrade the phytoalexins.

Albersheim: There are a number of publications in the literature by Hans
Van Etten of Cornell, Verna Higgins at Toronto, Burden & Bailey at Wye
College, also Francis Lyon at Dundee, that have shown that bacteria can grow
after sitting a while in the presence of phytoalexin, which initially
stopped them. All of these people have shown that phytoalexins are metabolized by some pathogens.

Brown: Microorganisms that produce antibiotics very often are not
susceptible to their own antibiotic because they set up permeability
barriers. With the phytoalexins, it is not enough that they are produced,
they also have to get into the pathogen. Perhaps a lot of the specificity
may lie in the development of permeability barriers.

Paxton: There is an inherent problem here because we have races, and as
far as we know the same phytoalexins are produced in the different cultivars.
So, you would have to invoke a mechanism in which in one cultivar, the
pathogen develops a permeability barrier to a particular phytoalexin, and
in another cultivar it does not develop a permeability barrier to the same
phytoalexin.

Brown: There are many things that will induce permeability barriers in
microorganisms that produce antibiotics and I could easily imagine the
situation in an interaction with the host which might cause differences in
permeability.

Albersheim: As far as I know, there is no evidence that there is specificity with regard to the action of the phytoalexins.

Bateman: Mode of action of phytoalexins really needs study. Van Etten demonstrated that phaseollin, a molecule very similar to glyceollin, readily lysed red blood cells immediately upon contact, also it causes a very rapid leakage in fungi. You can see this effect immediately if you treat the organism under the microscope. Organisms exposed to relatively low concentrations of phaseollin will lose up to 1/2 - 1/3 of their dry weight, but yet remain alive when transferred to a new medium free of the phytoalexin. In all respects except one, and I can't remember the exception, the physiological changes induced by the phytoalexins of this type are exactly the same as those induced by polyene antibiotics.

Paxton: Van Etten also showed that bean cells were susceptible to phaseollin. He has shown that in peas <u>Aphanomyces</u> induces very high levels of phytoalexin and yet the <u>Aphanomyces</u> keeps on growing and rots the pea. The mere fact that there are high levels of phytoalexin there obviously doesn't mean the fungus is inhibited in that tissue.

ELICITORS FROM FUNGAL WALLS TRIGGERING DEFENSE REACTIONS IN ANIMALS

Torgny Unestam, Institute of Physiological Botany, University of Uppsala, Sweden

Fungi are not common as parasites on vertebrates (Unestam, 1973). Fungal antigens in vertebrate immuno-systems are mostly very non-specific compared to bacterial and viral antigens and very often give cross reactions between different fungi in serological tests. Also, high antibody titer against a fungal parasite may be common in man also where the disease is not usually found (Hiloick-Smith et al., 1964). Apparently, fungal antigens (which are predominantly carbohydrates) are often genetically conservative fungal molecules. This seems to be true for most fungi. The high natural resistance in the vertebrate group against fungal infections may perhaps be partly explained this way.

In the different invertebrate groups fungal diseases are much more common (Unestam, 1973). Most fungi are, however, harmless also in these animals.

In arthropods, many harmless fungi are able to penetrate cuticle of different kinds. Therefore, active defence mechanisms are probably elicited upon hyphal penetration in vivo.

Crayfish phenoloxidases will be highly activated upon contact with the surface of purified walls of most fungi (Figure 1), but cell wall of other plants do not give the same response (Unestam & Beskow, 1976).

Small soluble fragments released from yeast, Saccharomyces cerevisiae, and other fungal walls specificly elicit this activation at extremely low concentrations (10^{-11} M or lower).

The yeast elicitor has been partly characterized. Its molecular weight varies between 10^4 and 10^6 dalton. Pre-treatment with a number of enzymes (Table 1) indicates that the eliciting effect is dependent upon β-1,3-glucan endgroups and possibly on protein in the molecule. A purified protein from

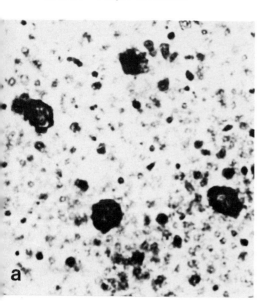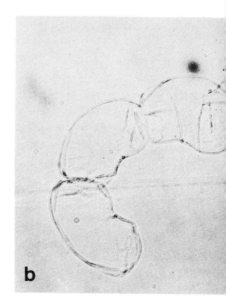

Figure 1. Phenol oxidase activity (dark zone) on the wall surface of
Saccharomyces walls (a) and lack of such activity on Haplopappus
walls (b). a: 1,300 x; b: 550 x.

Table 1. Effect of soluble Saccharomyces cell wall fragments and of
laminaran (β-1,3-; β-1,6-glucan) on crayfish phenol oxidase.

Pretreatment with	Activation of phenol oxidase	
	Saccharomyces	Laminaran
None	full	full
Papain	full	reduced - full
Pronase	full	double
Trypsin	reduced - full	small
Cellulase	full	full
Pronase + Cellulase	full	double
Endo-β-1,3-glucanase	full	full
Exo-β-1,3-glucanase	very small	very small

soybean also competitively inhibits the phenoloxidase activating capacity
of the elicitor. Surprisingly, endo-β-1,3-glucanase has no effect although

yeast wall glucans contain great amounts of β-1,3-glucan (Manners et al., 1973).

Purified β-1,3-glucan (from <u>Laminaria hyperborea</u>) has very similar, but not identical eliciting properties. This glucan and the yeast wall elicitor are similarly effected by enzyme pre-treatment (Table 1).

Recent work by Anderson-Prouty & Albersheim (1975) has characterized fungal wall fragments eliciting host response in bean tissues. The similarity with the fungal elicitor active in crayfish is striking and makes it tempting to speculate, that most fungi cannot avoid producing warning signals which are rather similar in different fungal groups and which are specificly recognized by plants as well as animals.

References

Anderson-Prouty, A.J. & Albersheim, P. (1975). Host-pathogen Interactions VIII. Isolation of a pathogen-synthesized fraction rich in glucan that elicits a defense response in the pathogen's host. Plant Physiology 56, 286-291.

Hiloick-Smith, G., Blank, H. & Karkany, I. (1964). Fungus Diseases and their Treatment. Churchill Ltd., London.

Manners, D.J., Masson, A.J. & Patterson, J.C. (1973). The structure of a β-(1-3)-D-glucan from yeast cell walls. Biochemical Journal 135, 19-30.

Unestam, T. (1973). Fungal diseases of Crustacea. Review of Medical and Veterinary Mycology 8, 1-20.

Unestam, T. & Beskow, S. (1976). Phenol oxidase in crayfish blood: Activation by and attachment on cells of other organisms. Journal of Invertebrate Pathology 27, 297-305.

DISCUSSION

Chairman: J. Paxton

Brown: Would you care to speculate what would happen if you used some of the black yeast, <u>Aureobasidium</u>, which presumably already has melanin on the outside of the wall?

Unestam: Do you mean the melanin might protect the cell walls from releasing elicitors?

Brown: There are some animal pathogens of which only the pigmentet black strains are pathogenic and the colorless ones are not pathogenic.

Unestam: Possibly the elicitor is not released or is in some way bound by the melanin and is not giving a response. I would like to have the resources to test all kinds of organisms to see whether they give that kind of effect.

Ballou: What appears to be the mechanism by which the presumed glucan activates the phenol oxidase?

Unestam: Compounds other than the elicitor will activate phenol oxidase, for example detergent in very high concentrations. We can also use proteases, so there is something going on in the molecules. Maybe the molecules change in one way or another to become active. It is very difficult to understand how these low concentrations can cause that effect. It cannot be nonspecific with these low concentrations. It is impossible to work with the purified enzyme because it looses its capacity to become activated. We have found that we have to treat the cell wall of live cells in some fungi, for instance by sonication, to get the response in crayfish. It might be that some cell walls are protected by something so that it doesn't release the elicitor.

GENERAL DISCUSSION ON FUNGAL ELICITORS

Chairman: J. Paxton

Goodman: I am a bit mystified by this specificity factor that is being postulated. I wonder whether or not we had ought to consider just a nutritional gap, the slowing down the pathogen may simply reflect the inability of the incompatible pathogen to grow because it doesn't have anything to grow on. In one case the compatible pathogen continues to grow, and the incompatible one is stopped. I believe we are going around Robin Hood's barn to come up with the possible answer as to why an incompatible pathogen doesn't really make it in the plant.

Paxton: Let me address your question. I think it demonstrates a beautiful case of serendipity. Dr. Gerdemann was aware of the fact that Phytophthora require thiamin for growth. His idea was that maybe the resistant plant lacked thiamin and the susceptible plant had enough thiamin and therefore the fungus grew in the susceptible plant and not in the resistant plant. In order to test this hypothesis he put strings through the stems of the plant (in what is now the famous wick technique) and found that indeed when he introduced thiamin into the plant we was able to make the plant susceptible. However, he ran a control and introduced water into the plant with his wicks and was able to induce susceptibility in a resistant plant. He looked at the string tips for what was coming out of the plant and found that material was coming out on these string tips that was inhibitory to the growth of the fungus. Later Dr. Chamberlain and I found that if you take these string tips, which have an inhibitory material on them, and put them back into the plants you can bestow resistance on what would have been a susceptible plant.

Goodman: I work with bacteria where we really haven't established the fact that inhibitors are produced in significant concentrations soon enough to suppress the growth of bacteria that give an incompatible response. In the inoculation of tobacco, with Pseudomonas pisi, which is an incompatible bacterium, there is reduced permeability and normal nonelectrolyte material does not leak from the plant. In the case of the compatible combination, let us say Pseudomonas tabaci, there is continued and increasing leakage of substrate to support the growth of the pathogen. These data suggest to me that there is more here than just an inhibitory substance that people would like to postulate. Substrate is terribly important, and substrate might be the limiting factor in the base of bacterial systems.

Kijne: Is there any evidence that these fungal elicitors enter the host cell, and if not, do they alter the permeability of the host cell plasma membrane?

Paxton: We haven't studied this and this is something that obviously needs to be studied.

Albersheim: We have been trying to study that. It is difficult because there are enzymes in the walls of the plants which modify the size of the elicitor and we want to look for the receptor. We haven't been successful.

Eriksson: I'm a little confused. It is claimed by your group, Peter, that most microorganisms you have tried can serve as elicitors, while Jack Paxton said that his group had found that Neurospora crassa and a Penicillium species did not serve as elicitors. I also would like to hear more about the similarity of extracellular enzymes, which are normally glycoproteins, to the cell walls and whether the carbohydrates on these glycoproteins can serve as elicitors.

Paxton: I am sure you would be able to find fungi that don't elicit phytoalexin production. We certainly did not extensively study the system we looked at. We simply picked out six or seven fungi to see if phytoalexin induction appeared to be uniform or if it was just specific to the fungus we were working with. It was by no means a study to see if all fungi had the inducers in their walls or not. Under different assay conditions, we may have detected elicitors in these fungi. Possibly if we had chosen a different set of assay conditions, we may have detected elicitors in these fungi. Possibly if we had chosen a different set of assay conditions, the elicitor activity we observed in these fungi may have gone undetected.

General discussion on fungal elicitors

Albersheim: Barbara Valent presented all our evidence and I would say that the question still is open. This field is new.

Eriksson: If any cell wall from any microorganism could serve as an elicitor, it would probably be easier to obtain suitable low molecular weight carbo- hydrates from Sporotrichum-pulverulentum, for instance. It's very easy to induce lytic enzymes that reduce the molecular weight of the polysaccharides in the cell wall of this fungus.

Albersheim: Glycoproteins secreted by fungi have structures which resemble some of the wall components, but we have not found any glycoproteins which are active elicitors. Glycoprotein secreted by Pms is not the elicitor component of the Pms.

Delmer: Peter, your group says that laminarin is not effective as an elicitor. In view of what Torgny Unestam has said, maybe you have been trying it at too high of a concentration? Have you tried this over a range of concentrations?

Albersheim: We have, but I don't remember what the exact concentrations were.

Paxton: I've tried it at tenth of a microgram per ml and find little activity up to 20 µg/ml. Laminarin is not very soluble. There is some activity there, but the elicitor activity disappears when you dilute the solution rather than increases.

Ballou: One of the properties of laminarin and such β-1,3-glucans is that they associate very strongly and if you go to dilute solutions, you really have to heat them up to dissociate the molecules and then keep it dilute so they don't reassociate. One of the exciting things about this work is that perhaps now people will try to determine the structures of fungal glucans. We shouldn't talk as if we have a molecule here, because essentially all that is known is that there are β-1,3 and β-1,6 linkages and even glucans with α-1,3 linkages in some of these fungi. Some of these carbohydrates are branched and some are linear. I think there will be a lot of stimulus to determine the structures. The chemical degradation you describe, periodate oxidation, reduction and then mild acid hydrolysis is the classic Smith degradation that removes terminal sugars. If you repeat that a second and

General discussion on fungal elicitors

a third time, do you still recover activity?

Valent: I do not know. I have not done it. I am working with very small amounts of material and after the first Smith degradation I just don't have much left.

Albersheim: Jürgen Ebel and I looked at the concentration of elicitors in the suspension system and their effect on accumulation of glyceollin. There is an optimum concentration for glyceollin accumulation and if you use too much (that is the elicitor and not the laminarin) you reduce the activity. We don't know if it involves association.

Raa: Do you imply that the elicitor mechanism also can account for resistance against saprophytes? You said all the fungi had the elicitor and the elicitor is unspecific, thus a saprophyte should also be able to induce phytoalexin production, if it could expost its cell wall to the plant.

Albersheim: That might be one reason why live plants are not attacked by saprophytes.

Raa: Are they saprophytes even without being excluded from the plant by the action of a phytoalexin?

Albersheim: No.

Paxton: I think the point is that saprophytes do indeed induce phytoalexin production and that's one of the reasons they can't attack the plant.

Raa: But, you said these elicitors were very firmly bound to the cell walls.

Paxton: No. They come off into the culture medium.

Raa: In nature, saprophytes don't penetrate the walls of host plant cells. You have to assume that these elicitors are released and pass through the wall and come in contact with the cell membrane and thus induce phytoalexin production. Why can't microorganisms which possess all the enzymes necessary to degrade the complete plant, attack a living plant?

Paxton: That's a very good question. Heat-treated plants temporarily become susceptible to other plant pathogens. Nonpathogens are still not able to attack, or at least visibly cause symptoms, on the heated plant.

General discussion on fungal elicitors

There appears to be a different mechanism which is preventing what we would consider true saprophytes from growing on the heated plant. We know that the heated plant's phytoalexin system has been damaged or even completely destroyed.

Raa: Albersheim said a healthy plant does not contain these phytoalexins.

Albersheim: Generally that's true. I can't say that is universal.

Raa: Then in spite of the fact that they are continuously exposed to saprophytes and thousands of different microorganisms, they are not reacting to them.

Albersheim: I don't think you can extend the argument that far. You are talking about one or two cells responding and the amount you have is very localized. If less than 1% of the cells respond, you just wouldn't detect it.

Bateman: Saprophytes and pathogens have distinct differences which I think we have to keep in mind. Olin Yoder and Bob Scheffer worked with Helminthosporium species which produced a host specific toxin. If they inoculated the plant with Neurospora, nothing happens. Neurospora does not form appresoria and does not form the penetration structure. However, if they inoculate the plant and supply the host specific toxin, then the plant would be degraded in the manner that the pathogen would degrade it. The pathogen has to have an infection structure, a mechanism of sealing itself to the plant surface and penetrate. Once beyond that point, there are other mechanisms which come into play, and these other mechanisms then, I think, may relate to the things we are discussing here. There is a distinct difference between saprophytes and plant pathogens in general with regard to their mode of ingress into plants. Unless the saprophyte actually is exposed to the plant cell surface, not much of a reaction will occur. I think this is a simple point we are overlooking in part.

Goodman: Give me the right to say that Pseudomonas fluorescens is a sapro-phyte and that 10^9 cells/ml is a significant number of bacterial cells. Pseudomonas fluorescens infiltrated at that concentration into tobacco, for example, does elicit a response from the plant. First, bacterial number increase stops. There are no apparent symptoms that one can ascribe to pathogenesis. Suppression of growth of microorganisms need not take a single form. With a number of different species of microorganisms, pathogens, there are systems of suppression that obviate the requirement of phytoalexins.

Albersheim: No argument!

Esquerré-Tugayé: Have you tried the elicitor on soybean protoplasts instead of a soybean suspension culture? Do you think that the plasmalemma is directly involved in the phytoalexin response?

Ayers: No, we haven't been dealing with cells which have had their cell walls removed. We don't believe, however, that a cell wall is required for activity, so we would expect that cells which have had their cell walls removed would still respond.

Esquerré-Tugayé: So you think that the receptor is on the plasmalemma?

Ayers: It's difficult to work with these molecules. There isn't evidence of a receptor in the cell membranes but we still believe that this is the most likely site.

Esquerré-Tugayé: What concentration of elicitor do you use in suspension culture? Are those concentrations as low as when you wound the seedlings?

Ayers: Jürgen Ebel worked with elicitor concentrations on the order of one microgram per ml. On cotyledons we can detect somewhere on the order of 1/10th this amount. They are within the same range.

Cooper: Before 1974 all phytoalexins were thought to come from necrotic cells, in other words, from cells resulting from a hypersensitive response. Obviously, now they don't. How does this fit in with the elicitor story? Obviously, you haven't got a hypersensitive response explaining this.

Paxton: My own feeling is that the hypersensitive response, to use Neville White's statement, is a monument to the battle rather than the reason the organism doesn't grow. Many of these organisms grow perfectly well on autoclaved plant material. So, the dead cell in and of itself certainly is not going to prevent the growth of many of these microorganisms. I don't think that a hypersensitive response, as reflected in the fact that you see dead cells, has very much to do with the fact that the plant is resistant.

Cooper: Yes, but often the dead cell was associated with disruption in redox, etc., so you have oxidation. Many of these phytoalexins were surely thought to be resulting from this release of oxidases of the phenol pathway.

Paxton: Burden & Bailey and other people have established that healthy cells can produce phytoalexin. Muller hypothesized phytoalexin production was part of the necrobiotic response and that has been shown to be incorrect.

Albersheim: There was no real evidence for that. It has been established now by Jack, Burden & Bailey, and by Mansfield that the live cells or those surrounding the wound are making the phytoalexins. But, the hypersensitive response can still be involved in determining race and varietal specificity. In fact, we believe it probably is, but there is no real evidence to establish that yet.

Cooper: Do your elicitors ever induce a hypersensitive response in treated plant material?

Albersheim: Pure elicitor apparently does not elicite a hypersensitive response.

Jarvis: There is some evidence from our group that the terpenoid phyto-alexins of potatoes are produced not by completely healthy tissue on its own, and especially not by necrotic cells, but by lieve cells which are in reasonable proximity, maybe four or five cell layers away from necrotic cell areas. Necrosis seems to be necessary for phytoalexin production. We don't have any definite evidence for or against elicitors in the system and it would be a little bit difficult to exclude it. But the fact that we can't show any phytoalexin production without necrosis has led us to wonder if the production of phytoalexins may be a response to the killing of neighbouring cells in a particular way, whatever that way happens to be.

Paxton: The phytoalexins themselves may be causing the cell death.

Albersheim: There is a group in Philadelphia that has a preparation of carbohydrate-like material from Phytophthora infestans which elicits pro-duction of all the phytoalexins in potatoes.

General discussion on fungal elicitors

<u>Delmer</u>: Where are the phytoalexins inside the cell? Are they leaking out of live cells? Is cell death required in order for them to be effective against the pathogen? Are they active when they are in living cells? Are they compartmentalized? Does the cell have to die or become leaky and might this, in fact, be the difference between resistance and susceptibility?

<u>Paxton</u>: The phytoalexins certainly leak out or we wouldn't be able to detect them in the droplets on these cotyledons. Whether they have to leak out in order to kill the fungus certainly hasn't been established.

<u>Albersheim</u>: In cell-suspension culture of soybeans you find about half the phytoalexin in the cells and half of it in the culture media. I don't think its location in the cell has been studied yet.

<u>Byrd</u>: Swinburne has shown that protease from a plant pathogenic fungus will elicit the formation of benzoic acid as a defense mechanism. Would you interpret this as perhaps reflecting the carbohydrate moiety in the pro-tease? He also claims protease from <u>Bacillus subtilis</u> will do exactly the same and will induce benzoic acid formation in Bramley seedling apples.

<u>Paxton</u>: I doubt that protease treatment of the cells is very subtle and I'm afraid that you might be approaching the situation you see when you dump mercuric chloride on the cells. It's obvious that wounding them mechanically will not induce phytoalexins but other types of injury will, and protease may be one of those other types of injury.

<u>Eriksson</u>: Could glyceollin serve as a substrate for phenol oxidases? Elicitors of the same kind may trigger the production of glyceollin. It might be that phenol oxidases are involved in this regulatory system. Are phytoalexins also polyphenolic compounds?

<u>Paxton</u>: Most of those that have been characterized are.

<u>Ayers</u>: We just haven't followed up on your suggestion of finding out whether they are substrates. I can't imagine that they aren't.

REGULATION OF SYNTHESIS OF CELL WALL-DEGRADING ENZYMES OF PLANT PATHOGENS

Richard M. Cooper, School of Biological Sciences, University of Bath, Bath, BA2 7AY, England

Degradation of host cell wall polysaccharides by extracellular enzymes of fungi and bacteria is a fundamental aspect of pathogenesis in a wide spectrum of plant diseases. Although the role of these enzymes may be often only secondary in terms of specificity or symptom induction, their initial or continued production may determine whether or not infection takes place. An understanding of the mechanisms regulating the synthesis of CWDE* is essential to our interpretation of pathogenesis, including specificity. Also such knowledge may be used to establish suitable culture conditions for selective enzyme biosynthesis, or for comparison of enzyme production by pathogenic or non-pathogenic mutants or isolates which in the past has often been carried out in unsuitable media.

This paper will deal with the two main controlling elements of polysaccharidase synthesis: induction and catabolite repression. The secretion and fate of enzymes in host tissue will be covered in an accompanying lecture by R.J.W. Byrde and S.A. Archer.

Much of our understanding of enzyme regulation is derived from studies on prokaryotes, in particular the lac operon of Escherichia coli; therefore some of the current concepts from this field and how they differ from gene regulation in eukaryotes, especially fungi, will be briefly outlined before considering the less clearly defined work on enzymes of plant pathogens.

Control of enzyme synthesis in prokaryotes and eukaryotes

Induction

Catabolic enzymes which attack exogenous substrates are, as a rule,

*cell wall-degrading enzymes

inducible by specific compounds structurally related to the enzyme's substrate or reaction products. Whereas specific repression is the rule for
anabolic enzymes involved in the synthesis of essential metabolites such as
amino acids and nucleotides (Richmond, 1968; Jacob & Monod, 1961).

Inducible enzymes are usually formed to a small extent by cells in the
absence of inducer when it is known as basal synthesis. Basal synthesis by
wild type E. coli results in up to 5 active molecules of β-galactosidase per
cell, whereas in the presence of galactoside inducers the level rises to
5000 molecules. This degree of inducibility, known as the induction ratio,
can vary from c. 1000 for β-galactosidase to < 10 for penicillin β-lactamases
of Gram-negative bacteria (Smith, 1963), which makes the distinction between
induced and constitutive synthesis marginal. Basal synthesis is not to be
confused with constitutive synthesis of inducible enzymes, which occurs in
certain mutants in the absence of inducers at rates similar to those of
induced wild type strains (Cohen-Bazire & Jolit, 1953). Basal, constitutive
and induced enzymes are identical as they are specified by the same structural gene.

The controlling elements of the lac operon are still being resolved
but have largely fitted into Jacob & Monod's (1961) original model. Since
then, the product of the regulatory gene, the lac repressor, has been
isolated as an allosteric tetrameric protein (MW 150,000), which can bind
to a specific site on the E. coli DNA (Gilbert & Müller-Hill, 1966, 1967;
Riggs, Suzuki & Bourgeois, 1970) and repress β-galactosidase transcription
in vitro (Zubay et al., 1967), but not in the presence of inducer which
binds at a second site on the repressor and liberates it from its complex
with DNA (de Crombrugghe et al., 1971; Gilbert & Müller-Hill, 1967; Oshima
et al., 1970).

The repressor acts at the operator site; mutations in this region
prevent binding of the repressor and allow constitutive synthesis from the
adjacent structural genes. Constitutive mutants can also arise from
mutations in the regulator gene.

Originally Jacob & Monod envisaged the operator as also 'setting the
rate' of expression of the structural genes. This role is now known to be
a function of a separate promoter region, which is the binding and initiation site for RNA polymerase (Ippen et al., 1968; Scaife & Beckwith, 1966).
Recently the promoter has been shown to contain a second site (Beckwith et
al., 1972; Dickson et al., 1975; Silverstone et al., 1970) which binds

cyclic adenosine 3', 5' monophosphate (cAMP) and its receptor protein (CRP) which are necessary factors for the transcription of inducible operons (de Crombrugghe et al., 1971; Emmer et al., 1970; Perlman et al., 1970; Zubay et al., 1970).

The structural genes for the 3 enzymes of the lac system are contiguous on the genome and the whole region may be transcribed as a single polycistronic mRNA molecule (Attardi et al., 1963).

In summary control of the lac operon is negative or repressive and induction is double negative or derepressive. However, there are exceptions to this, including the arabinose biosynthesis operon in E. coli which is under positive control (Sheppard & Englesberg, 1967).

Catabolite Repression

Synthesis of certain enzymes, particularly those of degradative metabolism, is reduced in the presence of glucose or other readily metabolized carbon sources. Known for many years as the 'glucose effect' its wider implications were discussed by Magasanik in 1961 when he coined the term catabolite repression (CR).

In effect, if the rate of catabolism exceeds that of anabolism then catabolites that accumulate intracellularly repress further synthesis of sensitive enzymes. This can apply to induced and constitutive synthesis.

In general carbon sources which support rapid growth, especially those which feed into the early reactions of glycolysis, are usually the best repressors. Studies with mutants, partly or wholly deficient in various catabolic pathways such as Kreb's cycle, suggest that the effector molecule lies at or below the level of fructose-6-phosphate and D-ribulose phosphate, and at or above pyruvate (Loomis & Magasanik, 1966; Paigen & Williams, 1970).

The extent of repression differs markedly depending on the enzyme, organism, and growth conditions, e.g. in E. coli, glucose represses expression of the lac operon by about 40% whereas the comparable figure for the structural gene for tryptophanase is about 95% (Peck et al., 1971). However, most inducible enzymes are subject to some degree of CR (Zubay et al., 1970).

The role of cAMP in transcription has already been stressed and many results suggest that CR may depend on a lowering of this key nucleotide. Pastan & Perlman (1970) demonstrated that addition of cAMP to cultures of a number of bacteria in which synthesis of various inducible enzymes was repressed by carbohydrates, largely overcame this repression. Similarly Makman & Sutherland (1965) found cAMP concentration in E. coli varied inversely with the concentration of glucose in the medium.

Sugars could influence cAMP levels by affecting the activity of adenyl cyclase (the enzyme responsible for its synthesis from ATP), cAMP phospho-diesterase (the enzyme responsible for its degradation) or the nucleotide's exit system from the cell. The latter mechanism is favoured in view of Makman & Sutherland's (1965) demonstration that sugars caused cells to excrete 99% of their internal cAMP into the medium, and results of Peter-kofsky & Gazdar (1971) showing that CR was not due to loss to cyclase or increase esterase acitivities. However, recent evidence has cast some doubt on whether CR can be fully explained by a decrease in cAMP levels (Haggerty & Schleif, 1975; Yudkin, 1976).

Cells can respond very rapidly to the presence of sugars, as in E. coli the initiation frequency of lac mRNA is reduced within 10 sec. of glucose addition (Haggerty & Schleif, 1975).

Sugars can also affect enzyme synthesis by inhibiting the entry of inducers. Although it is of minor significance for the lac operon, inhibi-tion of galactose entry contributes about half of the 90% repression exerted by glucose on the gal operon in E. coli (Adhya & Echols, 1966).

Gene regulation in eukaryotes

Although we now have a reasonable understanding of gene regulation in bacteria, knowledge of eukaryotes is relatively meagre because of the greatly increased size and structural complexity of the genetic apparatus (Tomkins et al., 1969). Certain features such as the nuclear membrane and DNA-protein association provide a possible basis for additional control (Johns, 1972; Marushige & Bonner, 1966); also much of the DNA comprises frequently repeated sequences which may fulfil a regulatory function (Britten & Kohne, 1968). Several models have been proposed for gene regulation in eukaryotes, including that of Britten & Davidson (1969) which is based on positive control, but details are beyond the scope of this paper.

Even in fungi the complexity of regulation has not provided a basis for a general theory (Burnett, 1975; Fincham & Day, 1971; Gross, 1969). Many of the systems examined do not conform to the bacterial model, although clusters of genes of related and possibly coordinated function have been found which are reminiscent of operons, such as the galactose utilization system in yeast (Douglas & Hawthorne, 1966). However, none so far have shown evidence of operators, and often several enzymes of related function are associated as aggregates rather than acting in sequence as in bacteria (Case & Giles, 1971; Creaser et al., 1967). Regulatory genes are common in fungi and their products are often diverse in their action by functioning as promoters or repressors of other genes, and in one case controlling enzyme activity by the formation of a complex with the enzyme (Messenguy & Wiame, 1969). That a considerable proportion of the eukaryotic genome may be responsible for determining regulation is illustrated by the methionine biosynthetic pathway in yeast, in which at least 15 regulatory genes are concerned with 7 structural genes (Cherest & Robichon-Szulmajster, 1973).

The kinetics of enzyme induction in fungi are often measured in hours rather than in minutes, which is probably a reflection of the increased cell size and genetic complexity.

Although CR occurs in fungi the mechanism of action may differ fundamentally from that of bacteria, e.g. glucose repression of α-glucosidase synthesis by Saccharomyces carlsbergensis acts at the level of translation not transcription (Wijk, 1968), and addition of cAMP did not affect protease synthesis of Neurospora crassa under repressed conditions (Drucker, 1972).

cAMP is thought to play a more complex role in eukaryotes, especially in animal cells where the nucleotide can mediate hormonal responses; it may effect derepression by activating specific protein kinases which modify DNA-histone interactions (Langan, 1968; Posternak, 1974; Rickenberg, 1974; Walsh et al., 1968).

Regulation of cell wall-degrading enzymes in vitro

In contrast to the critical studies on enzyme regulation in prokaryotes, much of the work reported on regulation of cell wall-degrading enzymes of plant pathogenic bacteria and fungi has involved long term trials in batch cultures under non-gratuitous conditions, in which effects of CR, growth rate, autolysis and changes in pH and medium composition have been ignored or insufficiently allowed for. Consequently interpretation of many results is difficult.

The criterion of inducibility is, generally, the detection of greater enzyme activity when a microorganism is grown on a polysaccharide substrate than when grown on a substrate such as glucose (e.g. Fuchs, 1965; Keen &

Table 1. Regulatory mechanisms of polysaccharide synthesis.

Enzyme	Organism	Regulation	Inducer	Reference
Polygalac-turonase	Verticillium albo-atrum (ex. cotton)	I.CR.b	Pga, Ga	Keen & Erwin (1971)
	Verticillium albo-atrum (ex. cotton)	C		Mussell & Strouse (1972)
	Verticillium albo-atrum (ex. lucerne)	C.CR		Heale & Gupta (1972)
	Verticillium albo-atrum (ex. potato and hop)	C		Talboys & Busch (1970)
	Pyrenochaeta terrestris	I.CR.b	Pga, Ga and analogues	Horton & Keen (1966)
	Penicillium chrysogenum	I	Pga, Ga and analogues	Phaff (1947)
	Sclerotinia fructigena isozymes PG I-III	I.CR.b	Pga, Ga	Archer (1973)
	PG IV	C		
Pectic transeliminase	Verticillium albo-atrum	I	Pga	Mussell & Strouse (1972)
	Verticillium albo-atrum	I & C	Pga	Talboys & Busch (1970)
	Erwinia carotovora	I.CR.b	Pga	Zucker & Hankin (1970)
	Erwinia carotovora &	C.CR		Moran & Starr (1969)
	Erwinia aeroideae	C.CR		
	Aeromonas liquefaciens	C.CR		Hsu & Vaughn (1969)
Pectin methylesterase	Penicillium chrysogenum	I	Pga, Ga and analogues	Phaff (1947)
	Sclerotinia fructigena	C		Cole (1956), Archer (1973)
	Fusarium oxysporum	C		Waggoner & Dimond (1955)
	Erwinia carotovora	I	Pga	de Herrera (1972)
Cellulase	Verticillium albo-atrum	I.CR	cellulose, cellobiose	Gupta & Heale (1971)
	Verticillium albo-atrum	I.CR	cellulose, cellobiose	Talboys (1958)

Enzyme	Organism	Regulation	Inducer	Reference
Cellulase	Pyrenochaeta terrestris	I.CR.b	cellulose, cellobiose	Horton & Keen (1966)
	Pyrenochaeta lycopersici	I.CR	cellulose, cellobiose	Goodenough & Maw (1975)
	Trichoderma viride	I.CR.b	cellobiose sophorose lactose	Mandels & Reese (1960, 1962), Nisizawa et al. (1971)
	Pseudomonas fluorescens	C.CR		Yamane et al. (1970)
	Myrothecium verrucaria	C.CR		Hulme & Stranks (1971)
	Cladosporium cucumerinum	C		Strider & Winstead (1961)
	Pythium aphanidermatum	C		Winstead & McCombs (1961)
Hemicellulases				
AF	Penicillium digitatum	C	Pga	Cole & Wood (1970)
AF	Corticium rolfsii	I	araban arabinose	Kaji & Yoshira (1969)
AF	Pathogenic and saprophytic fungi	See text		Fuchs et al. (1965)
AF	Sclerotinia fructigena	I.CR.b	Ga, ara-binose & analogues	Archer (1973)
Galactanase	Fusarium roseum	I	galactan	Mullen & Bateman (1975)
Xylanase	Fusarium roseum	I	xylan	Mullen & Bateman (1975)
Arabanase	Fusarium roseum	I	araban	Mullen & Bateman (1975)
Proteases	Colletotrichum lindemuthianum	I	cell walls collagen	Ries & Albers-heim (1973)
	Verticillium albo-atrum	I	proteins	Mussell & Strouse (1971)
	Monilinia fructicola	I	gelatin	Hall (1971)

I - inducible: C - constitutive; CR - repressible; b - basal synthesis

Pga - polygalacturonides; Ga - galacturonic acid; AF - α-arabinofuranosidase

Erwin, 1971). However, repressible constitutive enzymes would also appear
in higher yields on polysaccharides as the gradual rate of polymer breakdown
often fails to effect CR. Similarly when screening potential inducers (e.g.
monosaccharides) in batch cultures, enzyme synthesis often remains at a
basal level because inducing sugars also repress synthesis.

It is unlikely that polysaccharides per se are the inducers because
their molecular size would probably prevent entry into cells, and many are
inherently insoluble such as cellulose and xylan. Although the mono- or
oligosaccharides which result from polymer breakdown are the likely
effectors, this has been shown in comparatively few instances. It would
be tedious to review all the recorded cases of polysaccharidase synthesis,
but some of the more recent or significant work covering regulation of the
main groups of CWDE will be briefly discussed.

As previously mentioned most enzymes with exogenous substrates are
inducible (Jacob & Monod, 1961) and most inducible enzymes are subject to
CR (Zubay et al., 1970), therefore it is not surprising that polysaccharides
are also regulated by this dual control. However, constitutive synthesis
has been found in most groups although sometimes this has been confused with
basal synthesis when the assay method was sensitive (Table 1).

Pectic enzymes*

This group illustrates the great variation found in enzyme regulation
not only between but within species, e.g. some isolates of cotton strains
of Verticillium albo-atrum (Vaa) are induced and others are constitutive
for PG synthesis (Keen & Erwin, 1971; Mussell & Strouse, 1972); similarly
with PGTE production by strains of Erwinia carotovora (Moran & Starr, 1969;
Zucker & Hankin, 1970). However, control of PG of Pyrenochaeta terrestris
by induction-repression appears to be a species characteristic as it was
similar in 10 isolates from various locations (Horton & Keen, 1966).

Different pectic enzymes of a single isolate may be controlled by
different means, such as the constitutive PG and inducible PTE of some
cotton, hop and potato strains of Vaa (Mussell & Strouse, 1972; Talboys
& Busch, 1970). Even different forms of the same enzyme type may be under
separate control, as with PG isozymes of Sclerotinia fructigena (Archer

*Abbreviations: PG - polygalacturonase; PGTE - pectate trans-eliminase;
PTE - pectin trans-eliminase.

1973).

Most pectic enzymes from the 3 main groups (polygalacturonide hydrolases, trans-eliminases and pectin methylesterases) have been found to be
inducible, almost invariably by polygalacturonides and in some cases by the
monomer D-galacturonic acid or its structural derivatives dulcitol, mucic
acid, tartronic acid and L-galactonic acid, which suggests induction by a
common metabolite derived from these analogues (Horton & Keen, 1966; Phaff,
1947).

CR has been demonstrated usually by the addition of glucose to cultures,
but Horton & Keen (1966) demonstrated the non-specific nature of the phenomenon by the wide range of sugars and metabolites able to repress PG of
P. terrestris.

Cellulases

Apart from a few cases of constitutive synthesis, the majority of
cellulases are induced by forms of cellulose. Cellobiose, the end product
of cellulose degradation, is considered to be the natural inducer. However,
as with other low MW inducers, because of its ability to repress synthesis,
cellulase production may not occur until the disaccharide is virtually
depleted as in cultures of Vaa (Gupta & Heale, 1971). However, induction
of several fungal cellulases by cellobiose has been clearly demonstrated by
Mandels & Reese (1960) who restricted growth rate and CR either by imposing
mineral deficiencies, slow-feeding of cellobiose to cultures, or using
chemically modified inducers such as insoluble cellobiose octaacetate from
which the disaccharide was gradually released by esterase action.

Trichoderma viride is unusual in that it is induced by several other
disaccharides characterized by a β-glycosidic linkage, including lactose,
and especially sophorose. Although these are not normal products of cellulose degradation and may be considered unimportant in natural induction,
some cellulases (including that of T. viride) also show transglycosylating
activity whereby other disaccharides such as sophorose can be formed from
cellobiose (Mandels & Reese, 1960; Toda et al., 1968; Buston & Jabbar,
1954). This creates an interesting analogy with induction of E. coli β-
galactosidase by allolactose. This isomer of lactose has recently been
shown as the true inducer, and is formed from lactose by transferase
activity of β-galactosidase (Jobe & Bourgeois, 1972).

Hemicellulases

In spite of numerous reports on the ability of plant pathogens to produce hemicellulases there is a paucity of critical information on the control of their synthesis. Most information has come from Byrde's group on factors controlling AF production by S. fructigena and suggests a typical induction repression mechanism as with other polysaccharidases. To date, the constitutive AF production by P. digitatum appears to be exceptional (Cole & Wood, 1970).

A particularly interesting study was made by Fuchs, Jobson & Wouts (1965) on arabinanases of pathogenic and saprophytic fungi; only pathogens produced the enzyme on glucose, whereas both groups synthesized arabanase on groundnut meal. From the methods used no distinction can be made between induced or constitutive and nonrepressible or repressible synthesis, but the difference in mode of arabanase production by the two groups may be an important factor in pathogenesis. Especially as highest activities were found in fungal pathogens of legumes, which are known to contain a high proportion of arabinan.

Proteases

In view of the possible structural importance of protein in the cell wall (Lamport, 1965; Keegstra et al., 1973) some consideration should be given to proteases as CWDE.

The few reports all suggest that proteases are induced by proteins, and in the case of Colletotrichum lindemuthianum, by plant cell walls. However, regulation of synthesis may differ markedly from that of poly-saccharidases as -

(i) the inducible protease of C. lindemuthianum is not subject to CR, in fact production was optimal in the presence of 5% glucose (although protease of Vaa is repressible (Mussell, pers. comm.)).

(ii) specific induction by amino acids is unlikely in view of the hetero-geneity of monomer composition of proteins compared with polysaccharides, and the non-specific action of proteases compared with polysaccharides. In fact Mussell & Strouse (1971) found amino acids were ineffective in inducing a protease of V. albo-atrum.

Although a general picture of polysaccharidase regulation by induction/repression emerges from these numerous studies, some recent results have cast doubt on this generalisation. Thus, some workers have avoided CR by supplying potential inducers at low rates to cultures so that sugars never exceeded certain low levels; under these conditions several enzymes previously considered inducible were shown to be constitutive, e.g. Hsu & Vaughn (1969) provided a continuous but restricted supply of various sugars to Aeromonas liquefaciens which allowed constitutive synthesis of PGTE to c. 500 times higher than when sugars were in excess. Similarly synthesis of cellulases of Myrothecium verrucaria and P. fluorescens occurred in the absence of cellulose or cellobiose when CR was avoided (Hulme & Stranks, 1971; Yamane et al., 1970). Use of an inferior carbon/energy source for growth, such as glycerol, can have a similar effect as shown by Moran & Starr (1969) for constitutive PGTE production by Erwinia spp.

These conflicting observations prompted us to attempt a critical investigation into the control of synthesis of fungal CWDE (Cooper & Wood, 1973, 1975).

The vascular wilt pathogens of tomato, Vaa and F. oxysporum f.sp. lycopersici (Fol) were studied, as part of an existing programme on wilt diseases, and because both produce high levels of extracellular polysaccharidases and are easily handled in shake culture as bud cells and conidia.

Cultures were established of the same physiological age (as this can markedly affect enzyme production (Bull, 1972)), and were subjected to a period of starvation because of the persistence of endogenous catabolite repressors which can delay the response of microorganisms to added inducers (Kollar, 1958).

Many of the technical problems associated with supplyting potential inducers at low, constant rates, such as the use of pumps or elaborate culture apparatus, were ocercome by using diffusion capsules recently developed by Pirt (1971). These are simply filled with sugar solutions and added to shake cultures, into which sugars diffuse at linear rathes through a semi-permeable membrane; diffusion rates are determined by the internal sugar concentration and number of membranes in the capsules. This enables the manipulation of growth rates in culture.

Using this technique, sugars or uronic acids found in plant cell walls were supplied to cultures as potential inducers at constant rates close to those at which they were used by the fungi, so that sugar levels in culture fluids remained between 0-10 μg ml^{-1}. Under these conditions growth rates were between 1/8 and 1/30 of those in unrestricted cultures containing 1%

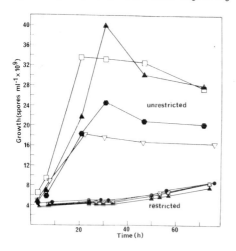 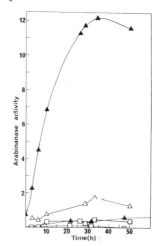

Figure 1(a). Growth of V. albo-atrum with carbon sources at initial concentrations of 1% (unrestricted cultures) or in restricted supply (restricted cultures to which sugars were supplied at 2-4 mg. 100 ml⁻¹h⁻¹ from diffusion capsules). Carbon sources were ($\triangledown$) glucose, ($\triangle$) galactose, ($\blacktriangle$) arabinose, ($\square$) xylose, ($\bullet$) galacturonic acid, ($\blacksquare$) cellobiose and ($\bullet$) arabinan.

Figure 1(b). Induction of arabinanase by cell wall sugars in restricted (---) and unrestricted (——) cultures of V. albo-atrum (after Cooper & Wood, 1975).

sugars (Figure 1a). Restricted feeding of sugars simulates the gradual release of monomers which results during growth on polysaccharides and more closely resembles conditions in vivo than batch cultures containing excess sugar. Mono- or disaccharides used in this way are gratuitous (sensu Benzer, 1953 and Monod & Cohn, 1952) in that they are not sensitive to the action of the enzyme they induces, and although they are metabolized as sole carbon source, inducer levels remain constant.

None of the various CWDE was produced above basal level on restricted supplies of glucose as would be expected if synthesis was constitutive with CR the main control mechanism. Synthesis of each polysaccharidase of both fungi was induced specifically by the monomer or dimer predominant in the enzymes specific polymeric substrate (Table 2).

Thus synthesis of pectic enzymes was induced by D-galacturonic acid, arabanase by L-arabinose, xylanase by D-xylose and cellulase by cellobiose.

An interesting exception was that for synthesis of galactanase and β-galactosidase, L-arabinose was a much more effective inducer than D-galactose;

Cooper - Synthesis of cell wall-degrading enzymes

Table 2. Induction of synthesis of polysaccharidases of Verticillium
albo-atrum by a restricted supply of cell-wall sugars.

Inducer	Enzyme activities*					
	Endo-PG	Endo-PTE	Arabina-nase	β-Galac-tosidase	Xylanase	Cellulase (Cx)
D-Galacturonic acid	100	100	15.8	0	16.5	0.30
L-Arabinose	0	0	100	100	3.5	0.25
D-Galactose	0	0	14.4	12.9	5.2	0.37
D-Xylose	0	0	3.7	0.26	100	0.08
D-Glucose	0-1.25	0	0-5.8	0.34	5.8	1.46
D-Mannose	0	0	0	0.34	0	0.20
L-Rhamnose	2.10	9.11	6.6	0	0	0.40
L-Fucose	1.38	0.16	0	0	0	0.25
Cellobiose	3.7	1.34	-	0	-	100

*Enzyme activities are expressed as a percentage of the maximum specific
activities attained in culture after identical periods of exposure to
different sugars.

Sugars were fed from diffusion capsules at linear rates of 2-4 mg per 100
ml per h to shake cultures of V. albo-atrum containing initially 4 x 10^9
spores ml^{-1}.

(after Cooper & Wood, 1973)

a degree of induction of arabanse synthesis by galactose also occurred.
The pentose L-arabinose is structurally identical to the hexose D-galactose
around carbons 1 to 4 which suggests that the inducer is a common metabolite. Similar examples of cross-induction are the greater synthesis of β-
galactosidase of Neurospora crassa with arabinose than with galactose as
inducer (Comp & Lester, 1971), and the induction of α-arabinofuranosidase of
S. fructigena with both L-arabinose and D-galacturonic acid which are
structurally identical around carbons 1 to 5 (Archer, 1973). However, L-
arabinose was ineffective in inducing pectic enzymes of S. fructigena, Vaa
and Fol.

The degree of specificity of polysaccharidase induction by cell wall
sugars was shown to be very high when the enzyme assay was specific (pectic
enzymes and cellulase) or where the substrate was pure (p-nitrophenyl-
glycosides). This is perhaps remarkable considering the many common inter-

mediates of carbohydrate metabolism, and in view of the structural similarities of many cell wall sugars, suggests that they are not usually metabolized to any extent before functioning as inducers.

Response to inducers was usually rapid; thus for Vaa after 3 hours xylanase and arabinanase activities had reached 1/3 and 1/5 respectively of the maximum specific activities detected throughout the growth periode of 36 hours (Figure 1b).

Only basal synthesis occurred on sugars other than the specific inducer and this synthesis was up to 7-fold greater in restricted than in unrestricted cultures, which suggests that basal synthesis is also subject to CR.

Results from this and other studies show that galacturonic acid is the common inducer for the 3 main types of pectic enzymes which can accumulate together under certain culture conditions. It would be of interest to ascertain whether the synthesis of a related group such as the pectic enzymes or cellulase complex are under co-ordinate control, i.e. controlled by a single regulatory gene (cf. Myers & Eberhart, 1966). This possibility is reinforced by recent observations of dual pectic enzyme acitivities associated with a single enzyme fraction even after extensive purification, e.g. PME activity remains in a homogeneous PG of B. cinerea and in a PGTE of Clostridium multifermentans (Urbanek et al., 1975; Miller & MacMillan, 1970); in these 2 examples dual activity was ascribed to enzyme aggregates which as previously mentioned are an indication of coordinated induction. Similarly Wang & Keen (1972) could not remove PGTE activity from a PG of Vaa, and we found 2 isozymes of PG and PTE of Vaa which retained almost identical activity profiles after wide and narrow range isoelectric focusing (Cooper, Rankin & Wood, 1976).

The advantage of certain complexes is obvious especially as the PME and PGTE of C. multifermentans exhibit similar means of substrate degradation, i.e. linear degradation from the reducing ends of pectic chains which might allow an alternation of de-esterification and cleavage of glycosidic bonds.

A more advanced system has been developed by Polyporus adjusta for degradation of wood polysaccharides, in which the induction of a group of enzymes - mannanase, xylanase and cellulase - appears to be under the control of a single regulatory gene (Eriksson & Goodell, 1974).

Even if coordinate induction of groups of enzymes does occur there are many exceptions, such as the different control mechanisms of PG and PTE of

Cooper - Synthesis of cell wall-degrading enzymes

some strains of Vaa and PME and PG of a Fol isolate (Table 1).

In summary our results show that with the exception of cellulase, monosaccharides are the true inducers of CWDE. In view of the ubiquity of glucose, both in the free state and as various polymers, the evolution of regulation of cellulase to a disaccharide inducer was a necessary step; similar amylase is induced by its corresponding glucose dimer, maltose (Gratzner et al., 1961). The only examples of polysaccharidases with larger inducers are those of Bacillus palustris which degrade the complex pneumococcal polysaccharides for which the inducer appears to be a tetrasaccharide or even higher (Torriani et al., 1962).

It follows from our studies with restricted cultures that when inducer levels are increased CR should operate when the supply exceeds the requirements of the microorganism. This was found to be the case with Vaa, as synthesis of both PG and PTE increased almost linearly with inducer supply up to certain low values after which synthesis was severely repressed: above 90% at only twice the optimum supply rate and complete repression at twice this rate (Figure 2a). Synthesis of arabanase and β-galactosidase followed similar patterns up to the same optimum rate of arabinose supply, but sensitivity to CR differed as synthesis then remained between 22-67% of the maximum.

Analysis of culture fluids showed that when repression started inducing sugar began to accumulate in cultures, indicating saturation of the fungal intra-cellular catabolite pool (Figure 2b).

Other observations are basically expressions of the same phenomenon, such as negative correlations between growth rate and enzyme production, correlation between the rates of catabolism of various carbohydrates and their abilities to repress enzyme synthesis, and for cellulases maximum levels attained in culture are inversely related to accessibility of cellulose substrates, i.e. the least accessible cellulases such an unmodified insoluble forms induce higher levels of cellulase than the readily degradable solubilised forms such as carboxymethylcellulose (Clarke et al., 1968; Okinaka & Dobrogosz, 1967; Horton & Keen, 1966; Mandels & Reese, 1960; Yamane et al., 1970).

These results illustrate the fine balance between induction and CR of CWDE synthesis in bacteria and fungi. CR of fungal polysaccharidases may be regulated by a similar mechanism to that previously described for bacteria, as addition of cAMP to cultures of Alternaria solani was shown by

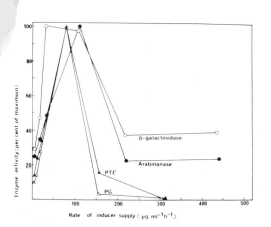

Figure 2(a). Induction and catabolite repression of synthesis of pectic enzymes and hemicellulases by V. albo-atrum in restricted cultures supplied with increasing rates of galacturonic acid and arabinose respectively. Enzyme activities were determined 6 h after addition of inducers.

Figure 2(b). Accumulation of inducing sugars in cultures of V. albo-atrum supplied with galacturonic acid (---) and arabinose (——) at the six different feed rates (1-6) shown in Figure 2(a) (adapted from Cooper & Wood, 1975).

Lukens & Sands (1973) to partially relieve PG and Cx synthesis from repression.

The contrasting sensitivities to CR of synthesis of pectic enzymes and hemicellulases of Vaa can be compared to the different degrees of repression exerted by glucose on certain E. coli operons (Peck et al., 1971). It has been proposed that this disparity results from different sensitivities of operons to cAMP, and may reflect the dissimilar functions of the enzymes (Lis & Schleif, 1973). Whether this has any bearing on the separate functions of endo-pectic enzymes which are primarily responsible for wall degradation, and exo-acting hemicellulases which probably fulfil a nutritional role, must remain speculative.

Regulation of synthesis of cell-degrading enzymes in vivo

Induction

In considering the part played by the host in regulating pathogen-synthesized enzymes, an indication of the central role of cell wall poly-saccharides can be obtained from studies with extracted plant cell walls. Thus recent evidence has shown that cell walls isolated from host plants are remarkably effective inducers of CWDE in cultures of fungal pathogens, especially when the relatively low proportions of individual sugars are considered, e.g. glc analysis of cell walls from tomato stems by Jones et al. (1972) showed a xylose content of < 7% and arabinose content of < 4%; however, we found activities of xylanase, arabanase and other enzymes of Vaa and Fol when grown on tomato cell walls were often of similar or greater levels than cultures containing specific inducing polysaccharides, in which amounts of monosaccharide inducers were 30-50 times greater (Cooper, 1974). The very high levels of PTE produced on cell walls are particularly interesting as this enzyme appears to be involved in the wilt syndrome.

Production of polysaccharidases by Vaa on host cell walls occurs in a sequential manner over several days (Figure 3); the first enzyme to be produced is PG followed by arabanase, xylanase and cellulase. This appears to be a general phenomenon as several other fungi including F. oxysporum, F. roseum, C. lindemuthianum and P. lycopersici synthesized CWDE in a similar sequence (Jones et al., 1972; Mullen & Bateman, 1975; English et al., 1971; Goodenough & Maw, 1976).

The sequential appearance of these enzymes presumably reflects the physicochemical susceptibility of cell wall polymers, resulting in induction as successive substrates become available during progressive degradation of the cell wall. This theory is reinforced by results of Albersheim's group showing that cell wall polysaccharides are not amenable to action of hemi-cellulases and cellulases of C. lindemuthianum and T. viride until poly-galacturonide of the cell wall has been degraded by endo-pectic enzymes (Karr & Albersheim, 1972; Keegstra et al., 1972; Talmadge et al., 1973). However, this is not the rule, as in contrast we found all polysaccharidases of Vaa were able to degrade tomato cell walls without the previous or simultaneous action of other enzymes (Cooper et al., 1976); similarly an AF isozyme of S. fructigena can degrade apple cell walls independently (Laborda et al., 1973), and a purified xylanase of T. pseudokoningii

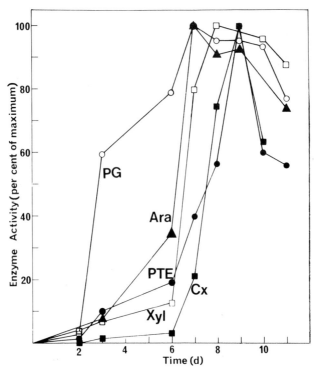

Figure 3. Production of polysaccharidases by V. albo-atrum grown on
tomato stem cell walls. PG - endopolygalacturonase, PTE -
endo-pectin trans-eliminase, Ara - arabinanase, Xyl -
xylanase, Cx - cellulase. Enzyme levels were similar on
walls from a resistant and susceptible cultivar (after
Cooper & Wood, 1975).

solubilizes xylose oligomers from cell walls of bean and corn (Bateman,
this symposium).

It seems likely that the order of polysaccharidase synthesis in
cultures containing cell walls also occurs in infected tissues, at least in
the absence of CR; analogously in vivo pectic enzymes are always the first
enzymes to be produced and cellulase the last, often only appearing when
tissues are becoming moribund. This sequence has been found in Pyrenochaeta-
infected onions, wheat infected with Gaeumannomyces graminis, leaf spots of
pea caused by Mycosphaerella pinodes and Ascochyta pinodella, and in Verticillium wilt of lucerne (Figure 4) (Horton & Keen, 1966b; Weste, 1970;
Heath & Wood, 1971; Heale & Gupta, 1972). Obviously the delay in cellulase

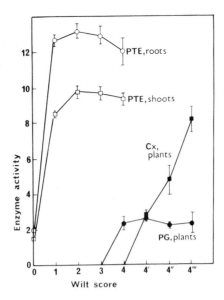

Figure 4. Levels of pectic and cellulolytic enzymes in relation to
wilting symptoms in lucerne infected with V. albo-atrum.
Wilt score: 0 - healthy; 4 - dead shoots; 4', 4", 4''' - as
4 but 1, 2 and 3 weeks later (after Heale & Gupta, 1972).

production may depend on factors other than the inaccessibility of cellulose
in the cell wall, e.g. Lisker et al. (1975) suggested that the delay in Cx
production as compared with PG by R. solani may be attributed partly to the
slower release of Cx although it was already formed and attached to cells.

In cell walls polysaccharides are insoluble and as such cannot induce
synthesis of enzymes. However, they probably provide inducers after slight
degradation by low levels of basal enzyme as has been demonstrated by Bull
(1972) for induction of a β-1,3-glucanase of Streptomyces.

In this connection, exo-acting enzymes would be far more effective in
releasing inducer than endo-enzymes. Some polysaccharidases such as
arabanases are invariably exo-acting, others may be produced as mixtures
of random and terminal types, such as PG complexes of S. fructigena,
S. rolfsii and Vaa, and PGTE of E. chrysanthemi (Archer, 1973; Bateman,
1972; Mussell & Strouse, 1972; Garibaldi & Bateman, 1971). Although it is
little appreciated some of the so-called 'endo'-pectic enzymes (as determined
by their ability to rapidly reduce substrate viscosity) appear to act in a

dual fashion by initially breaking the chain at random then progressing along the polymer releasing low MW residues. PG's of Vaa and C. lindemuth-ianum act in this way, and possibly those of B. cinerea and S. rolfsii (Cooper et al., 1976; English et al., 1972; Urbanek et al., 1975; Bateman, 1972). Such combined action would appear to benefit these pathogens as random cleavage effects extensive wall degradation and simultaneously releases low MW sugars for fungal nutrition and further enzyme induction.

Once sufficient inducer has been released by basal enzyme to initiate induced synthesis, the process should become autocatalytic. The significance of very low concentrations of inducers in these initial stages of polymer degradation is emphasized by Gupta & Heale's (1971) demonstration that cellulase synthesis by Vaa in liquid culture only occurs when hyphae are in physical contact with the insoluble substrate and not when separated by a membrane filter. In the transpiration stream of the host, localization of enzyme activity and inducers may occur within extracellular material which attaches hyphae of Vaa to xylem vessel walls (Figure 5a). It is interesting that Schmid (1966) found extracellular polysaccharides are a regular feature of wood-degrading fungi in which host cell wall degradation provides the main source of nutrition. Alternatively hyphae may form an intimate asso-ciation with the cell wall (Figure 5b).

I have already mentioned the possible importance of basal enzyme synthesis in pathogenesis as implied by the production of arabinanases by pathogenic but not saprophytic fungi in the absence of inducers (Fuchs' et al., 1965); the significance of basal activity is also suggested by Zucker & Hankin's (1970) comparison of E. carotovora, a pathogen of potatoes, and P. fluorescens, a common soil saprophyte; although both organisms produced an inducible PGTE, only the pathogen synthesized basal enzyme. Also the saprophyte showed long lag periods of many generations before enzyme induction when transferred to inducing media from glucose media or from natural sources; in contrast E. carotovora showed no such lag in induction. These results may explain some earlier significant work of Lapwood (1957) who compared the abilities of 2 pseudomonads and a Flavobacterium sp. with E. aroideae to rot potatoes and produce pectic enzymes in culture. Although they all produced pectic enzymes of similar activity in culture, the unique ability of Erwinia to rot potatoes under normal conditions was ascribed to a much more rapid rate of pectic enzyme production which enabled its establishment before suberization prevented further attack.

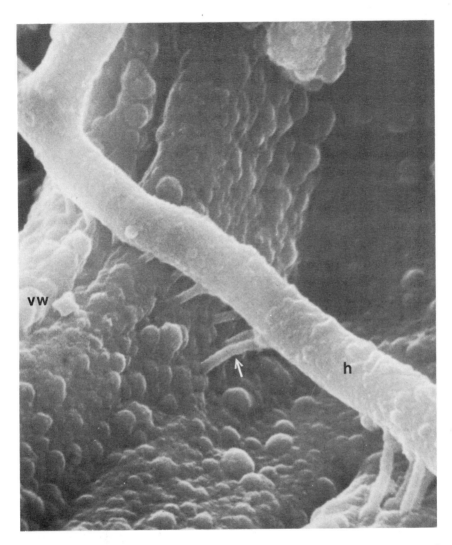

Figure 5 (a).   Attachment of <u>V. albo-atrum</u> hypha (h) to tomato xylem vessel
wall (VW) by extracellular material (→).   X 7500 (after
Cooper & Wood, 1976).

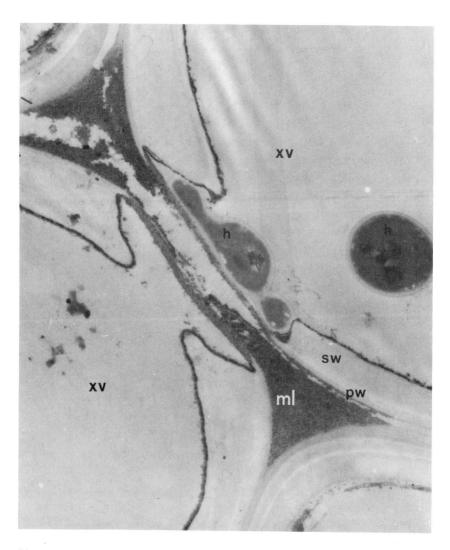

Figure 5 (b). V. albo-atrum hypha (h) closely adpressed to the membrane of a bordered pit pair between two xylem vessels (XV) of tomato. Degradation of the middle lamella (ml) has extended back to the intercellular zones; primary (pw) and secondary wall layers (SW) appear intact. X 12130.

Thus speed of induction, which may depend on basal synthesis, is of obvious advantage in pathogenesis enabling cell wall penetration before host defence mechanisms become effective. In this context cultures of Xanthomonas malvacearum and F. oxysporum cubense pre-adapted to pectic substrates were more pathogenic- to cotton and banana respectively than non-adapted cultures (Abo-El-Dahab, 1964; Deese & Stahmann, 1962). Although there is little critical information on the speed of polysaccharidase induction in other diseases, both appearance and action of enzymes can be rapid, e.g. histological and histochemical evidence showed degradation of pectin and cellulose in cotton cell walls by R. solani only 12 hours after host-pathogen contact (Weinhold & Motta, 1973); this substantiates analyses of bean cell walls by Bateman (1970) which indicated a 40% reduction in galacturonic acid content 24 hours after infection by R. solani. Similarly, during invasion of tomato roots Vaa causes significant cell wall degradation within 24 hours of inoculation (Bishop, C.D. & Cooper, unpublished results).

In view of these and other observations it would be of considerable interest to compare the pathogenicity of a range of mutants of a parasitic sp. differeing with respect to regulation of one or more CWDE's. In addition to the usual comparison of mutants, productive or deficient for enzyme synthesis, it is suggested that attention is also paid to inducibility, repressibility and basal synthesis.

It should be mentioned here that polysaccharidases produced by some pathogens in vivo can differ qualitatively from those produced in vitro. These include pectic enzymes of R. solani in bean, Botrytis spp. in onion, P. expansum and S. fructigena in apple and Hypomyces solani in squash (Bateman, 1963, 1967; Hancock et al., 1964; Cole & Wood, 1961; Byrde & Fielding, 1968; Hancock, 1976). Differences include variation in enzyme types and in isozymes of single enzymes. A common occurrence is the differential production of a PG in culture and a PTE in the host or vice versa. This failure of the 2 enzymes to appear simultaneously as would be expected with galacturonic acid as common inducer, is because PTE's of several fungal pathogens only accumulate above pH 7 whereas PG's are preferentially produced under acidic conditions (Bateman, 1966; Cooper & Wood, 1975; Hancock, 1965, 1968; Mullen & Bateman, 1975). This pH effect may influence some aspect of enzyme production (synthesis? secretion?), as differential inactivation of PG and PTE of Vaa and Fol at alkaline pH does not completely explain the phenomenon - stabilities of both enzymes are maximal around pH 5.

Other disparities may be attributable to several causes including insufficient extraction procedures from infected tissue, alteration of enzyme properties on purification (Harman & Corden, 1972; Swinburne & Corden, 1969), and inactivation by host constituents (cf. Byrde & Archer, this symposium).

Such variations are probably not due to differential induction in vivo and in vitro, as we found CWDE of Vaa appear identical whether induced on monosaccharides, polysaccharides or extracted cell walls. Also several recent investigations into various diseases have revealed identical forms of CWDE in culture and in infected tissue, which again suggests a common induction mechanism (Bateman, 1972; Byrde et al., 1971; Ferraris et al., 1974; Heale & Gupta, 1972; Laborda et al., 1974; Mullen & Bateman, 1975; Russel, 1975; Wang & Pinckard, 1971).

Catabolite Repression

The question of whether natural host sugar levels can affect disease development by repression of pathogen-synthesized enzymes remains largely unanswered.

Most CWDE are significantly repressed in culture by the addition of sugar levels to c. 10-50 mM (Table 3). The most notable exception is the extreme sensitivity to repression of cellulase synthesis by P. terrestris; this observation has led to the general concept that cellulases are only produced late in the disease cycle when host sugar levels are virtually exhausted. However, this may not be the case as other cellulases including that of the related P. lycopersici are repressed by similar sugar levels to other polysaccharidases.

It could be considered that if sugar levels in host tissues exceed that required to repress enzyme synthesis in vitro then infection could be prevented or delayed. However, detection of free sugar levels in plant tissues and extrapolation to in vitro studies is of limited value as such figures would only be applicable to diseases involving complete protoplast disruption, as in the maceration of parenchyma by pectic enzymes. In addition, intermediary metabolites and amino acids in plant cytoplasm should be considered as they can also effect CR (Nisizawa et al., 1971, Paigen & Williams, 1970).

Workers have studied the role of CR in vivo along several different

Cooper - Synthesis of cell wall-degrading enzymes

Table 3. Sensitivity of polysaccharidase synthesis to catabolite repression.

Enzyme	Organism	Level of carbon source to effect repression*	Effector	Reference
Cellulase	Trichoderma viride	0.01M	glucose, glycerol	Nisizawa et al. (1971)
Cellulase	Pyrenochaeta lycopersici	0.01M	glucose	Goodenough & Maw (1975)
Cellulase	Pyrenochaeta terrestris	>0.0005M	glucose	Horton & Keen (1966a)
PG	Pyrenochaeta terrestris	0.05M	many carbon compounds	Horton & Keen (1966b)
PG	Fusarium oxysporum f.sp. lycopersici	0.05M	glucose	Patil & Dimond (1968)
PG & PTE	Verticillium albo-atrum	$\geq$0.05M	glucose, galacturonic acid	Cooper (1974)
PG	Verticillium albo-atrum	0.1M	glucose	Keen & Erwin (1971)
PG & AF	Sclerotinia fructigena	0.01M	glucose	Archer (1978)

*to c. basal synthesis

lines. Thus Horton & Keen (1967) examined the effect of altering host sugar
levels on pathogenesis and enzyme production; they found that severity of
onion pink root rot and activities of PG and Cx in infected tissue were
inversely correlated with amount of sugar in host tissue, which were
increased either by spraying with glucose or maleic hydrazide, or decreased
by removing cotyledons. Natural sugar levels in onion roots were below
0.1%, i.e. sufficient to repress synthesis of Cx but not PG, which was
reflected by the sequential appearance of these two enzymes. Similarly
Patil & Dimond (1968) decreased PG synthesis and symptom severity in
Fusarium wilt of tomato by supplying glucose into infected vascular elements.
However, this is an academic point because natural conditions in xylem
vessels of tomato (and probably other species) should be favourable for
polysaccharidase production, as sugar levels range from zero (Dixon & Pegg,
1972) to only 12 ppm (Blackhurst, 1961) and total amino acids from 0.5 to
1 nM (Dixon & Pegg, 1972). Although such levels are too low to effect
repression, it should be noted that CR can be more severe when essential

nutrients such N, S and PO$_4$ are limiting, which could be the case in dilute vascular fluids (Clark & Marr, 1964).

Other workers have made a qualitative and quantitative determination of free sugars in host tissue and then tested their ability to repress enzyme synthesis of the pathogen in culture. Goodenough & Maw (1976) found that the major sugars in tomato roots were glucose (1.2%), fructose (1%) and inositol (0.7%); when added at these levels to cultures of the root pathogen P. terrestris growing on cell walls, synthesis of CWDE was repressed for 4 days. Prima facie this could affect the rate of root invasion, which in fact in this disease is comparatively slow. However, the tolerance of a Lycopersicon hirsutum var was not related to the concentration of these soluble sugars as levels in roots were lower than in a susceptible L. esculentum var.

In contrast Bugbee (1972) found a degree of negative correlation between sucrose levels of beet (Beta vulgaris) (which ranged from 6-18% fresh weight) and susceptibility to the storage rot pathogen Phoma betae, although some of the results were contradictory. Sucrose concentrations varied with cultivar, age and as a result of defoliation; it was suggested that the effect was due to repression of an endo-PGTE which is involved in the disease.

These findings bear on the hypothesis of Horsfall & Dimond (1957) on high and low sugar diseases; some of the diseases they described are associated with low sugar levels in host tissues such as Helminthosporium diseases of cereals, and Dutch elm disease, and may require the participation of CWDE (Bateman, Jones & Yoder, 1973; MacDonald et al., 1970), whereas certain diseases linked with high sugar levels such as rusts and powdery mildews do not appear to degrade host cell walls to a significant extent. However, some doubt is cast as to the expression of this phenomenon via CR, as Fusarium and Verticillium wilts were placed by them into different categories and these genera appear to have similar mechanisms of parasitism. Also the theory conflicts with that of Grainger's (1968) host C_p/R_s ratio which implies that plants are most susceptible to all types of infection when the ratio of carbohydrates to residual matter is high.

Role of polysaccharidase regulation in disease resistance/specificity

Factors which affect the susceptibility/resistance of cell wall polysaccharides to enzymic degradation can also influence regulation of inducible enzymes, as basal activity will be equally affected, and sustained enzyme production relies on a continuous release of inducer from the cell wall.

It is now established that both polysaccharidase induction by cell wall sugars and action on polysaccharides are of a high order of specificity (Cooper & Wood, 1973, 1975; Reese et al., 1968). This opens up the possibility of involvement of cell wall polysaccharides in specific resistance by influencing one or both of these interdependent phenomena as originally hypothesized by Albersheim, Jones & English (1969). However, there is now a considerable body of evidence against such a role in specific interactions between host and parasite, i.e. in those diseases which exhibit a gene-for-gene relationship, or even where the pathogen does not interact differentially with resistant host cultivars, such as Verticillium wilt of tomato.

Some of this opposing evidence comes from recent advances in analysis of plant cell walls by Albersheim's group; this has shown that polysaccharides of primary cell walls from different species are similar in terms of monosaccharide composition and glycosidic linkages (Albersheim et al., 1973; Burke et al., 1974; McNab et al., 1967; Wilder et al., 1973). It follows that cell walls from disease, resistant and susceptible cultivars are almost identical in this respect as shown for vars. of bean (P. vulgaris), soybean, corn (Z. mays) and tomato by Nevins et al. (1967, 1968) and Jones et al. (1972).

These similarities in structure have been confirmed by the ability of cell walls of different cvs. to induce similar levels of polysaccharidases when included in cultures as sole carbon source. Thus in cultures of Vaa containing cell walls from tomato cvs. with susceptibility or monogenic resistance to wilt, we found activities of PG, PTE and Cx were almost identical throughout the growth period (Cooper & Wood, 1975). Other workers (Jones et al., 1972; Langcake & Drysdale, 1975) have obtained similar results with pectic enzymes of Fol. Using the same technique English et al. (1969, 1971) could not distinguish races of C. lindemuthianum by their abilities to produce several glycosidases on cell walls from resistant or susceptible bean vars. Disease tolerance also appears to depend on factors other than cell wall polysaccharides as P. terrestris synthesized similar amounts of polysaccharidases on walls from a tolerant and susceptible tomato var. (Goodenough et al., 1975).

These observations suggest that differential induction by cell wall polymers is not responsible for the presence of higher enzyme levels in infected tissue of susceptible vars. than in resistant vars.; such as the demonstration by Deese & Stahmann (1962) and Mussell & Green (1970) of greater pectic enzyme production by F. oxysporum in susceptible cvs. of banana and tomato respectively. However, this could be determined by other interactions between CWDE and host cell walls as described by Mussell in this symposium.

An alternative and probably more accurate comparison of cell wall polysaccharides in different vars. can be obtained by controlled enzymic degradation of extracted cell walls. Using 8 tomato cvs. we found no correlation between their susceptibility or resistance to Verticillium or Fusarium wilt and the susceptibility of their cell walls to degradation by (i) pectic enzymes, (ii) a range of CWDE; living tissue disks from stems were also macerated at similar rates by pectic enzymes (Figure 6) (Cooper et al., 1976).

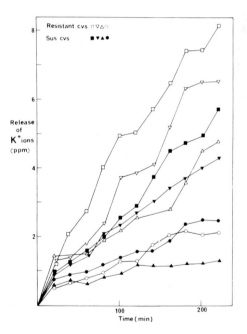

Figure 6. Maceration* of tissue from resistant and susceptible tomato cultivars by F. oxysporum pectin trans-eliminase.

*Assessed by determination of K$^+$ion release as an indicator of protoplast disruption.

Studies based on isolated cell walls are open to criticism because of inevitable physicochemical changes incurred during extraction. However, an unique study of the regulatory role of the host cell wall in situ is made possible by the unusual parasitic habit of vascular wilt fungi; as they exist in contact with cell walls in the relatively inert xylem vessels, interference from reactions of host cytoplasm should be minimal in the early stages of infection. We exploited this situation by infiltrating spores of Vaa into vascular tissue of stem segments from tomato cvs. and subsequently extracting for enzyme activity from vascular bundles. This technique confirmed previous results obtained in vitro using extracted cell walls, as identical levels of PTE were produced after 3 days growth in vessels of a resistant and a susceptible cv. (Cooper, 1974).

Further evidence against the involvement of polysaccharidase regulation in specificity is based on a consideration of some aspects of enzyme induction and action:

(i) As monosaccharides appear to be the natural inducers of these enzymes, induction does not depend on the presence of specific glycosidic linkages except in the case of cellulase, which generally is not considered to be involved in the critical initial stages of infection. Albersheim & Anderson-Prouty (1975) made the point that as walls of all higher plants are composed of the same 10 or so monosaccharides, there is insufficient information to account for varietal specificity.

(ii) This information is reduced even further by the previously mentioned phenomenon of cross-induction (i.e. the induction of one enzyme by structurally similar sugars such as arabinose, galacturonic acid and galactose).

(iii) Initial wall degradation by pectic enzymes may also result in the uncontrolled induction of several polysaccharidases, irrespective of the polysaccharide niche of the pathogen. This is because pectic enzymes not only release galactoronides from cell walls, but also a significant propor- tion of the sugars covalently linked to pectic polysaccharides such as rhamnose, arabinose, galactose and xylose; these could function as inducers if the pathogen is capable of degrading the sugar-uronide links. Extensive wall degradation has been shown with a highly purified PG of C. lindemuthia- num and a PGTE isozyme of E. chrysanthemi which released c. 65% of the rhamnose, arabinose and galactose from tobacco cell walls (English et al., 1972, Basham & Bateman, 1975).

No attempt will be made to involve CR in specificity as it is a non-specific phenomenon effected by a vast array of carbon-energy sources. Minor differences in host sugar levels are unlikely to be under sufficiently fine control to determine varietal resistance, especially as synthesis of most CWDE is repressed by similar sugar levels. Even a role in general resistance or tolerance has yet to be clearly demonstrated.

In contrast, cell wall polysaccharides can determine general resistance as exhibited by mature tissues or certain morphological parts of plants, e.g. the development of resistance with age by bean hypocotyls to R. solani and C. lindemuthianum has been well documented and ascribed to cell wall changes, including the conversion of pectin to calcium pectate (Bateman & Lumsden, 1965), major alteration in monosaccharide composition (Nevins et al., 1968) and lignification (Griffey & Leach, 1965). This is reflected by the ability of fungal enzymes to cause extensive degradation to walls from susceptible (4-5 day) hypocotyls but only limited breakdown of walls from resistant (18-20 day) hypocotyls (Bateman et al., 1969; English et al., 1969). It is interesting that cell walls appeared to affect tissue suscep-tibility by influencing enzyme activity rather than induction, as similar levels of CWDE's appeared in R. solani cultures after 9 days growth on resistant or susceptible walls.

However, a time course study may have revealed significant differences in induction, as found by English et al. (1971) for production of α-galactosidase by C. lindemuthianum grown on cell walls extracted from different parts of bean plants. Walls from tissues susceptible to infection (leaves, first internodes or hypocotyls) induced much higher levels than walls from roots which are resistant to infection (Figure 7). An earlier analysis of cell walls from these 4 regions by Nevins et al. (1967) showed that roots differed markedly from other tissues by their much higher arabinose content, but the significance of this in resistance is not obvious.

The recently discovered fundamental differences between cell wall polymers of dicots and monocots (Burke et al., 1974), in particular the high proportion of arabinoxylan in the latter, may also play a part in determining the general resistance of one group to pathogens of the other group. Especially in view of an earlier report by McNab et al. (1967) that enzymes of C. lindemuthianum were unable to degrade cell walls of several monocots.

In conclusion, the exciting prospect that host cell wall polymers can

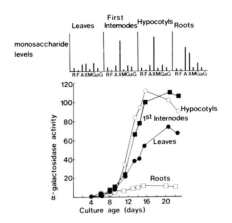

Figure 7. Monosaccharide composition of cell walls from different parts of Red Kidney bean and relative abilities to induce α-galactosidase of Colletotrichum lindemuthianum. R - rhamnose, F - fucose, A - arabinose, X - xylose, M - mannose, Ga - galactose, G - glucose (adapted from Nevins et al., 1967, and English et al., 1971).

determine specificity now seems remote, in view of recent advances in our understanding of polysaccharidase regulation and plant cell wall structure. However, this does not detract from the central role of induction-repression control of CWDE during parasitism, enabling pathogens to make rapid quali- tative and quantitative changes in response to host polysaccharides.

References

Abo-El-Dahab, M.K. (1964). Production of pectic and cellulolytic enzymes by Xanthomonas malvacearum. Phytopathology 54, 597-601.

Adhya, S. & Echols, H. (1966). Glucose effect and the galactose enzymes of Escherichia coli: correlation between glucose inhibition of induction and inducer transport. Journal of Bacteriology, 601-608.

Albersheim, P. & Anderson-Prouty, A. (1975). Carbohydrates, proteins, cell surfaces, and the biochemistry of pathogenesis. Annual Review of Plant Physiology 26, 31-52.

Albersheim, P., Bauer, W.D., Keegstra, K. & Talmadge, K.W. (1973). The structure of the wall of suspension cultured sycamore cells. pp. 117-147 in Biogenesis of plant cell wall polysaccharides, ed. by F. Loweus. Academic Press.

Albersheim, P., Jones, T.M. & English, P.D. (1969). Biochemistry of the cell wall in relation to infective processes. Annual Review of Phytopathology 7, 171-194.

Archer, S.A. (1973). Physiological aspects of brown rot of apple. Ph.D. thesis, University of Bristol.

Attardi, G., Naono, S., Rouviere, J., Jacob, F. & Gros, F. (1963). Production of messenger RNA and regulation of protein synthesis. Cold Spring Harbour Symposium on Quantitative Biology 28, 363-372.

Basham, H.G. & Bateman, D.F. (1975). Relation of cell death in plant tissue treated with a homogeneous endo-pectate lyase to cell wall degradation. Physiological Plant Pathology 5, 249-262.

Bateman, D.F. (1963). Pectolytic activities of culture filtrates of Rhizoctonia solani and extracts of Rhizoctonia-infected tissues of bean. Phytopathology 53, 197-204.

Bateman, D.F. (1966). Hydrolytic and trans-eliminative degradation of pectic substances by extracellular enzymes of Fusarium solani f. phaseoli. Phytopathology 56, 238-244.

Bateman, D.F. (1967). Alteration of cell wall components during pathogenesis by Rhizoctonia solani. pp. 58-79 in The dynamic role of molecular constituents in plant parasite interactions, ed. by C.J. Mirocha & I. Uritani.

Bateman, D.F. (1970). Depletion of galacturonic acid content in bean hypocotyl cell walls during pathogenesis by Rhizoctonia solani and Sclerotium rolfsii. Phytopathology 60, 1846-1847.

Bateman, D.F. (1972). The polygalacturonase complex produced by Sclerotium rolfsii. Physiological Plant Pathology 2, 175-184.

Bateman, D.F., Jones, T.M. & Yoder, O.C. (1973). Degradation of corn cell walls by extracellular enzymes produced by Helminthosporium maydis Race T. Phytopathology 63, 1523-1529.

Bateman, D.F. & Lumsden, R.D. (1965). Relation of calcium content and nature of the pectic substances in bean hypocotyls of different ages to susceptibility to an isolate of Rhizoctonia solani. Phytopathology 55, 734-738.

Bateman, D.F., Van Etten, H.D., English, P.D., Nevins, D.I. & Albersheim,

P. (1969). Susceptibility to enzymatic degradation of cell walls from bean plants resistant and susceptible to Rhizoctonia solani. Plant Physiology 44, 641-648.

Beckwith, J., Grodzicker, T. & Arditti, R. (1972). Evidence for two sites in the lac promoter region. Journal of Molecular Biology 69, 155-160.

Benzer, S. (1953). Induced synthesis of enzymes in bacteria analysed at the cellular level. Biochimica et Biophysica Acta 11, 383-395.

Blackhurst, F.M. (1961). Induction of Verticillium wilt disease symptoms in detached shoots of resistant and susceptible tomato plants. Annals of Applied Biology 52, 79-88.

Britten, R.J. & Davidson, E.H. (1969). Gene regulation for higher cells: a theory. Science 165, 349-357.

Britten, R.J. & Kohne, D.E. (1968). Repeated sequences in DNA. Science 161, 529-540.

Bugbee, W.M. (1972). Sucrose and cell walls as factors affecting Phoma storage rot of sugar beet. Phytopathology 63, 480-484.

Bull, A. (1972). Environmental factors influencing the synthesis and excretion of exocellular molecules. Journal of Applied Chemistry and Biotechnology 22, 261-292.

Burke, D., Kaufman, P., McNeil, M. & Albersheim, P. (1974). The structure of plant cell walls VI. A survey of the walls of suspension cultured monocots. Plant Physiology 54, 109-115.

Burnett, J.H. (1975). Mycogenetics. John Wiley, 375 pp.

Buston, H.W. & Jabbar, A. (1954). Synthesis of β-linked glucosaccharides by extracts of Chaetomium globosum. Biochimica et Biophysica Acta 15, 543-548.

Byrde, R.J.W. & Fielding, A.H. (1968). Pectin methyl trans-eliminase as the maceration factor of Sclerotinia fructigena and its significance in brown rot of apple. Journal of General Microbiology 52, 287-297.

Byrde, R.J.W., Fielding, A.H., Archer, S.A. & Davies, E. (1971). The role of extracellular enzymes in the rotting of fruit tissues by Sclerotinia fructigena. pp. 39-54 in Fungal pathogenicity and the plant's response, ed. by R.J.W. Byrde & C.V. Cutting. Academic Press.

Case, M.E. & Giles, N.H. (1971). Partial enzyme aggregates formed by pleiotropic mutants in the arom gene cluster of Neurospora crassa. Proceedings of the National academy of Sciences, U.S.A. 68, 58-62.

Cherest, H. & Robichon-Szulmajster, H. (1973). The roles of 2 independent genes in one of the regulatory systems involved in methionine biosynthesis in Saccharomyces cerevisiae. pp. 165-178 in Genetics of industrial microorganisms, ed. by Z. Vanek, Z. Hostalek & J. Cudlin. Vol. II. Elsevier, Amsterdam.

Clarke, P.H., Houldsworth, M.A. & Lilley, M.D. (1968). Catabolite repression and the induction of amidase synthesis by Pseudomonas aeruginosa 8602 in continuous culture. Journal of General Microbiology 51, 225-234.

Clark, D.J. & Marr, A.G. (1964). Studies on the repression of β-galactosidase in Escherichia coli. Biochimica et biophysica Acta 92, 85-98.

Cohen-Bazire, G. & Jolit, M. (1953). Isolement par sélection du mutants d'Escherichia coli synthétisant spontanément l'amylomaltose et la β-galactosidase. Annls. Inst. Pasteur, Paris 84, 937-945.

Cole, A.L.J. & Wood, R.K.S. (1970). Production of hemicellulases by Penicillium digitatum. Phytochemistry 9, 695-699.

Cole, J.S. (1956). Studies in the physiology of parasitism XX. The pathogenicity of Botrytis cinerea, Sclerotinia fructigena, and Sclerotinia laxa, with special reference to the part played by pectolytic enzymes. Annals of Botany 20, 16-38.

Cole, M. & Wood, R.K.S. (1961). Pectic enzymes and phenolic substances in apples rotted by fungi. Annals of Botany 25, 435-452.

Comp, P.C. & Lester, G. (1971). Properties of an extracellular β-galactosidase secreted by Neurospora crassa. Journal of Bacteriology 107, 162-167.

Cooper, R.M. (1974). Cell wall-degrading enzymes of vascular wilt fungi. Ph.D. thesis, University of London.

Cooper, R.M. & Wood, R.K.S. (1973). Induction of synthesis of extracellular cell wall-degrading enzymes in vascular wilt fungi. Nature 246, 309-311.

Cooper, R.M. & Wood, R.K.S. (1974). Scanning electron microscopy of Verticillium albo-atrum in xylem vessels of tomato plants. Physiological Plant Pathology 4, 443-446.

Cooper, R.M. & Wood, R.K.S. (1975). Regulation of synthesis of cell wall-degrading enzymes by Verticillium albo-atrum and Fusarium oxysporum f.sp. lycopersici. Physiological Plant Pathology 5, 135-156.

Cooper, R.M., Rankin, B. & Wood, R.K.S. (1976). Cell wall-degrading enzymes

of vascular wilt fungi II. Properties and modes of action of polysaccharides of Verticillium albo-atrum and Fusarium oxysporum f.sp.
lycopersici (in preparation).

Creaser, E.H., Bennett, D.J. & Drysdale, R.B. (1967). The purification and
properties of Histidinol dehydrogenase from Neurospora crassa.
Biochemical Journal 103, 36-41.

de Cromburgghe, B., Chen, B., Anderson, W., Nissley, P., Gottesman, M. &
Pastan, I. (1971). Lac DNA, RNA polymerase and cyclic AMP receptor
protein, cyclic AMP, lac repressor and inducer are the essential
elements for controlled lac transcription. Nature, New Biology 231,
139-142.

de Herrera, E.C. (1972). Production of pectin methyesterase by Erwinia
carotovora. Phytopathologische Zeitschrift 74, 48-54.

Deese, D.C. & Stahmann, M.A. (1962). Pectic enzyme and cellulase formation
by Fusarium oxysporum cubense on stem tissues from resistant and
susceptible banana plants. Phytopathology 52, 247-255.

Dickson, R.C., Abelson, J., Barnes, W.M. & Reznikoff, W.S. (1975). Genetic
regulation: The lac control region. Science 187, 27-34.

Dixon, G.R. & Pegg, G.F. (1972). Changes in amino-acid content of tomato
xylem sap following infection with strains of Verticillium albo-atrum.
Annals of Botany 36, 147-154.

Douglas, H.C. & Hawthorne, D.C. (1966). Regulation of genes controlling
synthesis of the galactose pathway enzymes in yeast. Genetics 54,
911-916.

Drucker, H. (1972). Regulation of extracellular proteases in Neurospora
crassa. Journal of Bacteriology 110, 1041-1049.

Emmer, M., de Crombrugghe, B.D., Pastan, I. & Perlman, R. (1970). Cyclic
AMP receptor protein of E. coli: its role in the synthesis of inducible
enzymes. Proceedings of the National Academy of Sciences, U.S.A. 66,
480-487.

English, P.D. & Albersheim, P. (1969). Host-pathogen interactions: I. A
correlation between α-galactosidase production and virulence. Plant
Physiology 44, 217-224.

English, P.D., Jurale, J.B. & Albersheim, P. (1971). Host-pathogen interactions. II. Parameters affecting polysaccharide-degrading enzyme
secretion by Colletotrichum lindemuthianum grown in culture. Plant
Physiology 47, 1-6.

English, P.D., Maglothin, A., Keegstra, K. & Albersheim, P. (1972). A cell wall degrading endo-PG secreted by Colletotrichum lindemuthianum. Plant Physiology 49, 293-297.

Eriksson, K.-E. & Goodell, E.W. (1974). Pleiotropic mutants of the wood-rotting fungus Polyporus adustus lacking cellulase, mannanase, and xylanase. Canadian Journal of Microbiology 20, 371-378.

Ferraris, L., Garibaldi, A. & Matta, A. (1974). Polygalacturonase and poly-galacturonate trans-eliminase production in vitro and in vivo by Fusarium oxysporum f.sp. lycopersici. Phytopathologische Zeitschrift 81, 1-14.

Fincham, J.R.S. & Day, P.R. (1971). Fungal Genetics. Blackwell Scientific Publications, 402 pp.

Fuchs, A. (1965). The trans-eliminative breakdown of Na-polygalacturonate by Pseudomonas fluorescens. Antonie Van Leewenhoek 31, 323-340.

Fuchs, A., Jobson, J.A. & Wouts, W.M. (1965). Arabanases in phyto-pathogenic fungi. Nature 206, 714-715.

Garibaldi, A. & Bateman, D.F. (1971). Pectic enzymes produced by Erwinia chrysanthemi and their effects on plant tissue. Physiological Plant Pathology 1, 25-40.

Gilbert, W. & Müller-Hill, B. (1966). Isolation of the lac repressor. Proceedings of the National Academy of Sciences, U.S.A. 56, 1891-1898.

Gilbert, W. & Müller-Hill, B. (1967). The lac operator is DNA. Proceedings of the National Academy of Sciences, U.S.A. 58, 2415-2421.

Goodenough, P.W., Kempton, R.J. & Maw, G.A. (1976). Studies on the root rotting fungus Pyrenochaeta lycopersici: extracellular enzyme secretion by the fungus grown on cell wall material from susceptible and tolerant plants. Physiological Plant Pathology 8, 243-251.

Goodenough, P.W. & Maw, G.A. (1975). Studies on the root rotting fungus Pyrenochaeta lycopersici: the cellulase complex and regulation of its extracellular appearance. Physiological Plant Pathology 6, 145-157.

Grainger, J. (1968). C_p/R_s and the disease potential of plants. Horticultural Research 8, 1-40.

Gratzner, H. & Sheenan, D.N. (1969). Neurospora mutants exhibiting hyperpro-duction of amylase and invertase. Journal of Bacteriology 97, 544-549.

Griffey, R.T. & Leach, J.G. (1965). The influence of age of tissue on the development of bean anthracnose lesions. Phytopathology 55, 915-918.

Gross, S.R. (1969). Genetic regulatory mechanisms in the fungi. Annual Review of Genetics 3, 395-425.

Gupta, D.P. & Heale, J.B. (1971). Induction of cellulase (Cx) in

Verticillium albo-atrum. Journal of General Microbiology 63, 163-173.

Haggerty, D.M. & Schleif, R.F. (1975). Kinetics of the onset of catabolite repression in Escherichia coli as determined by lac messenger ribonucleic acid initiations and intracellular cyclic adenosine 3',5'-monophosphate levels. Journal of Bacteriology 123, 946-953.

Hall, R. (1971). Pathogenicity of Monilinia fructicola I. Hydrolytic enzymes. Phytopathologische Zeitschrift 72, 245-254.

Hancock, J.G. (1965). Relationship between induced changes in pH and production of polygalacturonate trans-eliminase by Colletotrichum trifolii. Phytopathology 55, 1061 (Abst.).

Hancock, J.G. (1968). Degradation of pectic substances during pathogenesis by Fusarium solani f.sp. cucurbitae. Phytopathology 58, 62-69.

Hancock, J.G. (1976). Multiple forms of endo-pectate lyase formed in culture and in infected squash hypocotyls by Hypomyces solani f.sp. cucurbitae. Phytopathology 66, 40-45.

Hancock, J.G., Millar, R.L. & Lorbeer, J.W. (1964). Pectolytic and cellulolytic enzymes produced by Botrytis allii, B. cinerea and B. squamosa in vitro and in vivo. Phytopathology 54, 928-931.

Harman, G.E. & Corden, M.E. (1972). Purification and partial characterisation of the polygalacturonase produced by Fusarium oxysporum f.sp. lycopersici. Biochimica et Biophysica Acta 264, 328-338.

Heale, J.B. & Gupta, D.P. (1972). Mechanism of vascular wilting induced by Verticillium albo-atrum. Transactions of the British Mycological Society 58, 19-28.

Heath, M.C. & Wood, R.K.S. (1971). Role of cell wall-degrading enzymes in the development of leaf spots caused by Ascochyta pisi and Mycosphaerella pinodes. Annals of Botany 35, 451-474.

Horsfall, J.G. & Dimond, A.E. (1951). Interactions of tissue sugar, growth substances and disease susceptibility. Zeitschrift Planzenbau Planzenshutz 64, 415-421.

Horton, J.C. & Keen, N.T. (1966a). Regulation of induced cellulase synthesis in Pyrenochaeta terrestris by utilizable carbon compounds. Canadian Journal of Microbiology 12, 209-220.

Horton, J.C. & Keen, N.T. (1966b). Sugar repression of endo-polygalacturonase and cellulase synthesis during pathogenesis by Pyrenochaeta terrestris as a resistance mechanism in onion pink root rot. Phytopathology 56, 908-916.

Hsu, E.J. & Vaughn, R.H. (1969). Production and catabolite repression of the constitutive polygalacturonic acid trans-eliminase of Aeromonas liquefaciens. Journal of Bacteriology 98, 172-181.

Hulme, M.A. & Stranks, D.W. (1971). Regulation of cellulase production by Myrothecium verrucaria grown on non-cellulosic substrates. Journal of General Microbiology 69, 145-155.

Ippen, K., Miller, J.H., Scaife, J. & Beckwith, J. (1968). New controlling element in the lac operon of E. coli. Nature 217, 825-827.

Jacob, F. & Monod, J. (1961). Genetic regulatory mechanisms in the synthesis of proteins. Journal of Molecular Biology 3, 318-356.

Jobe, A. & Bourgeois, S. (1972). The natural inducer of the lac operon. Journal of Molecular Biology 69, 397-408.

Johns, E.W. (1972). Histones, chromatin structure and RNA synthesis. Nature, New Biology 237, 87-88.

Jones, T.M., Anderson, A.J. & Albersheim, P. (1972). Host-pathogen interactions IV. Studies on the polysaccharide-degrading enzymes secreted by Fusarium oxysporum f.sp. lycopersici. Physiological Plant Pathology 2, 153-166.

Kaji, A. & Yoshira, O. (1969). Production and properties of α-L-arabinofuranosidase from Corticium rolfsii. Applied Microbiology 17, 910-913.

Karr, A.L. & Albersheim, P. (1970). Polysaccharide-degrading enzymes are unable to attack plant cell walls without prior action by a "wall-modifying enzyme". Plant Physiology 46, 69-80.

Keegstra, K., Talmadge, K.W., Bauer, W.D. & Albersheim, P. (1973). The structure of plant cell walls. III. A model of the walls of suspension-cultured sycamore cells based in the interconnections of the macromolecular components. Plant Physiology 51, 188-196.

Keen, N.T. & Erwin, D.C. (1971). Endopolygalacturonase: evidence against involvement in Verticillium wilt of cotton. Phytopathology 61, 198-203.

Keen, N.T. & Horton, J.C. (1966). Induction and repression of endopolygalacturonase synthesis by Pyrenochaeta terrestris. Canadian Journal of Microbiology 12, 443-453.

Kollar, G. (1958). Biochemical studies on the synthesis of streptomycin. II. Formation of and role played in the biosynthesis of streptomycin by Streptomyces griseus α-mannosidase. Acta Microbiologica Acedemia Scientiarum Hungarica 5, 19-34.

Laborda, F., Archer, S.A., Fielding, A.H. & Byrde, R.J.W. (1974). Studies on the α-L-arabinofuranosidase complex from Sclerotinia fructigena in relation to brown rot of apple. Journal of General Microbiology 81, 151-163.

Langan, T.A. (1968). Histone phosphorylation: stimulation by adenosine 3',5'-monophosphate. Science 162, 579-580.

Lamport, D.T.A. (1965). The protein component of primary cell walls. Advances in Botanical Research 2, 151-218.

Langcake, P. & Drysdale, R.B. (1975). The role of pectic enzyme production in the resistance of tomato to Fusarium oxysporum f.sp. lycopersici. Physiological Plant Pathology 6, 247-258.

Lapwood, D.H. (1957). Studies in the physiology of parasitism. XXIII. On the parasitic vigour of certain bacteria in relation to their capacity to secrete pectolytic enzymes. Annals of Botany 21, 167-184.

Lis, J.T. & Scheif, R. (1973). Different cyclic AMP requirements for induction of the arabinose and lactose operons of Escherichia coli. Journal of Molecular Biology 79, 149-162.

Lisker, N., Katan, J., Chet, I. & Henis, Y. (1975). Release of cell-bound polygalacturonase and cellulase from mycelium of Rhizoctonia solani. Canadian Journal of Microbiology 21, 521-526.

Loomis, W.F. & Magasanik, B. (1966). Nature of the effector of catabolite repression in Escherichia coli. Journal of Bacteriology 92, 170-177.

Lukens, R.J. & Sands, D.C. (1973). Alternaria solani, catabolite repression of pectinase and cellulase. Proceedings of the 2nd International Congress of Plant Pathology (Abst.).

MacDonald, W.L. & McNabb, H.S. (1970). Fine-structural observations on the growth of Ceratocystis ulmi in elm xylem tissue. Bioscience 20, 1060-1061.

Magasanik, B. (1961). Catabolite repression. Cold Spring Harbour Symposium on Quantitative Biology 26, 249-256.

Makman, R.S. & Sutherland, E.W. (1965). Adenosine 3',5'-phosphate in Escherichia coli. Journal of Biological Chemistry 240, 1309-1314.

Mandels, M., Parrish, F.W. & Reese, E.T. (1962). Sophorose as an inducer of cellulase in Trichoderma viride. Journal of Bacteriology 83, 400-408.

Mandels, M. & Reese, E.T. (1960). Induction of cellulase in fungi by cellobiose. Journal of Bacteriology 79, 816-826.

Marushige, K. & Bonner, J. (1966). Template properties of liver chromatin.

Journal of Molecular Biology 15, 160-174.

McNab, J.M., Nevins, D.J. & Albersheim, P. (1967). Differential resistance of cell walls of Acer pseudoplatanus, Triticum vulgare, Hordeum vulgare, and Zea mays, to polysaccharide-degrading enzymes. Phytopathology 57, 625-631.

Messenguy, F. & Wiame, J.M. (1969). The control of ornithine transcarbamylase activity by arginase in Saccharomyces cerevisiae. FEBS letters 3, 47-49.

Miller, L. & MacMillan, J.D. (1970). Mode of action of pectic enzymes II. Further purification of exopolygalacturonate lyase and pectinesterase from Clostridium multifermentans. Journal of Bacteriology 102, 72-78.

Monod, J. & Cohn, M. (1952). La biosynthese induite des enzymes (adaptation enzymatique). Advances in Enzymology 13, 67-119.

Moran, F. & Starr, M.P. (1969). Metabolic regulation of polygalacturonic acid trans-eliminase in Erwinia. European Journal of Biochemistry 11, 291-295.

Mullen, J.M. & Bateman, D.F. (1975). Polysaccharide degrading enzymes produced by Fusarium roseum "Avenaceum" in culture and during pathogenesis. Physiological Plant Pathology 6, 233-246.

Mussell, H.W. & Green, R.J. (1970). Host colonization and polygalacturonase production by two tracheomycotic fungi. Phytopathology 60, 192-195.

Mussell, H.W. & Strouse, B. (1971). Proteolytic enzyme production by Verticillium albo-atrum. Phytopathology 61, 904 (Abst.).

Mussell, H.W. & Strouse, B. (1972). Characterization of two polygalacturonases produced by Verticillium albo-atrum. Canadian Journal of Biochemistry 50, 625-632.

Myers, M.G. & Eberhart, B. (1966). Regulation of cellulase and cellobioase in Neurospora crassa. Biochimica et Biophysica Acta 24, 782-785.

Nevins, D.J., English, P.D. & Albersheim, P. (1967). The specific nature of plant cell wall polysaccharides. Plant Physiology 42, 900-906.

Nevins, D.J., English, P.D. & Albersheim, P. (1968). Changes in cell wall polysaccharides associated with growth. Plant Physiology 43, 914-922.

Nisizawa, T., Suzuki, H., Nakayama, M. & Nisizawa, K. (1971). Inductive formation of cellulase by sophorose in Trichoderma viride. Journal of Biochemistry, Tokyo 70, 375-385.

Okinaka, R.T. & Dobrogosz, W.J. (1967). Catabolite repression and pyruvate metabolite in Escherichia coli. Journal of Bacteriology 93, 1644-1650.

Oshima, Y., Horiuchi, T., Iida, Y. & Kameyama, T. (1970). Transcription and repression of the lac operon in vitro. Cold Spring Harbour Symposium on Quantitative Biology 35, 425-432.

Paigen, K. & Williams, B. (1970). Catabolite repression and other control mechanisms in carbohydrate utilization. Advances in Microbial Physiology 4, 251-323.

Pastan, I. & Perlman, R. (1970). Cyclic adenosine monophosphate in bacteria. Science N.Y. 169, 339-344.

Patil, S.S. & Dimond, A.E. (1968). Repression of polygalacturonase synthesis in Fusarium oxysporum f.sp. lycopersici and its effect on symptom reduction in infected tomato plants. Phytopathology 58, 676-682.

Peck, R.M., Markey, F. & Yudkin, M.D. (1971). Effect of 5-fluoroacil on β-galactosidase synthesis in an Escherichia coli mutant resistant to catabolite repression of the lac operon. FEBS letters 16, 43-44.

Perlman, R., Chen, B., de Crombrugghe, B., Emmer, M., Gottesman, M., Varmus, H. & Pastan, I. (1970). The regulation of lac operon transcription by cyclic adenosine 3',5'-monophosphate. Cold Spring Harbour Symposium on Quantitative Biology 35, 419-423.

Peterkofsky, A. & Gazdar, C. (1971). Glucose and the metabolism of adenosine 3',5'-cyclic monophosphate in Escherichia coli. Proceedings of the National Academy of Sciences, U.S.A. 68, 2794-2798.

Phaff, H.J. (1947). The production of exocellular pectic enzymes by Penicillium chrysogenum. I. On the formation and adaptive nature of polygalacturonase and pectinesterase. Archives of Biochemistry 13, 67-81.

Pirt, S.J. (1971). The diffusion capsule, a novel device for the addition of a solute at a constant rate to a liquid medium. Biochemical Journal 121, 293-297.

Posternak, T. (1974). Cyclic AMP and cyclic GMP. Annual Review of Pharmacology 14, 23-33.

Reese, E.T., Maguire, A.H. & Parrish, F.W. (1968). Glucosidases and exoglucanases. Journal of Biochemistry 46, 25-34.

Richmond, M.H. (1968). Enzymatic adaptation in bacteria: its biochemical and genetic basis. Essays in Biochemistry 4, 106-154.

Rickenberg, H.V. (1974). Cyclic AMP in prokaryotes. Annual Review of Microbiology 28, 353-369.

Ries, S.M. & Albersheim, P. (1973). Purification of a protease secreted by Colletotrichum lindemuthianum. Phytopathology 63, 625-629.

Cooper - Synthesis of cell wall-degrading enzymes

5555555555

Riggs, A.D., Suzuki, H. & Bourgeois, S. (1970). Lac repressor-operator interactions. I. Equilibrium studies. Journal of Molecular Biology 48, 67-83.

Russell, S. (1975). The role of cellulase produced by Verticillium albo-atrum in Verticillium wilt of tomatoes. Phytopathologische Zeitschrift 82, 35-48.

Scaife, J. & Beckwith, J.R. (1966). Mutational alteration of the maximal levels of lac operon expression. Cold Spring Harbour Symposium on Quantitative Biology 31, 403-408.

Schmid, R. (1966). Studies on extracellular structures of fungal hyphae XI. International Congress for Electron Microscopy 2, 785-786.

Sheppard, D.E. & Englesberg, E. (1967). Further evidence for positive control of the L-arabinose system by gene ara C. Journal of Molecular Biology 25, 443-454.

Silverstone, A.E., Arditti, R.R. & Magasanik, B. (1970). Catabolite-insensitive revertants of lac promoter mutants. Proceedings of the National Academy of Sciences, U.S.A. 66, 773-779.

Smith, J.T. (1963). Penicillinase and ampicillin resistance in a strain of Escherichia coli. Journal of General Microbiology 30, 299-306.

Strider, D.L. & Winstead, N.N. (1961). Production of cell-wall dissolving enzymes by Cladosporium cucumerinum in cucumber tissue and in artificial media. Phytopathology 51, 765-768.

Swinburne, T.R. & Corden, M.E. (1969). A comparison of the polygalacturo-nases produced in vivo and in vitro by Penicillium expansum. Journal of General Microbiology 55, 75-87.

Talboys, P.W. (1958). Degradation of cellulose by Verticillium albo-atrum. Transactions of the British Mycological Society 41, 242-248.

Talboys, P.W. & Busch, L.V. (1970). Pectic enzymes produced by Verticillium species. Transactions of the British Mycological Society 55, 367-381.

Talmadge, K.W., Keegstra, K., Bauer, W.D. & Albersheim, P. (1973). The structure of plant cell walls. I. The macromolecular components of the walls of suspension-cultured sycamore cells with a detailed analysis of the pectic polysaccharides. Plant Physiology 51, 158-173.

Toda, S., Suzuki, H. & Nisizawa, K. (1968). The mode of action of Trichoderma cellulases toward normal and reduced cello-oligosaccharides. Journal of Fermentation Technology 46, 711-718.

Tomkins, G.M., Gelehrter, T.D., Granner, D., Martin, D., Samuels, H.H. & Thompson, E.B. (1969). Control of specific gene expression in higher

Cooper - Synthesis of cell wall-degrading enzymes

organisms. Science 166, 1474-1480.

Torriani, A. & Pappenheimer, A.M. (1962). Inducible polysaccharide depoly-
merases of Bacillus palustris. Journal of Biological Chemistry 237,
3-13.

Urbanek, H. & Zalewska-Sobczak, J. (1975). Polygalacturonase of Botrytis
cinerea E-2000 PERS. Biochimica et Biophysica Acta 377, 402-409.

Van Wijk, R. (1968). α-glucosidase synthesis, respiratory enzymes and
catabolite repression in yeast. Proc. Konoklyke Akad Wet 71, 302-313.

Waggoner, P.E. & Dimond, A.E. (1955). Production and role of extracellular
pectic enzymes of Fusarium oxysporum f.sp. lycopersici. Phytopathology
45, 79-87.

Walsh, D.A., Perkins, J.P. & Krebs, E.G. (1968). An AMP dependent protein
kinase from rabbit skeletal muscle. Journal of Biological Chemistry
243, 3763-3774.

Wang, M.C. & Keen, N.T. (1970). Purification and characterization of endopolygalacturonase from Verticillium albo-atrum. Archives of Biochemistry
and Biophysics 141, 749-757.

Wang, S.C. & Pinckard, J.A. (1971). Pectic enzymes produced by Diplodia
gossypina in vitro and in infected cotton bolls. Phytopathology 61,
1118-1124.

Weinhold, A.R. & Motta, J. (1973). Initial host responses in cotton to
infection by Rhizoctonia solani. Phytopathology 63, 157-162.

Weste, G. (1970). Extracellular enzyme production by various isolates of
Ophiobolus graminis var avenae. Phytopathologische Zeitschrift 67,
327-336.

Wilder, B.M. & Albersheim, P. (1973). The structure of plant cell walls.
IV. A structural comparison of the wall hemicellulose of cell
suspension cultures of sycamore (Acer pseudoplatanus) and of red kidney
bean (Phaseolus vulgaris). Plant Physiology 51, 889-893.

Winstead, N.N. & McCombs, C.L. (1961). Pectinolytic and cellulolytic
enzyme production by Pythium aphanidermatum. Phytopathology 51,
270-273.

Yamane, K., Suzuki, H., Hirotani, M., Ozaza, H. & Nisizawa, K. (1970).
Effect of nature and supply of carbon sources on cellulase formation
in Pseudomonas fluorescens var cellulose. Journal of Biochemistry,
Tokyo 67, 9-18.

Yudkin, M.D. (1976). Mutations in Escherichia coli that relieve catabolite

Cooper - Synthesis of cell wall-degrading enzymes

repression of tryptophanase synthesis. Mutations distant from the
tryptophanase gene. Journal of General Microbiology 92, 125-132.

Zubay, G., Schwartz, D. & Beckwith, J. (1970). Mechanism of activation of
catabolite sensitive genes. A positive control system. Proceedings
of the National Academy of Sciences 66, 104-110.

Zucker, M. & Hankin, L. (1970). Regulation of pectate lyase synthesis in
Pseudomonas fluorescens and Erwinia carotovora. Journal of Bacteriology
104, 13-18.

DISCUSSION

Chairman: D.F. Bateman

Kauss: Is the appearance of extra-cellular enzymes produced by pathogens
attributed to derepression of enzyme formation, in other words, is the action
of metabolites at the level of the genome? I wish to draw your attention to
an alternate mechanism of regulating enzyme levels, namely, catabolite
inactivation. In this case the catabolite binds with the enzyme directly
and when such binding occurs, a built-in suicide mechanism inactivates the
enzyme. This mechanism is believed to operate for a number of enzymes in
yeast and in bacteria (H. Holzer, Trends in Biochem. Sci. 1, 178-180, 1976).

Cooper: There was a limitation on the amount of material I could cover. In
many of the cases I reviewed, workers have tested the effects of repressing
catabolites on enzyme activities and have found no effects at all. With the
exception of the effects of cellobiose on the action of cellulase, I am
unaware of any effects of monosaccharides on polysaccharidases.

Albersheim: I would like to refer to your comments about enzyme aggregates
or multienzyme particles. Over the years, we have labored to purify such
diverse enzymes as endopolygalacturonase, endopolygalacturonic acid lyase,
pectinesterase, galactanases, arabinases, and others. Every one of these
enzymes was difficult to purify and at times we were sure that we were dealing
with multi-enzyme particles. The fact is, that when we labored long enough,
we have always been able to succeed in separating distinct enzyme activities.
I think that Eriksson and Bateman have had similar experiences. Would you

describe for us the evidence for the multi-enzyme particles you mentioned?

Cooper: There is not enough evidence from my work to be dogmatic about the existence of multi-enzyme particles but my comments were based on a purification scheme involving ammonium sulphate precipitation and double isometric focusing, using both wide range and narrow range Ampholines. In our work the purification has not been carried any further than this. I think it is well established, however, that enzyme aggregates do exist. For example, Miller & MacMillan (J. Bacteriol. 102, 72-78, 1970), after use of an extensive purification scheme, were not able to separate pectin esterase and polygalacturonate lyase activities. I believe that they ended up with a supposedly homogeneous preparation.

Albersheim: This may be. We have had particles which could go through all kinds of purifications and different enzyme activities remain superimposed and you would say that you are never going to get them apart, but if you keep working at it, eventually something happens and you find the trick and they come apart.

Cooper: I wonder if one could disassociate aggregates anyway?

Albersheim: Perhaps Eriksson, Bateman, or Bauer would like to comment because I know they have had similar experiences in purifying plant cell wall degrading enzymes.

Bateman: I am well aware of the problem. We have spent a great deal of time attempting to purify the individual polysaccharide-degrading enzymes produced by Sclerotium rolfsii without a great deal of success in some cases. In our work we have used a number of different enzyme purification procedures. We have never been able to completely separate, for example, the endopolygalacturase and exopolygalacturonase activities produced by this pathogen. A major problem in attempting to purify the polysaccharidases produced by Sclerotium rolfsii probably resides in the fact that the enzymes produced by this organism have similar molecular weights as well as similar isoelectic points. Because of this, these enzymes are difficult to separate from each other. Based on our studies, however, I feel that the individual activities produced by this organism represent individual proteins rather than multi-enzyme particles.

Cooper: Enzyme aggregates have been found in critical work on certain fungal pathways. Enzyme aggregates are a reflection of coordinate induction. I thought it would be interesting to look at results from work on cell wall degrading enzymes to see if they are produced in a coordinate fashion ...??

Eriksson: We have always found that isoelectric focusing is the most sensitive method you can apply for separation of enzymes. There are of course a number of different other possibilities. You can try to separate aggregates under different concentrations of urea for instance. You can run the enzymes through columns and analyze very carefully the adsorption curve for different activities and see if you can obtain any signs of skewness. But we have found that isoelectric focusing is probably superior to any other separation method.

Jarvis: You have been emphasizing the importance of basal level of cell wall degrading enzymes and the rapid buildup of these enzymes before host defense mechanisms are called into play. I would like to raise a question about the host's own enzymes and their degradating effect on its own walls. We know that pectin methylesterase is widely distributed in plants and other polysaccharidases are known to occur in tissues.

Cooper: I think we should bear in mind that we are talking about penetration of very localized areas of the plant cell wall. Such as the bore hole around the hyphae. I find it difficult to see how a host enzyme is going to participate in that. Penetration is a very rapid, localized process. How does a host regulate its own enzyme activities, particularly pectic enzyme activity? I would dearly like to know. You can barely detect polygalacturonase activity in tomato stems but it is not difficult to find this enzyme in tomato fruit.

Goodman: Based on some points raised by you, and others raised by Dr. Pegg earlier, relative to ethylene, I would like to raise a question about the advisability of using excised plant parts in experiments. I wonder if those who have used detached plant parts in experiments have considered the possibility of avoiding problems like ethylene or wound responses by aging detached tissues for a period prior to use. Dr. Novacky in my department, who does work in the area of electrophysiology, uses strips of tissue for implanting micropipettes into individual cells. He and other people who work in this area age detached tissues 24 hours or longer in a minimal salt and buffer. I wonder if people, for example those in Dr. Albersheim's laboratory and others, who have worked with plant parts have experienced any difficulti with this sort of thing. If they haven't, would they think that aging tissue for a while would avoid problems inherent for something like ethylene which is so volatile?

Ayers: We have found in our elicitor studies, if we do these type of incubations or washes of wounded cotyledon tissue for example, that we lose our response. So I don't know if this answers or complicates your question.

Ballou: One of your earlier slides dealt with the relative merits of cell walls vs. isolated polysaccharides for inducing enzyme formation. One of the more dramatic differences was associated with induction of polygalacturonate trans-eliminase. It occurs to me that a possible artifact may result from using isolated pectic acid in vitro studies. The pectic constituents of cell walls would be expected to be more highly methylated than the isolated polysaccharides. Do you have any control experiments that would eliminate the possibility that the eliminase is more effectively induced by the more natural polysaccharide as it exists in the wall?

Cooper: Yes, I mentioned briefly a minute ago that studies have been carried out by infiltrating stem segments which contain cell walls in situ. In such experiments, the trans-eliminase is produced quickly and polygalacturonase does not appear. The trans-eliminase is the only pectic enzyme that can be picked up in vascular fluids. It is also produced in greater quantity than any other enzyme when the fungus is grown on isolated cell walls. One thing I would like to know is the relative abilities of methyl-galacturonic acid and galacturonic acid to serve as inducers for pectin methylesterase. I don't know if the methylester is required for induction of this enzyme or not. This would be very difficult to show anyway because the basal level pectin methylesterase may cleave off the methyl group so you really won't know what you're looking at.

Bateman: It may be good to keep in mind that when discussing the methoxyl content of pectin in tissue, it can be expected that the methoxyl group content will vary with tissue age, at least in bean. In our studies it was demonstrated that about 70% of the uronic acid carboxyls are methylated in young bean hypocotyl tissues. Whereas after the hypocotyls have matured, the methoxyl content is reduced to about 25 to 30%. Also, one has to keep in mind that in host parasite interactions there is a tendency to get a stimulation of host pectin methylesterase activity in the vicinity of the infection site. All of these factors must be kept in mind as we consider induction phenomena in vivo.

Pegg: In your scanning electron microscope picture of the hyphae in vessels, you said that the sheath encapsulating the hyphae was a polysaccharide. Have you got any evidence for that?

Cooper: None whatsoever, except that transmission micrographs show that there is an amorphous coating around the hyphae. It is similar to what other investigators are describing as polysaccharide. This material, whatever it is, may well be the medium for the enzyme inducers, enzyme localization, etc. I really don't know what it is at the moment but certainly it is extra cellular.

Pegg: It occurs to me that this might be an important deposit. I was interested yesterday in Dr. Unestam's paper which showed that a similar deposit may result from phenoloxidase activity. Why a hyphae in a vessel has this coating, you know, we don't really know. It could be either of host or fungal origin. Would you want to speculate on this?

Cooper: Having looked at many infected vessels I would suggest that the coating around the vessels and that around the hyphae are similar. You can see similar electron dense deposits in the vacuoles of the host which other investigators have described as polyphenol. But I have never seen such deposits moving through to the vessels, but I think there is a better chance of this. We are, at the moment, in a state of ignorance as to what this material is and what it is doing is a puzzle.

Pegg: We have found this material around hyphae in susceptible but not around hyphae in colonized vessels of resistant plants.

Cooper: I haven't made this type of comparison yet. How did you get the fungus into resistant plants?

Pegg: It goes in.

Cooper: Our observation with this coating around the hyphae of Verticillium may not be a general phenomenon because in plants infected with Fusarium wilt we have not observed any equivalent deposits.

Pegg: Your transmission electron micrographs did not show all hyphae to be encapsulated.

Cooper: I think that the degree of encapsulation observed depends upon the age of infection in the particular vessel. I think it occurs perhaps in all vessels.

HOST INHIBITION OR MODIFICATION OF EXTRACELLULAR ENZYMES OF PATHOGENS

R.J.W. Byrde and S.A. Archer, Long Ashton Research Station and Department of Botany, University of Bristol, U.K.

Introduction

Soluble extracellular products are the first pathogen-synthesised metabolites to be encountered by host tissues, and in some diseases the outcome of the host-pathogen encounter may well be determined by interactions involving such extracellular products of pathogens, especially wall-degrading enzymes. If the host influences the pathogen's enzymes it may thus modify the extent of infection. Enzymes secreted by the pathogen may be activated, inhibited, immobilised or denatured, and the process of secretion itself may be affected. This paper reviews what we know of such effects on the extracellular enzymes of pathogens and attempts to relate such processes to the outcome of the host-pathogen interaction.

1. Secretion

When a pathogen has synthesized its extracellular enzymes by processes described by Cooper in this symposium, they must be secreted to act on the host. This involves transport from the site of synthesis to, and through, the cytoplasmic membrane. In an important review of secretory processes, Lampen (1974) described the transport of enzymes in eukaryotic cells from the membranebound ribosomes of the rough endoplasmic reticulum via the smooth endoplasmic reticulum, and often a golgi apparatus, into vesicles. Subsequently, these vesicles fuse with the plasma membrane and release their proteins (Figure 1). He pointed out that the essential step in secretion is not this exocytosis but the passage of the polypeptide chain through the membrane of the rough endoplasmic reticulum, since the resulting enzyme is then extracytoplasmic.

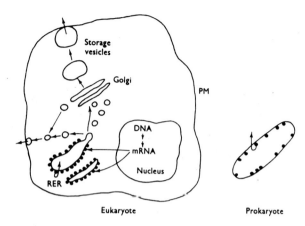

Eukaryote   Prokaryote

Figure 1. Schematic representation of the secretory process in eukaryotic
and prokaryotic cells. In the eukaryote, the path of the polypeptide chain from its synthesis on the bound ribosomes of the
rough endoplasmic reticulum (RER) and transfer into the cisternal
space to its eventual release from the plasma membrane (PM) by
exocytosis, is shown. Two modes of transport via smooth
membrane vesicles are illustrated, either directly or by way of
a golgi apparatus and storage vesicles. The proposed synthesis
on membrane-bound ribosomes in the prokaryote and transfer
through the membrane to the periplasmic space is also presented.
In gram-positive bacteria many of the secreted enzymes will pass
readily through the cell wall (not shown) into the extracellular
fluid. Gram-negative organisms tend to trap the enzymes between
the cytoplasmic membrane and the outer, more lipophilic cell wall.
(After Lampen, 1974).

In plant pathogenic fungi, there is evidence for these processes. Thus
Hislop, Barnaby, Shellis & Laborda (1974), following preliminary studies by
Calonge, Fielding & Byrde (1969), demonstrated histochemically in mycelium of
Monilinia (= Sclerotinia) fructigena the presence of sphcrosome-like bodies
with acid phosphatase or α-L-arabinofuranosidase activity. Other electron
micrographs showed lomasomes with enzyme activity that seemed to have arisen
by exocytosis (Figure 2). Such lomasomes seem more plentiful in host tissues
than in rich culture media (Calonge, Fielding & Byrde, 1969), but this
probably relfects an effect of the host on enzyme synthesis rather than on
secretion.

Maxwell, Williams & Maxwell (1972), by contrast, found no evidence to
correlate the presence of multivesicular bodies in hyphal tips with the

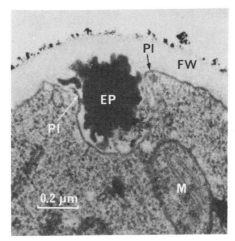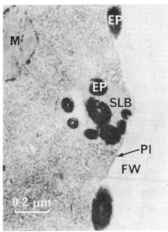

Figure 2. α-L-Arabinofuranosidase localisation in de-repressed mycelium of <u>Monilinia fructigena</u>. EP - enzyme product, M - mitochondrion, Pl - plasmalemma, SLB - spherosome-like body, FW - fungal wall (after Hislop et al., 1974).

secretion or production of extracellular hydrolase. However, there were most crystal-containing microbodies in hyphal tips grown on media supporting the greatest production of extracellular enzyme.

Once through the plasmalemma, the enzyme must traverse the wall before it can encounter host tissue. Binding to the fungal wall (Lisker, Katan, Chet & Henis, 1975) occurs (see Section 6); molecular sieving may also be involved, and it is at this stage that we believe the host may influence secretion in fungi, as distinct from synthesis. Chang & Trevithick (1974) proposed that the passage of extracellular enzymes through fungal walls occurs most easily at the apex of fungal hyphae (cf. Brown, 1917), where the wall is porous. As this wall becomes less porous with further growth, some extracellular enzymes are trapped 'in transit', and become wall-bound. Chang & Trevithick showed that three enzymes of high mol. wt - aryl-β-glucosidase, invertase (heavy) and trehalase - had a greater proportion of activity in the wall fraction than another three of lower mol. wt - ribonuclease, acid protease and amylase. The mean mol. wt of the former group

was approximately ten times greater than that of the latter. They also found
that during periods of vigorous culture growth, such as conidial germination
and early log phase, characterised by a high hyphal tip/total wall ratio,
the proportion of heavy invertase (mol. wt 210,000) which leaked into the
culture filtrate was maximal. Later, when hyphal tips comprised a lower
proportion of wall surface, a light form of invertase (mol. wt 51,000)
reached its greatest proportion in culture filtrates. If this hypothesis is
correct, any inhibition of hyphal growth by host defence mechanisms, e.g.
phytoalexins, would reduce the proportion of young, actively secreting hyphal
tips and thus alter the proportions of secreted enzymes or enzyme forms, in
favour of an accumulation of low mol. wt types.

In prokaryotes, by contrast, it is believed that enzyme synthesis
occurs on membrane-bound ribosomes followed by direct transfer through the
membrane to the periplasmic space.

There is good evidence (Lampen, 1974; Sanders & May, 1975) that the
bacterial enzymes are secreted as modified polypeptides which take on the
enzymic tertiary structure after passage through the membrane. Significantly
for our theme, Sanders & May showed that proteolytic enzymes, either of
bacterial or exogenous origin, could inactivate emerging α-amylase polypeptides which were more susceptible to proteolysis than the complete
molecule. They suggested that this was evidence that a conformational
change occurred during emergence and that the full tertiary structure was
assumed only after secretion. Just as in eukaryotes, the wall is also
recognised as a barrier to the liberation of bacterial enzymes (Lampen,
1974).

2. Non-specific Binding to Host Walls

After secretion the activity of the pathogen's enzymes may be modified
by various factors. The first barrier encountered is the host cell wall,
some of the constituent polymers of which are substrates for some of the
pathogen's enzymes.

There is microscopic evidence that the degradative effects of some
pathogen-secreted enzymes are very localised. This is especially true of
diseases caused by biotrophic pathogens, but is also noted in diseases caused
by some facultative parasites (Bracker & Littlefield, 1973). Brown rot of
apple is characterised by total colonisation of the host tissue, but chemical
(Cole & Wood, 1961a) and microscopic evidence (Calonge, Fielding, Byrde &

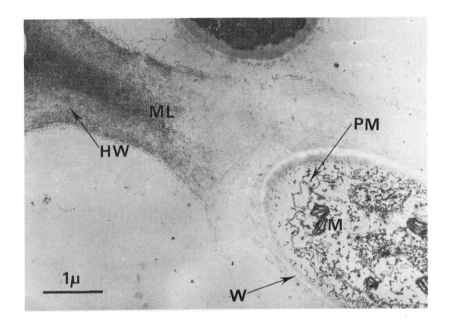

Figure 3. Electron micrograph of immature pear fruit tissue infected by
Monilinia fructigena, showing an intercellular hypha, and
adjacent localised host wall degradation. The host cell
contents have been lost from the field of view. The hypha at
top centre, with electron-dense contents, probably has a
storage function. HW - host wall, M - mitochondrion, ML -
middle lamellar region, PM - plasma membrane, W - fungal wall.
(Micrograph: R.J. Pring)

Akinrefon, 1969) reveals that the wall is degraded only in regions surrounding hyphae (Figure 3). Much of this restriction could be caused by the
physical constraints of the plant cell wall which, like the fungal wall, may
act as a molecular sieve. Thus enzymes such as phospholipase and protease
which kill naked higher plant protoplasts have no necrotic effect when
applied to tissue sections or callus (Tseng & Mount, 1974; Stephens & Wood,
1975). The most rational explanation for this phenomenon is that the plant
cell wall prevents these enzymes reaching the host plasmalemma, the presumed
site of action. This may be an over-simplification, as under the same
conditions host cytoplasmic proteins are able to diffuse in the reverse

direction (Stephens & Wood, 1975).

Other mechanisms besides molecular sieving can account for the observed immobilisation of pathogen enzymes. Non-covalent binding of both host and pathogen enzymes to plant cell walls is a well-documented phenomenon (Jansen, Jang & Bonner, 1960; Blackhurst & Wood, 1963; Parr & Edelman, 1975; Skare, Paus & Raa, 1975). Such binding is generally reversed by a high ionic concentration in the aqueous environment and, for this reason, a salt-based extraction medium is frequently employed when examining lesions for pecto-lytic enzyme content (Blackhurst & Wood, 1963; Keen & Erwin, 1971). For example, the extraction of polygalacturonase (PG) from Rhizopus stolonifer infected strawberries increases as the ionic strength is raised to approxi-mately 0.2M (Table I).

Table 1. Effect of salt concentration on the extraction of polygalac-turonase (PG) from strawberries infected with Rhizopus stolonifer.

Extractant	PG (relative)
Distilled water	100
0.05M K phosphate buffer, pH 7.0	113
Buffer + NaCl, 0.05M	173
Buffer + NaCl, 0.1M	282
Buffer + NaCl, 0.15M	298
Buffer + NaCl, 0.2M	296

20 g of chilled infected tissue was homogenized with 20 ml of extractant in an M.S.E. 'Atomix' homogenizer at full speed for 2 min. Extracts were centrifuged at 2,000 x g for 5 min at 4^0 and assayed as described by Calonge, Fielding, Byrde & Akinrefon (1969).

Surfactants can assist in the solubilisation of host wall-bound enzymes (Copping & Street, 1972) and we have noted a similar effect in M. fructigena lesions of apple fruit. Polygalacturonase extraction was increased by some of the cationic surfactants tested at 0.1% (w/v) (see also Section 3).

Once host cells have been killed, the interactions between host consti-tuents and pathogen enzymes described here will be augmented and modified by reactions involving cytoplasmic constituents. For example, the solute concentration may become sufficient to desorb enzymes from the walls. In

tissues rich in phenolic compounds (see Section 3), the enzyme-wall association may become irreversibly complexed by oxidized phenolic polymers.

3. The Role of Phenolic Compounds

Polyphenolic compounds occur in many plants and frequently influence
the relationship between host and pathogen. In healthy cells, the polyphenols
are generally believed to be localized in the vacuole, though there is
evidence of them in chloroplasts (Monties, 1975; Siedow & San Pietro, 1974),
and of enzymes involved in their biosynthesis (Gestetner & Conn, 1974).
Polyphenol oxidases have been detected in vesicles and amyloplasts and also
in the thylakoids of chloroplasts (see Czaninski & Catesson, 1974). As
uncontrolled oxidation does not occur in healthy tissue, enzyme and substrate, if present in the same organelle, are clearly not in contact.

When a necrotrophic pathogen invades tissue there is an increase in
permeability of the host membranes (Wheeler & Hanchey, 1968), so that polyphenol oxidases come into contact with their substrates, oxidizing them to
quinones. Sometimes the enzymes occur in a latent form, which can become
activated by a number of compounds, including the enzymic degradation
products of pectic substances (Deverall & Wood, 1961). The quinones formed
subsequently polymerize to compounds of high mol. wt and low aqueous
solubility. Components of this reaction inactivate polyphenolase, and many
other enzymes, including extracellular wall-degrading enzymes secreted by
pathogens.

Four mechanisms have been suggested to account for this inhibition.
Firstly, the quinone end-product may have a direct effect on the enzyme
protein, e.g. by the 1,4-addition of quinones to an amino or imino group
on the enzyme (Patil & Dimond, 1967) or reaction with thiol groups (Pierpoint,
1969). However, on the basis of the half-lives of the quinones, which are
less than 1 min for $\underline{o}$-quinone and DOPA-quinone, Kosuge (1969) considered
that the time required for inhibition in Patil & Dimond's experiments was
too long for this mechanism to be operative. Similar conditions presumably
apply to the other mechanisms for direct action quinones listed by Kosuge,
who also pointed out that highly active polygalacturonase has been isolated
from necrotic tissue with a high polyphenol oxidase and peroxidase content.

Secondly, the polyphenol oxidase may act directly on the protein of
the pathogen's enzyme. In large concentrations, phenolase can catalyse the
oxidation of tyrosine groups in proteins (Kosuge, 1969). This author also

Byrde & Archer - Extracellular enzymes of pathogens

speculated that phenolase might affect the membranes of plants and microorganisms. Such a mechanism does not seem relevant in the inhibition of
the PG of M. fructigena by oxidized apple extracts (Byrde, Fielding &
Williams, 1960), since autoxidized polyphenols were at least as effective
as those oxidized enzymically; also, the polyphenol oxidase preparation
alone was inactive (see also Tripathi, this symposium), and the synthetic
compound pentagalloyl glucose was strongly inhibitory (Williams, 1963).

Thirdly, the polymerized end-products of the polyphenol oxidase system
may inhibit the enzymes by a relatively non-specific protein precipitation
reaction, which can also be brought about by pre-formed polyphenols with
mol. wts of about 700 and above (Williams, 1963). The enzymes affected,
the inhibitory plant extracts and compounds involved were reviewed by Byrde
(1963), Williams (1963) and, for cellulases, by Mandels & Reese (1965).
The characterized polyphenols most often implicated as oxidized polymers
are catechins and leucoanthocyanins.

The mechanism of enzyme inactivation by plant polyphenols is not fully
known, but an insoluble enzyme-polyphenol complex seems to be formed.
Goldstein & Swain (1965) studied the non-competitive inhibition of a non-
fungal β-glucosidase by tannic acid (an octa- and nona-galloyl derivative
of glucose) and a condensed tannin. Precipitate formation depended on pH,
ionic strength and the concentration of enzyme and tannin. The enzyme could
be reactivated from the precipitate by cationic or non-ionic (but not
anionic) surfactants, by caffeine (cf. Mejbaum-Katzenellenbogen, Dobryszycka,
Boguslawska-Jaworska & Morawiecka, 1959), and also by such polymers as
polyvinyl pyrollidone and some polyethylene glycols. Goldstein & Swain
concluded that non-covalent linkages were involved in complex formation.
Enzymes have also been reactivated by treating tannic acid precipitates
with acetone (Hathway & Seakins, 1958), a method used also for enzyme
concentration (Shibata & Nisizawa, 1965).

The kinetics of the inhibition of cellulase by a melanin derived by
autoxidation of the phenolic compound DL-3,4-dihydroxyphenylalanine are
also relevant. Bull (1970) demonstrated that his inhibition data were
best fitted by the theoretical curve for a melanin:enzyme ratio of 3.
Lineweaver-Burk plots of the reaction showed a 'mixed-type' inhibition,
involving both competitive and irreversible non-competitive effects. An
earlier report had implicated competitive inhibition (of pectinase) by a
grape leaf component (Bell & Etchells, 1958), subsequently shown to be

tannin in nature (Porter & Schwartz, 1962). In Bull's experiments the enzyme-inhibitor complex was partly reactivated by simple bases, protamine and the cationic detergent hexadecyltrimethyl ammonium bromide (Cetavlon), reinforcing the view that covalent linkages were not involved in the enzyme-melanin reaction. Electrostatic attraction was suggested as an important factor in complex formation. The improved extraction of PG from infected fruit tissue mentioned in Section 2 of our review may thus largely reflect the reactivation of enzyme from complexes with oxidized host polyphenols.

Not all extracellular enzymes are inhibited by oxidized polyphenols to the same extent (e.g. Pollard, Kieser & Sissons, 1958; Hathway & Seakins, 1958). Multiple forms of an enzyme produced by a given pathogen may also vary in susceptibility (Laborda, Archer, Fielding & Byrde, 1974).

Fourthly, the oxidized polyphenols may act to 'protect" the substrate. For example, Beckman, Mueller & Mace (1974) showed that membranes of calcium pectate infused with polyphenols were more resistant to oxalic acid and pectolytic enzyme attach and postulated that phenolic infusion is an important defence reaction in vascular wilts. Lignification, suggested as a defence mechanism by Hijwegen (1963), can protect cell walls from degradation (Friend, 1973; Friend, Reynolds & Aveyard, 1973). Lignin may act as both a physical and chemical barrier. Friend (1976) suggested that esterification of wall polymers, especially galactan and galacturonan, at C_2 and C_3 by hydroxy-cinnamic acids may render them more resistant to enzymolysis. Other barriers to wall-degrading enzymes include plant gums and the often neglected wound response compound suberin (Wood, 1967, p. 149; Zucker & Hankin, 1970).

Although oxidation of polyphenols can result in inhibition of pectolytic enzymes, a role for oxidized phenols in disease resistance, first postulated by Cook , Bassett, Thompson & Taubenhaus (1911), cannot be considered proven (Wood, 1967). Hunter (1974) has since suggested that oxidized d-catechin is a factor limiting the spread of Rhizoctonia solani in older leaves of cotton, and pointed to the presence of larger quantities of inhibitor precursors in these leaves than in the more susceptible younger ones. Significantly, Cole & Wood (1961b) noted that pectolysis proceeded to a much greater extent in Bramley's Seedling apples rotted by Penicillium expansum, which seemed able to prevent phenolic oxidation, than in those rotted by some other pathogens (see also Walker, 1969, 1970). Byrde (1957) had earlier suggested that polyphenol oxidation might be particularly important on wound surfaces, and

it is in such places that oxidation can occur before the pathogen arrives;
within the tissues oxidation often occurs only after the pathogen has
damaged the cells and progressed further (Brown, 1965).

4. Effect of low Molecular Weight, Non-phenolic Cytoplasmic Constituents

When tissues are disrupted a range of low mol. wt host metabolites leak
into the intercellular environment where some may reach very high concentrations. This is true of certain organic acids, citric and malic acids
reaching concentrations of 0.1 M in certain fruits. The most obvious effect
of organic acids leaking into the intercellular space will be to alter the
pH which will modify the activity of any pathogen enzymes present and may
inactivate those which are acid-labile. The occurrence of multimolecular
forms of enzymes may also be pH-dependent, as has been shown for grape
catechol oxidase (Dubernet & Ribéreau-Gayon, 1974).

Irreversible loss of activity is usually limited to extremes of pH
such as are encountered in some fruit tissues. The extent to which
inactivation occurs is minimised by the 'robust' nature of extracellular
enzymes in general and pectolytic enzymes in particular. Enzymes, and even
multiple forms possessing the same activity, differ greatly in their
stability under extremely acid conditions; thus the α-L-arabinofuranosidase
from Corticium rolfsii is much more acid-tolerant (Kaji & Yoshihara, 1969)
than the two extracellular forms from M. (S.) fructigena (Laborda, Fielding
& Byrde, 1973). In addition, catalytic activity of the enzyme from C.
rolfsii is affected less by low pH than those from M. fructigena, which also
differ between themselves in this respect. The intracellular form of the
enzyme in M. fructigena is even less acid-tolerant.

Not all diseased tissues become more acid, and the changes may result
from the pathogen's activities as well as the host's. Upon infection of
Phaseolus bean by Sclerotium rolfsii, the pH of the host tissue decreases
from 6.0 to 4.0, close to the optimum for the PG secreted by this pathogen
(Bateman & Beer, 1965). A similar situation occurs in Sclerotinia sclerotiorum infection of sunflower and tomato stems (Hancock, 1966). On the
other hand, Colletotrichum trifolii infection of lucerne (Hancock & Millar,
1965), Fusarium solani f. phaseoli infection of bean (Bateman, 1965) and
F. solani f. cucurbitae infection of squash (Hancock, 1968) all result in
an increase in the pH to 7.0-8.0, conditions under which the activity of
pectin or pectate lyase, the principal wall-degrading enzyme present in

each disease, is enhanced.

Most host metabolites probably occur in sufficient concentration to affect a pathogen's extracellular enzymes. There are rare exceptions: thus some polyols are glucosidase and galactosidase inhibitors (Keleman & Whelan, 1966), while myoinositol and D-xylose are reported to inhibit α- but not β-galactosidases (Sharma, 1971; Sharma & Sharma, 1976). However, the importance of such inhibitors in plant pathogenesis is unknown.

Ethylene is of particular interest among host compounds of low mol. wt since, as a gas, its diffusion is rapid and it can act as an efficient messenger. Lund & Mapson (1970) showed that polygalacturonate lyase preparations from Erwinia carotovora could bring about release of ethylene from tissue of cauliflower florets. But more relevant to this paper are the stimulatory effects of ethylene on the action of preparations containing pectolytic enzymes detected by Cronshaw & Pegg (1976); ethylene can also act indirectly in the opposite direction by stimulating host defence mechanisms (Stahmann, Clare & Woodbury, 1966).

Of the mineral constituents of plants the only elements likely to affect pathogens' extracellular enzymes (few extracellular enzymes have co-factor requirements) are the divalent cations calcium and magnesium. Many pectate lyases, for example that of Erwinia carotovora (Starr & Moran, 1962), require calcium ions, whereas the enzyme from some other organisms, e.g. Bacillus polymyxa (Nagel & Wilson, 1970) is merely increased in activity. In practice, only severely nutrient-deficient plants are likely to contain insufficient calcium to affect a pathogen's pectate lyase measurably, unless some other effect such as calcium chelation by organic acids is also operative.

Calcium may also affect a pathogen's extracellular enzymes (in particular pectolytic) by an indirect mechanism via the substrate. A high calcium content in plant tissues has often been associated with disease resistance (Bateman & Lumsden, 1965; Edgington, Corden & Dimond, 1961; Skare et al., 1975) and is usually attributed to the formation of an insoluble calcium pectate complex resistant to the action of fungal PG (Bateman, 1964; Wallace, Kuĉ T Draudt, 1962). Bateman (1964) accounted for induced resistance of bean hypocotyls to Rhizoctonia solani by suggesting that in the early stages of disease, metallic cations - mainly calcium - accumulate at the infection site activating host pectin esterase which results in de-esterification of pectic substances. The resulting free pectic acids form calcium salts which are more resistant to degradation by PG. This hypothesis agreed with the observations that bean tissue bearing lesions is more difficult to macerate by enzyme than is healthy tissue, and that the middle lamella of cells adjacent to lesions is especially resistant. Corden (1965) demonstrated that there is sufficient calcium in the vascular tissue of tomato plants to decrease PG activity by 58%. By way of contrast, calcium has been shown to increase the maceration of potato slices by Colletotrichum trifolii (Hancock & Millar, 1965), presumably due to the stimulation of a calcium-dependent pectate lyase.

5. Specific Interactions with Host Constituents of High Molecular Weight

In Section 2 we have described non-specific binding of the pathogen's enzymes to plant cell walls, but a more specific type of interaction may exist. Plant cell walls and plasma membranes contain proteins or glyco-proteins, some of which possess carbohydrate-binding activity (Kauss & Glaser, 1974; Clarke, Knox & Jermyn, 1975; Jermyn & May Yeow, 1975; Bowles & Kauss, 1975; Kauss & Bowles, 1976). Such molecules (lectins) interact in a specific antibody-antigen like fashion (Lis & Sharon, 1973) with carbo-hydrate groups which may form part of a polysaccharide or glycoprotein molecule. Many, but certainly not all, fungal extracellular enzymes are glycoproteins (see Table 2), and might be expected to precipitate or interact in some other way with lectins possessing the appropriate sugar specificity. Such interactions will at least immobilise pathogens' extra-cellular enzymes and may lead to irreversible loss of activity.

The only experimental evidence for this hypothesis concerns inter-actions between plant wall proteins and fungal pectolytic enzymes, none of which have been shown to be glycoproteins. Of the pathogen-secreted enzymes tested by Albersheim & Anderson (1971), only PGs were inhibited, a significant result in view of the postulated importance of these enzymes in 'opening up' the cell wall to digestion by other enzymes. The evidence available suggests that such interactions, not necessarily of a lectin-carbohydrate nature, are not involved in host-parasite specificity, as each of several inhibitors from different plants reacted with the PG from a variety of unrelated pathogens (Albersheim & Anderson-Prouty, 1975). For example, an apparently homogeneous preparation of the inhibitor protein from Phaseolus vulgaris hypocotyls inhibited almost equally endo-PGs secreted by <u>Colletotrichum lindemuthianum</u>, <u>Helminthosporium maydis</u> and <u>Aspergillus niger</u>. Furthermore specificity of interactions involving races of <u>C. lindemuthianum</u> and cultivars of <u>P. vulgaris</u> showed no correlation with the abilities of the inhibitor proteins to inactivate the PGs from the various races (Anderson & Albersheim, 1972). It was concluded that the endo-PGs from the three races studied were identical, as were the three inhibitor proteins. Inhibition of the pectinase of <u>Cladosporium cucumerinum</u> by a cucumber hypocotyl wall component was reported by Skare et al. (1975). Although the inhibitor was not further characterised, it could be washed from walls by molar NaCl solution, a treatment stated to remove ionic bound proteins. The occurrence in plants of such inhibitors may provide some background resistance against weak pathogens, allowing only those organisms that are able to secrete sufficient PG to saturate the host defence system, to initiate an infection. The wider applicability of this model awaits verification.

6. Possible Effects of Host Enzymes on Pathogen Enzymes

It is possible that pathogen enzymes may be affected directly or indirectly by host enzymes. Thus, host proteases might degrade pathogen enzymes, although the latter have usually evolved a structure resistant to such an obvious hazard (e.g. Fielding & Byrde, 1969). However, the precursors of extracellular bacterial enzymes may be especially vulnerable as they emerge from the membrane (see Section 1).

Many extracellular fungal enzymes are glycoproteins, and examples of these are listed in Table 2. We have indications that a PG (pI 4.6) of <u>M. fructigena</u>, known to be a very thermostable enzyme (Archer & Fielding, 1975), is also a glycoprotein. Such enzymes would be liable to lose their carbohydrate moiety if they encounter a host glycosidase, in situ or following liberation of the latter after wall degradation. Relevant enzymes that occur in plants include α- and β-glucosidases, α-galactosidase (Dey & Pridham, 1969), β-galactosidase (Bartley, 1974; Wallner & Walker, 1975), β-1,3-glucanase (Albersheim & Valent, 1974; Wallner & Walker, 1975), <u>N</u>-acetyl-glucosaminidase (Li & Li, 1970), and α-mannosidase (Li, 1966). Mannosidases may be particularly important in deglycosylation because of the widespread occurrence of mannose in the fungal glyco-enzymes (Table 2). However, the mere presence of mannose, or any other sugar, in the molecule

Table 2. Sugars present in some extracellular fungal glyco-enzymes.

Sugars	Enzyme	Fungus	Reference
Glucosamine, glucose, mannose	α-glucosidase	Aspergillus fumigatus	Rudick & Elbein (1974)
Glucosamine, mannose	β-glucosidase I and II	A. fumigatus	Rudick & Elbein (1973, 1975)
Galactose, glucose, mannose	Glucoamylase	A. niger	Pazur et al. (1963)
Mannose	α-amylase		
Glucose, mannose	Pectin lyase I and II	A. niger	Van Houdenhoven (1975)
Glucosamine, mannose	Invertase	Neurospora crassa	Meachum et al. (1971)
N-Acetylglucosamine, mannose	Amylase	Rhizopus javanicus	Watanabe & Fukimbara (1975)
Glucose, mannose	Endo-1,4-β glucanase T_1	Sporotrichum pulverulentum	Eriksson & Pettersson (1975)
Arabinose, galactose, mannose	Endo-1,4-β glucanase T_{2b}		

affords no proof of its availability as a substrate.

The removal of the carbohydrate component would probably decrease stability (Pazur, Knull & Simpson, 1970) and resistance to dehydration (Darbyshire, 1974) and increase susceptibility to proteolytic attack (see Section 7).

Polyhydroxyl compounds can stabilise proteins in solution (Bradbury & Jakoby, 1972; Heitefuss, Buchanan-Davidson & Stahmann, 1959), and we have evidence (Table 3) that the polysaccharide mucilage of M. fructigena will help 'protect' its enzymes from inactivation by low pH and by oxidized phenolic compounds. Microbial enzymes may also be 'protected' to some extent by being embedded in the pathogen cell wall (see Section 1). Then such enzymes as chitinase, endo-β-1,3-glucanase and lysozyme which have been suggested as host defence factors (Albersheim & Anderson-Prouty, 1975), might expose and release microbial enzymes as their parent wall is lysed. This may temporarily increase the activity of tne newly-released enzyme, but will later expose it to the deleterious effects that this review describes.

Table 3.   Inactivation of pectin lyase (PL), pectin esterase (PE) and
           polygalacturonase (PG) by low pH and oxidized 0.01 M-catechin,
           and protection by extracellular polysaccharide (EP) of
           M. fructigena.   Results expressed as % of control.

Treatment	PL	PE*	PG (pI 4.6)
pH 2.5	47	100	47
ditto + 0.2% EP	75	100	100
oxidized catechin	-	71	63
ditto + 0.2% EP	-	100	100

*mixture of enzyme forms.

Exposure to low pH (citrate-phosphate buffer) and oxidized
catechin (at pH 5.5 in 0.05 M-citrate buffer) was for 30
min, except for PG (24 h).

For assay procedures see Calonge, Fielding, Byrde &
Akinrefon (1969).

7.   Inhibitors (High Molecular Weight) of Specific Pathogen Enzymes

There are reports of the occurrence in plants of enzyme inhibitors of
high mol. wt which are themselves considered to be proteinaceous.   Their
relevance to plant pathology is uncertain because most of the reports
concern other biological systems.

There are well-known and widespread protease inhibitors (Ryan, 1973)
which characteristically inhibit enzymes of animal or microbial origin
having specificities similar to trypsin or chymotrypsin, for example
pronase and subtilisin.   Inhibition of plant proteases has been less
frequently reported, but some inhibitors have been implicated in resistance
to insect damage and it is feasible that this role may extent also to
microbial invaders.   Further support for the 'protective agent' theory
came from the finding that inhibitor concentrations in potato and tomato
leaves increased rapidly following insect damage (Green & Ryan, 1972).
The effect, which could be simulated by wounding leaves, was systemic,
resulting in a rapid accumulation of protease inhibitor throughout the
foliage.

Following earlier work on cereals (Kneen & Sandstedt, 1946; Miller &

Kneen, 1947), wheat has been reported to contain an α-amylase inhibitor of macromolecular dimensions which is presumably a protein (O'Donnell & McGeeny, 1976). The inhibitory effect against a fungal α-amylase was not particularly strong when compared with the effect on, enzyme from some mammalian sources, but this is probably of little significance as the fungus tested, Aspergillus oryzae, is not a pathogen of wheat.

The possibility of a range of plant proteins inhibiting fungal and bacterial extracellular enzymes clearly demands further study. Some may prove to be identical to the cell wall glycoproteins already discussed in Section 5: others located in the cytoplasm may have other primary functions and only become involved in the plant-parasite interaction following death of the host cell.

Conclusion

Modification to pathogen enzymes that occurs early in the invasion processes may change the whole course of infection, whereas effects after cell death only modify symptoms rather than the balance of the host-pathogen relationship. Thus, effects on enzyme secretion, and interactions of the pathogen's enzymes with host glycoproteins, may be expected to be of more importance to disease development than those of oxidized polyphenols and other cytoplasmic or vacuolar constituents.

The multiple forms common among pathogen enzymes (Garibaldi & Bateman, 1971; Laborda et al., 1974; Hancock, 1976) may confer an advantage on the pathogen, not least because specialized forms may be particularly stable under adverse conditions encountered in the host. The presence of carbohydrate components leads to the stabilization of the three-dimensional structure of the protein (Pazur et al., 1970), so glycoproteins are likely to be more resistant to proteolytic attack and extremes of temperature, although susceptible to binding by lectins.

The inhibition of the wall-degrading enzymes of a pathogen is not necessarily a disadvantage (Skare et al., 1975) but may be a characteristic of a well-balanced host-pathogen relationship. Thus, any localisation of these enzymes by the host represents a form of control over them, and hence an evolutionary step towards a more biotrophic relationship (cf. Lewis, 1974). Also, necrotrophic fungi which have some degree of specialisation towards their hosts often rely on them for overwintering. Thus, if the Monilinia spp. were as effective as Penicillium expansum in destroying wall tissue (Cole & Wood, 1961a), their ability to survive on mummified fruits would be greatly if not completely impaired (Byrde & Willetts, 1977).

Clearly, the types of interplay between host and pathogen outlined in this review do not provide a basis for specificity in 'gene-for-gene' interactions. However, many of the processes we have described may have a delaying effect on the establishment or spread of a pathogen. They may thus contribute to the accumulation of factors constituting non-race-specific ('field') resistance, the importance of which in both natural plant communities and also in crop situations is being increasingly realized.

Acknowledgements

We wish to thank Mr. R.J. Pring for permission to use his unpublished electron micrograph; also Professor J.O. Lampen and Dr. E.C. Hislop for permission to use published illustrations.

References

Albersheim, P. & Anderson, A.J. (1971). Proteins from plant cell walls inhibit polygalacturonases secreted by plant pathogens. Proceedings of the National Academy of Science of the U.S.A. 68, 1815-1819.

Albersheim, P. & Anderson-Prouty, A.J. (1975). Carbohydrates, proteins, cell surfaces and the biochemistry of pathogenesis. Annual Review of Plant Physiology 26, 31-52.

Albersheim, P. & Valent, B.S. (1974). Host-pathogen interactions. VII. Plant pathogens secrete proteins which inhibit enzymes of the host capable of attacking the pathogen. Plant Physiology 53, 684-687.

Anderson, A.J. & Albersheim, P. (1972). Host-pathogen interactions. V. Comparison of the abilities of proteins isolated from three varieties of Phaseolus vulgaris to inhibit the endopolygalacturonases secreted by three races of Colletotrichum lindemuthianum. Physiological Plant Pathology 2, 339-346.

Archer, S.A. & Fielding, A.H. (1975). Thermostable polygalacturonase secreted by Sclerotinia fructigena. Journal of Food Science 40, 423-424.

Bartley, I.M. (1974). β-Galactosidase activity in ripening apples. Phytochemistry 13, 2107-2111.

Bateman, D.F. (1964). An induced mechanism of tissue resistance to polygalacturonase in Rhizoctonia-infected hypocotyls of bean. Phytopathology 54, 438-445.

Bateman, D.F. (1966). Hydrolytic and trans-eliminative degradation of pectic substances by extracellular enzymes of Fusarium solani f. phaseoli. Phytopathology 56, 238-244.

Bateman, D.F. & Beer, S.V. (1965). Simultaneous production and synergistic action of oxalic acid and polygalacturonase during pathogenesis by Sclerotium rolfsii. Phytopathology 55, 204-211.

Bateman, D.F. & Lumsden, R.D. (1965). Relation of calcium content and nature of the pectic substances in bean hypocotyls of different ages to susceptibility to an isolate of Rhizoctonia solani. Phytopathology 55, 734-738.

Beckman, C.H., Mueller, W.C. & Mace, M.E. (1974). The stabilization of artificial and natural cell wall membranes by phenolic infusion and its relation to wild disease resistance. Phytopathology 64, 1214-1220.

Bell, T.A. & Etchells, J.L. (1958). Pectinase inhibitor in grape leaves. Botanical Gazette 119, 192-196.

Blackhurst, F.M. & Wood, R.K.S. (1963). Verticillium wilt of tomatoes - further experiments on the role of pectic and cellulolytic enzymes. Annals of Applied Biology 52, 89-96.

Bowles, D.J. & Kauss, H. (1975). Carbohydrate binding proteins from cellular membranes of plant tissue. Plant Science Letters 4, 411-418.

Bracker, C.E. & Littlefield, L.J. (1973). Structural concepts of host-pathogen interfaces. pp. 159-317 in Fungal pathogenicity and the plant's response, ed. by R.J.W. Byrde & C.V. Cutting. Academic Press, London & New York.

Bradbury, S.B. & Jakoby, W.B. (1972). Glycerol as an enzyme stabilizing agent: effects on aldehyde dehydrogenase. Proceedings of the National Academy of Sciences of the U.S.A. 69, 2373-2376.

Brown, W. (1917). Studies in the physiology of parasitism. IV. On the distribution of cytase in cultures of Botrytis cinerea. Annals of

Botany 31, 489-498.

Brown, W. (1965). Toxins and cell-wall dissolving enzymes in relation to plant disease. Annual Review of Phytopathology 3, 1-18.

Bull, A.T. (1970). Kinetics of cellulase inactivation by melanin. Enzymologia 39, 333-347.

Byrde, R.J.W. (1957). The varietal resistance of fruits to brown rot. II. The nature of resistance in some varieties of cider apples. Journal of Horticultural Science 32, 227-238.

Byrde, R.J.W. (1963). Natural inhibitors of fungal enzymes and toxins in disease resistance. Connecticut Agricultural Experiment Station Bulletin 663, 31-47.

Byrde, R.J.W., Fielding, A.H. & Williams, A.H. (1960). The role of oxidized polyphenols in the varietal resistance of apples to brown rot. pp. 95-99 in Phenolics in plants in health and disease, ed. by J.B. Pridham. Pergamon Press, Oxford.

Byrde, R.J.W. & Willetts, H.J. (1977). The brown rot fungi of fruit: their biology and control. 171 pp. Pergamon Press, Oxford.

Calonge, F.D., Fielding, A.H. & Byrde, R.J.W. (1969). Multivesicular bodies in Sclerotinia fructigena and their possible relation to extracellular enzyme secretion. Journal of General Microbiology 55, 177-184.

Calonge, F.D., Fielding, A.H., Byrde, R.J.W. & Akinrefon, O.A. (1969). Changes in ultrastructure following fungal invasion and the possible relevance of extracellular enzymes. Journal of Experimental Botany 20, 350-357.

Chang, P.L.Y. & Trevithick, J.R. (1974). How important is secretion of exoenzymes through apical walls of fungi? Archives of Microbiology 101, 281-293.

Clarke, A.E., Knox, R.B. & Jermyn, M.A. (1975). Localisation of lectins in legume cotyledons. Journal of Cell Science 19, 157-167.

Cole, M. & Wood, R.K.S. (1961a). Types of rot, rate of rotting and analysis of pectic substances in apples rotted by fungi. Annals of Botany 25, 417-434.

Cole, M. & Wood, R.K.S. (1961b). Pectic enzymes and phenolic substances in apples rotted by fungi. Annals of Botany 25, 435-452.

Cook, M.T., Bassett, H.P., Thompson, F. & Taubenhaus, J.J. (1911). Protective enzymes. Science 33, 624-629.

Copping, L.G. & Street, H.E. (1972). Properties of the invertases of cultured sycamore cells and changes in their activity during culture growth. Physiologia Plantarum 26, 346-354.

Corden, M.E. (1965). Influence of calcium nutrition on Fusarium wilt of tomato and polygalacturonase activity. Phytopathology 55, 222-224.

Cronshaw, D.K. & Pegg, G.F. (1976). Ethylene as a toxin synergist in Verticillium-wilt of tomato. Physiological Plant Pathology 9, 33-44.

Czaninski, Y. & Catesson, A.-M. (1974). Polyphenol oxidases (plants). Vol. 2, pp. 66-78 in Electron microscopy of enzymes: Principles and Methods, ed. by M.A. Hayat. van Nostrand Reinhold, New York.

Darbyshire, B. (1974). The function of the carbohydrate units of three fungal enzymes in their resistance to dehydration. Plant Physiology 54, 717-721.

Deverall, B.J. & Wood, R.K.S. (1961). Chocolate spot of beans (Vicia faba L.) - interactions between phenolase of host and pectic enzymes of the pathogen. Annals of Applied Biology 49, 473-487.

Dey, P.M. & Pridham, J.B. (1969). Substrate specificity and kinetic properties of α-galactosidases from Vicia faba. Biochemical Journal 115, 47-54.

Dubernet, M. & Ribéreau-Gayon, P. (1974). Isoelectric point changes in Vitis vinifera catechol oxidase. Phytochemistry 13, 1085-1087.

Edgington, L.V., Corden, M.E. & Dimond, A.E. (1961). The role of pectic substances in chemically induced resistance to Fusarium wilt of tomato. Phytopathology 51, 179-182.

Eriksson, K.E. & Petterson, B. (1975). Extracellular enzyme system utilized by the fungus Sporotrichum pulverulentum (Chrysosporium lignorum) for the breakdown of cellulose. I. Separation, purification and physico-chemical characterisation of five endo-1,4-β glucanases. European Journal of Biochemistry 51, 193-206.

Fielding, A.H. & Byrde, R.J.W. (1969). The partial purification and properties of endopolygalacturonase and α-L-arabinofuranosidase secreted by Sclerotinia fructigena. Journal of General Microbiology 58, 73-84.

Friend, J. (1973). Resistance of potato to Phytophthora. pp. 383-399 in Fungal pathogenicity and the plant's response, ed. by R.J.W. Byrde & C.V. Cutting. Academic Press, London & New York.

Friend, J. (1976). Lignification in infected tissues. pp. 291-304 in .

Biochemical aspects of plant-parasite relationships, ed. by J. Friend & D.R. Threlfall. Academic Press, London & New York.

Friend, J., Reynolds, S.B. & Aveyard, M.A. (1973). Phenylalanine ammonia lyase, chlorogenic acid and lignin in potato tuber tissue inoculated with Phytophthora infestans. Physiological Plant Pathology 3, 495-507.

Garibaldi, A. & Bateman, D.F. (1971). Pectic enzymes produced by Erwinia chrysanthemi and their effects on plant tissue. Physiological Plant Pathology 1, 25-40.

Gestetner, B. & Conn, E.E. (1974). The 2-hydroxylation of trans-cinnamic acid by chloroplasts from Melilotus alba Desr. Archives of Biochemistry and Biophysics 163, 617-624.

Goldstein, J.L. & Swain, T. (1965). The inhibition of enzymes by tannins. Phytochemistry 4, 185-192.

Green, T.R. & Ryan, C.A. (1972). Wound-induced proteinase inhibitor in plant leaves: a possible defence mechanism against insects. Science, New York 175, 776-777.

Hancock, J.G. (1966). Degradation of pectic substances associated with pathogenesis by Sclerotinia sclerotiorum in sunflower and tomato stems. Phytopathology 56, 975-979.

Hancock, J.G. (1968). Degradation of pectic substances during pathogenesis by Fusarium solani f.sp. cucurbitae. Phytopathology 58, 62-69.

Hancock, J.G. (1976). Multiple forms of endo-pectate lyase formed in culture and in infected squash hypocotyls by Hypomyces solani f.sp. cucurbitae. Phytopathology 66, 40-45.

Hancock, J.G. & Millar, R.L. (1965). Relative importance of polygalacturonate trans-eliminase and other pectolytic enzymes in southern anthracnose, spring black stem and Stemphylium leaf spot of alfalfa. Phytopathology 55, 346-355.

Hathway, D.E. & Seakins, J.W.T. (1958). The influence of tannins on the degradation of pectin by pectinase enzymes. Biochemical Journal 70, 158-163.

Heitefuss, R., Buchanan-Davidson, D.J. & Stahmann, M.A. (1959). The stabilization of extracts of cabbage leaf proteins by polyhydroxyl compounds for electrophoretic and immunological studies. Archives of Biochemistry and Biophysics 85, 200-208.

Hijwegen, T. (1963). Lignification, a possible mechanism of active resistance against pathogens. Netherlands Journal of Plant

Pathology 69, 314-317.

Hislop, E.C., Barnaby, V.M., Shellis, C. & Laborda, F. (1974). Localization of α-L-arabinofuranosidase and acid phosphatase in mycelium of Sclerotinia fructigena. Journal of General Microbiology 81, 79-99.

Hunter, R.E. (1974). Inactivation of pectic enzymes by polyphenols in cotton seedlings of different ages infected with Rhizoctonia solani. Physiological Plant Pathology 4, 151-159.

Jansen, E.F., Jang, R. & Bonner, J. (1960). Binding of enzymes to Avena coleoptile cell walls. Plant Physiology 35, 567-574.

Jermyn, M.A. & May Yeow, Y. (1975). A class of lectins present in the tissues of seed plants. Australian Journal of Plant Physiology 2, 501-531.

Kaji, A. & Yoshihara, O. (1969). Production and properties of α-L-arabinofuranosidase from Corticium rolfsii. Applied Microbiology 17, 910-913.

Kauss, H. & Bowles, D.J. (1976). Some properties of carbohydrate-binding proteins (lectins) solubilized from cell walls of Phaseolus aureus. Planta (Berlin) 130, 169-174.

Kauss, H. & Glaser, C. (1974). Carbohydrate binding proteins from plant cell walls and their possible involvement in extension growth. FEBS Letters 45, 304-307.

Keen, N.T. & Erwin, D.C. (1971). Endopolygalacturonase: evidence against involvement in Verticillium wilt of cotton. Phytopathology 61, 198-203.

Keleman, M.V. & Whelan, W.J. (1966). Inhibitions of glucosidases and galactosidases by polyols. Archives of Biochemistry and Biophysics 117, 423-428.

Kneen, E. & Sandstedt, R.M. (1946). Distribution and general properties of an amylase inhibitor in cereals. Archives of Biochemistry 9, 235-249.

Kosuge, T. (1969). The role of phenolics in host response to infection. Annual Review of Phytopathology 7, 195-222.

Laborda, F., Fielding, A.H. & Byrde, R.J.W. (1973). Extra- and intra-cellular α-L-arabinofuranosidase of Sclerotinia fructigena. Journal of General Microbiology 79, 321-329.

Laborda, F., Archer, S.A., Fielding, A.H. & Byrde, R.J.W. (1974). Studies on the α-L-arabinofuranosidase complex from Sclerotinia fructigena in relation to brown rot of apple. Journal of General Microbiology 81, 151-163.

Lampen, J.O. (1974). Movement of extracellular enzymes across cell membranes. In Transport at the cellular level, ed. by M.A. Sleigh & D.H. Jennings. Symposia of the Society of Experimental Biology 28, 351-374.

Lewis, D.H. (1974). Micro-organisms and plants: the evolution of parasitism and mutualism. In Evolution in the microbial world, ed. by M.J. Carlile & J.J. Skehel. Symposia of the Society for General Microbiology 24, 367-392. University Press, Cambridge.

Li, S.-C. & Li, Y.-T. (1970). Studies on the glycosidases of jack bean meal. III. Crystallization and properties of β-N-acetylhexosaminidase. Journal of Biological Chemistry 245, 5153-5160.

Li, Y.-T. (1966). Presence of α-D-mannosidic linkage in glycoproteins. Liberation of D-mannose from various glycoproteins by α-mannosidase isolated from jack bean meal. Journal of Biological Chemistry 241, 1010-1012.

Lis, H. & Sharon, N. (1973). The biochemistry of plant lectins (phytohaemagglutinins). Annual Review of Biochemistry 42, 541-574.

Lisker, N., Katan, J., Chet, I. & Henis, Y. (1975). Release of cell-bound polygalacturonase and cellulase from mycelium of Rhizoctonia solani. Canadian Journal of Microbiology 21, 521-526.

Lund, B.M. & Mapson, L.W. (1970). Stimulation by Erwinia carotovora of the synthesis of ethylene in cauliflower tissue. Biochemical Journal 119, 251-263.

Mandels, M. & Reese, E.T. (1965). Inhibition of cellulases. Annual Review of Phytopathology 3, 85-102.

Maxwell, D.P., Williams, P.H. & Maxwell, M.D. (1972). Studies on the possible relationships of microbodies and multivesicular bodies and oxalate, endopolygalacturonase and cellulase (Cx) production by Sclerotinia sclerotiorum. Canadian Journal of Botany 50, 1743-1748.

Meachum, Z.D., Colvin, H.J. & Braymer, H.D. (1971). Chemical and physical studies of Neurospora crassa invertase: molecular weight, amino acid and carbohydrate composition, and quarternary structure. Biochemistry 10, 326-332.

Mejbaum-Katzenellenbogen, W., Dobryszycka, W., Boguslawska-Jaworska, J. & Morawiecka, B. (1959). Regeneration of protein from insoluble protein-tannin compounds. Nature, London 184, 1799-1800.

Miller, B.S. & Kneen, E. (1947). The amylase inhibitor of Leoti sorghum. Archives of Biochemistry 15, 251-264.

Monties, B. (1975). Compartimentation des polyphénols présents dans les feuilles et les chloroplastes d'angiospermes: mise en évidence par extraction progressive. Compte rendu de l'Academie des Sciences, Paris. Série C, 280, 1331-1334.

Nagel, C.W. & Wilson, T.M. (1970). Pectic acid lyases of Bacillus polymyxa. Applied Microbiology 20, 374-383.

O'Donnell, M.D. & McGeeny, K.F. (1976). Purification and properties of an α-amylase inhibitor from wheat. Biochimica et Diophysica Acta 422, 159-169.

Parr, D.R. & Edelman, J. (1975). Release of hydrolytic enzymes from cell walls of intact and disrupted carrot callus tissue. Planta (Berlin) 127, 111-119.

Patil, S.S. & Dimond, A.E. (1967). Inhibition of Verticillium polygalac-turonase by oxidation products of polyphenols. Phytopathology 57, 492-496.

Pazur, J.H., Kleppe, K. & Ball, E.M. (1963). The glycoprotein nature of some fungal carbohydrases. Archives of Biochemistry and Biophysics 103, 515-516.

Pazur, J.H., Knull, H.R. & Simpson, D.L. (1970). Glycoenzymes: a note on the role for the carbohydrate moieties. Biochemical and Biophysical Research Communications 40, 110-116.

Pierpoint, W.S. (1969). o-Quinones formed in plant extracts. Their reaction with bovine serum albumin. Biochemical Journal 112, 619-629.

Pollard, A., Kieser, M.E. & Sissons, D.J. (1958). Inactivation of pectic enzymes by fruit phenolics. Chemistry and Industry 1958, 952.

Porter, W.L. & Schwartz, J.H. (1962). Isolation and description of the pectinase inhibiting tannins of grape leaves. Journal of Food Science 27, 416-418.

Rudick, M.J. & Elbein, A.D. (1973). Glycoprotein enzymes secreted by Aspergillus fumigatus. Purification and properties of β-glucosidase. Journal of Biological Chemistry 248, 6506-6513.

Rudick, M.J. & Elbein, A.D. (1974). Glycoprotein enzymes secreted by Aspergillus fumigatus. Purification and properties of α-glucosidase. Archives of Biochemistry and Biophysics 161, 281-290.

Rudick, M.J. & Elbein, A.D. (1975). Glycoprotein enzymes secreted by
Aspergillus fumigatus. Purification and properties of a second β-
glucosidase. Journal of Bacteriology 124, 534-541.
Ryan, C.A. (1973). Proteolytic enzymes and their inhibitors in plants.
Annual Review of Plant Physiology 24, 173-196.
Sanders, R.L. & May, B.K. (1975). Evidence for extrusion of unfolded
extracellular enzyme polypeptide chains through membranes of Bacillus
amyloliquefaciens. Journal of Bacteriology 123, 806-814.
Sharma, C.B. (1971). Selective inhibition of α-galactosidases by
myoinositol. Biochemical and Biophysical Research Communications
43, 572-579.
Sharma, T.N. & Sharma, C.B. (1976). D-xylose, an anomer-specific inhibitor
of α-galactosidase. Phytochemistry 15, 643-646.
Shibata, Y. & Nisizawa, K. (1965). Microheterogeneity of β-glycosidases
in apricot emulsin. Archives of Biochemistry and Biophysics 109,
516-621.
Siedow, J.N. & San Pietro, A. (1974). Studies on photosystem I. Characteristics of "310 material" isolated from spinach chloroplasts.
Archives of Biochemistry and Biophysics 164, 145-155.
Skare, N.H., Paus, F. & Raa, J. (1975). Production of pectinase and
cellulase by Cladosporium cucumerinum with dissolved carbohydrates
and isolated cell walls of cucumber as carbon sources. Physiologia
Plantarum 33, 229-233.
Stahmann, M.A., Clare, B.G. & Woodbury, W. (1966). Increased disease
resistance and enzyme activity induced by ethylene and ethylene production by black rot infected sweet potato tissue. Plant Physiology
41, 1505-1512.
Starr, M.P. & Moran, F. (1962). Eliminative split of pectic substances by
phytopathogenic soft-rot bacteria. Science, New York 135, 920-921.
Stephens, G.J. & Wood, R.K.S. (1975). Killing of protoplasts by soft-rot
bacteria. Physiological Plant Pathology 5, 165-181.
Tseng, T.C. & Mount, M.S. (1974). Toxicity of endopolygalacturonate trans-
eliminase, phosphatidase and protease to potato and cucumber tissue.
Phytopathology 64, 229-236.
Van Houdenhoven, F.E.A. (1975). Studies on pectin lyase. Mededelingen
Landbouwhogeschool Wageningen, 75-13.
Walker, J.R.L. (1969). Inhibition of the apple phenolase system through

infection by Penicillium expansum. Phytochemistry 8, 561-566.

Walker, J.R.L. (1970). Phenolase inhibitor from cultures of Penicillium expansum which may play a part in fruit rotting. Nature, London 227, 298-299.

Wallace, J., Kuč, J. & Draudt, H.M. (1962). Biochemical changes in the water-insoluble material of maturing apple fruit and their possible relationship to disease resistance. Phytopathology 52, 1023-1027.

Wallner, S.J. & Walker, J.E. (1975). Glycosidases in cell wall-degrading extracts of ripening tomato fruits. Plant Physiology 55, 94-98.

Watanabe, K. & Fukimbara, T. (1975). The structure of the carbohydrate moiety of a glycopeptide Gp-1-b from Rhizopus saccharogenic amylase. Agricultural and Biological Chemistry 39, 1711-1717.

Wheeler, H. & Hanchey, P. (1968). Permeability phenomena in plant disease. Annual Review of Phytopathology 6, 331-350.

Williams, A.H. (1963). Enzyme inhibition by phenolic compounds. pp. 87-95 in Enzyme chemistry of phenolic compounds, ed. by J.B. Pridham. Pergamon Press, Oxford.

Wood, R.K.S. (1967). Physiological Plant Pathology. Blackwell Publications Ltd., Oxford & Edinburgh. 570 pp.

Zucker, M. & Hankin, L. (1970). Physiological basis for a cycloheximide-induced soft rot of potatoes by Pseudomonas fluorescens. Annals of Botany 34, 1047-1062.

Late insertion: For a comprehensive review of proteinase inhibitors (Section 7), see Richardson, M. The proteinase inhibitors of plants and micro-organisms. Phytochemistry (in press).

DISCUSSION

Chairman: D.F. Bateman

Bateman: Dr. Pegg, would you comment on the role of ethylene?

Pegg: The work referred to on Verticillium concerns the interaction of
ethylene with a polysaccharide from the culture filtrate. The polysaccharide is most probably a fructosan with a molecular weight in the region of
30,000. In the purification procedure used it is completely free from
pectolytic enzyme activity. The Verticillium wilt syndrome has been
attributed to several different agents based on a bioassay of culture
filtrate components; a lipopolysaccharide (Keen, Long & Erwin, Physiological Plant Pathology 2, 317-331, 1972), endo polygalacturonase (Mussell,
Phytopath. 63, 62-70, 1973), and pectin transeliminase (Heale & Gupta,
Trans. Brit. Mycol. Soc. 58, 19-28, 1972). The results we have obtained
suggest that all these groups of workers could be correct. Enzyme fractions
containing endo PG and PTE induced wilting of excised tomato leaves in 6
hours and also stimulated endogenous ethylene release from the leaves. The
protein-free polysaccharide fractions induced neither wilting nor increased
ethylene release. When leaves which had been treated with 5.0 ppm ethylene
were exposed to the polysaccharide, wilting, chlorosis and necrosis developed in a short time. A point of further interest was that ethylene pretreatment of tomato plants markedly increased the sensitivity of detached
leaves to pectolytic enzymes.

    It is very clear to us that Verticillium is producing more than one
metabolite in the infected host and that these molecules are interacting.
If our results are right, they have implications for other host-pathogen

interactions where ethylene is produced.

Could I ask Dr. Byrde how the mucilage he mentioned might protect the enzymes from inhibitors.

Byrde: The suggestion was made on the basis that other workers have shown that polyhydroxyl substances can protect proteins.

Mussell: I am very curious to know if you or S.A. Archer have carried out any experiments in which you have put together the pathogen enzyme and isolated host cell wall in order to see if you get an inhibition or destruction of the pathogen enzyme. I am thinking in terms of a real disease situation in which the pathogen enzyme has host cell wall to inter- act with, bind to, hide behind, etc. In some of our work, at least, binding of enzymes to cell walls seems to cut down on enzyme activity. Perhaps binding of enzymes on cell walls can bring about changes in the steric configuration of the enzyme which might protect it from some of the inhibitors you are postulating.

Byrde: No, I don't think so. We haven't carried out such studies. In our work with the macerating enzyme of course, this has been carried out in infected tissues. In this case the oxidized phenolics are effective inhibitors.

Mussell: I have been trying to rationalize the work published by Hunter which you mentioned. From the studies carried out at the National Cotton Pathology Laboratory one would think that there is no way a pectic enzyme would have any chance of survival in a cotton plant because of all the quinones and phenolics that are present. Yet, with Verticillium and Rhizoctonia infections as well as a number of other diseases, there is obvious pectic enzyme activity on the tissue. So, I wonder if the cell walls could protect the enzyme since the substrate is located in this structure.

Byrde: Do you think perhaps the enzyme action occurs in advance of the pathogen in these tissues? The phenolics may not get oxidized until after the enzyme action has occurred.

Mussell: I haven't a clue.

Byrde: That's how we see it. We believe that the phenolics and the phenoloxidase are not getting together in the tissue until after the pectic enzymes have acted. The fungal hyphae produces enzyme which acts in advance of actual tissue colonization.

Mussell: We don't have a lot of hard data but it is our feeling that pectic enzymes and the vascular wilt may, in fact, be leading the way far in advance of the fungus. We visualize that as the fungus grows along the vessel walls, the enzyme is released into solution and moves with the translocation stream. In cotton, there is a possibility that some of the materials released by enzyme action into the vascular system is also being translocated. Possibly some of the symptoms we see in the foliage are resulting from release of materials within the vascular system. So maybe we are observing the effects of enzyme action which take place prior to oxidation of phenolics in the tissue. On the other hand, if you look at some of the analytical data on phenolics and quinones in cotton leaves, one might expect that the enzymes may be precipitating out because of the great concentration of these materials.

Byrde: We feel that certain of the unoxidized phenolics may be inhibitory. Certainly a few phenolics with a high fat to water ratio are toxic to the fungus, but most of the naturally occurring are not toxic per se, i.e. they don't stop fungal growth. In general, it is not until they are oxidized, unless they have a very high molecular weight, that they have any inhibitory effect on enzymes.

Bateman: Would you comment on where within the fungal hyphae excretion of enzymes like α-L-arabinofuranosidase are excreted? Does excretion take place near the hyphal tips or does it occur at other locations?

Byrde: Dr. Hislop expected to find excretion to take place near the hyphal tips but, in fact, his data indicate that it is more or less at random. In some hyphae excretion appears to be a little more common near the hyphal tips than elsewhere, but not necessarily so. One weak point in our work, and Dr. Hislop is always pointing this out, is that we don't really know which isoenzyme we are detecting in the vesicles. Also, it is possible that some enzymes may be more affected than others by the fixation procedures employed.

Eriksson: I would like to make a short comment on the protective power that lignin may offer to plants infected by pathogens. If one has a suspension of spruce fibers and treats this preparation with a mixture of polysaccharide-degrading enzymes capable of degrading the polysaccharides in wood, you will observe that the enzymes have very little effect. If you have holo-cellulose containing very little lignin it is possible to delete as much as 60 to 80% of the xylan in this holo-cellulose with very highly specific xylanase. Of course, you can also ballmill a highly lignified material to very small particles and then it is possible to degrade the polysaccharides with a mixture of polysaccharide-degrading enzymes only.

Byrde: Thank you. That seems to be very good evidence for the protective effect of lignin.

Cooper: I would like to raise a technical point in connection with your comments about ballmilling. In the process of preparing cell walls, we have used a procedure that gives a very finely divided product. I feel that results obtained using extracted cell walls are obviously always going to be suspect because of breaking lignin carbohydrate linkages and increasing the area exposed within such walls. I don't really see any way around this particular problem.

Jarvis: We have done a few controlled experiments on the effect of ballmilling for different lengths of time under different conditions. It is customary to talk about ballmilling as though there is just one process. In fact, you can ballmill under all sorts of different conditions and get all sorts of different results. I am not sure of the conditions Dr. Eriksson used in his studies but it is quite common for wood chemists to ballmill wood until it is actually colloidal. I think this is a bit smaller than any of the rest of us would normally prepare our plant tissue. The actual breakage of polymers within the wall by ballmilling, when it is carried out to a moderate extent, so far as I can make out is not a significant problem.

Byrde: May I ask a question? We are always fascinated by the great stability of cell wall degrading enzymes in the presence of proteolytic enzymes. Does anyone know the mechanism by which they are so robust?

Eriksson: We have tried to study the structure of endo as well as exo cellulases with the aid of proteolytic enzymes. Most of the cellulases appear to be glycoproteins. It seems as if the carbohydrate part plays an impoltant role in the protection of these enzymes from degradation by proteolytic enzymes. It turned out to be impossible to inactivate the cellulase, a glycoprotein, with protease when this cellulase is in its native state. On the other hand, if the enzyme is denatured it is readily attacked by the protease. So, I suggest that the carbohydrate portion of the cellulase serves as a protective mechanism which renders the enzyme in its native state resistant to proteolytic enzymes.

Byrde: I believe that there are some polysaccharidases which are not, in fact, glycoproteins or at least have very few sugar residues. One of the enzymes we have worked with seems to have less than 4 hexose units per molecule. Also, I believe that the polygalacturonase studied by Keen has less than 4 sugar residues per mole. When we say less than 4 sugar residues per molecule it must be remembered that this is reaching limits of detection.

Mussell: A very few sugar residues may be quite important. I wish to point out that the Verticillium we work with is capable of producing 2 inducible proteases. One of these enzymes is capable of hydrolyzing enzyme proteins produced by any other organism except Verticillium. The proteases of Verticillium do not degrade the endo or exo-polygalacturonase produced by itself, nor will it degrade any of the multiple forms of cellulase in their native state which we have examined. On the other hand, our work with endo-polygalacturonase indicates that if you take the enzyme out of solution and then put it back into solution something happens to the carbohydrate fraction. The re-hydrated enzyme is in some way a different enzyme and exhibits a slightly reduced activity. The re-hydrated enzyme also is susceptible to digestion by the protease produced by Verticillium. So I think that the carbohydrate portion of these enzymes do protect them from protease action.

Cooper: I would like to raise a question about protein inhibitors of pectic enzymes in plant cell walls. I believe that such inhibitors have been reported only for polygalacturonase. Has anyone found such inhibitors that affect the trans-eliminases? In our work with the tomato/Verticillium system we have found inhibitors of the trans-eliminase.

Albersheim: In our work we did test these inhibitors against trans-
eliminase. We did not find any inhibition of these enzymes. The inhibitor
that we worked with is a glycoprotein and inhibits the polygalacturonase on
a 1:1 basis. This inhibitor did not work against exopolygalacturonase or
any other enzymes that we know of.

Cooper: I wish to raise a question regarding the enzymes produced by
Erwinia in potato rot. A trans-eliminase with a pH optimum of about 9 is
involved. I think the pH of the potato tuber is about 6. In vitro, one
does not obtain any trans-eliminase activity at pH 6 yet, the potato tissue
is readily broken down. Also, optimum production of the enzyme by Erwinia
occurs under alkaline conditions and again, the potato tuber in which it
is growing is approximately pH 6. I believe the pH of the rot remains
similar to that in the uninfected host. Whether in this case we're dealing
with a microenvironment situation or something else, I don't know. But in
any event, the enzyme is supposed to fuse well ahead of the bacterium.

Raa: Well, with animal proteases the pH optimum changes depending upon the
substrate being attacked. The pH optimum may very well be different in an
intact tissue or cell wall from that when a pure pectin substrate is used.

Cooper: There has been a good deal of work using potato disks as a
substrate and the pH optimum for maceration follows very closely that which
has been obtained for optimum breakdown of isolated pectin. I believe the
optimum pH in both instances is around pH 9, with the trans-eliminase.

Kimmins: I wonder what a pH of 6 means. If you stick an electrode into a
potato tuber what relationship does this bear to the hydrogen ion concentration at the site of the substrate and the enzyme. I suspect nothing.

Raa: I could add something from the animal world to this. If you isolate
enzymes from an animal tissue or digestive tract and look at the pH curve
for the enzymes involved in tissue degradation, you will find that none of
these enzymes which you may think are involved in tissue degradation have a
pH optimum which lies close to the optimum for tissue autolysis. The tissue
is degraded at a maximum rate near a neutral pH. This is in the lower range
of activity of the isolated enzymes, proteases and glycosidases, involved
in degradation of that tissue. I think the problem is related to how the
substrates and the enzyme become exposed to the endogenous enzyme or enzymes

directly injected or exposed in the tissue. Perhaps in the case of potato tuber where you add the extracellular enzyme, the situation could be different from that where the pathogen enters into the tissue and delivers the enzyme at the proper place at the proper time.

XYLANASE FROM <u>TRICHODERMA PSEUDOKONINGII</u>: EFFECTS ON ISOLATED PLANT
CELL WALLS

<u>D.F. Bateman</u>, C.J. Baker and C.H. Whalen, Department of Plant Pathology,
Cornell University, Ithaca, New York 14853, USA

Sources of pure polysaccharide degrading enzymes would serve as
powerful tools in the further elucidation of plant cell wall structure and
facilitate studies to determine the role of specific polysaccharidases
produced by plant pathogens in plant cell wall breakdown during tissue
invasion and pathogenesis. Here we report the purification and characterization of an endo-β-1,4 xylanase produced by <u>Trichoderma pseudokoningii</u>
Rifai as well as some preliminary experiments on the release of cell wall
sugars from isolated corn (<u>Zea mays</u> L.) and bean (<u>Phaseolus vulgaris</u> L.)
cell walls by this enzyme.

Material and Methods

Available commercial sources of xylan contained only 30 to 40%
xylose as determined by the procedure for analyzing polysaccharides as
described by Jones & Albersheim (1972). A xylan preparation containing
more than 90% xylose was prepared from a commercial xylan (Nutritional
Biochemicals Corp.) preparation by adapting the procedure used by Meier
(1958). This purified xylan was used to prepare O-hydroxyethyl xylan, a
soluble derivative, using a procedure similar to that used for production
of hydroxyethyl ethers of cellulose (Klug, 1963). All other chemicals were
reagent grade unless otherwise noted.

Crude xylanase was produced by <u>T. pseudokoningii</u> (isolate C-3) when
grown in shake culture (80 oscillations/min.) for 4 days at $25^{o}C$ on a
medium containing/liter: 181 mg $MgSO_4$, 149 mg KCl, 1 g NH_4NO_3, 650 mg
KH_2PO_4, 3.5 mg $ZnSO_4 \cdot H_2O$, 6.9 mg $MnSO_4 \cdot H_2O$, 3.3 mg $FeCl_3 \cdot 6H_2O$, 3.1 mg
$CuSO_4 \cdot 5H_2O$, 1 g yeast extract (Difco) and 1 g xylan.

Enzyme purification was accomplished by ion exchange chromatography on CM-Sephadex (C-25) and gel filtration in Bio-Gel P-10. The isoelectric point of the enzyme was determined by isoelectric focusing (Bateman et al., 1973) using pH range 8-11 Ampholine carriers (LKB). Molecular weight estimates were based on gel filtration in Bio-Gel P-10 and SDS slab gel electrophoresis (Maizel, 1971), using appropriate reference proteins.

Reaction mixtures contained 0.3 ml 0.5% xylan or 0.2 ml 1.0% hydroxy-ethyl xylan, 0.1 ml 0.2 M Na acetate buffer (pH 5.0), 0.1 ml enzyme, and glass distilled water for a final volume of 0.5 ml. Activity was determined by the release of reducing groups with time as determined by the procedure of Nelson (1944) using xylose as the standard. Reaction products were examined by paper chromatography (Mullen & Bateman, 1975).

Protein was determined by the procedure of Lowry et al. (1951); bovine serum albumin was used as the reference.

Cell walls were prepared from the shoots and hypocotyls, respectively, of 5-day old corn and 7-day old bean seedlings using the procedure of Barnett (1974), and were further treated with organic solvents (Bateman et al., 1969). The release of cell wall constitutents by endo-β-1,4 xylanase was determined in reaction mixtures containing 10 mg cell wall, 1.5 ml enzyme and 0.5 ml 0.2 M Na acetate buffer (pH 5.0). Reactions in the presence of 0.005% Thimerosal were allowed to proceed for 18 hr at $25^{\circ}C$ with constant stirring. Cell wall residues and supernatants of reaction mixtures were assayed for noncellulosic carbohydrate constituents by the procedure of Jones & Albersheim (1972).

Results

A time course study of xylanase production by T. pseudokoningii revealed that maximum activity occurred on day 4 under the described culture conditions. Crude dialyzed filtrates of 4-day old cultures con-tained about 140 μg protein/ml and had a xylanase specific activity (μmoles/min/mg protein) of approximately 5. Paper chromatography of aliquots of xylan reaction mixtures after 0, 30, 60 and 120 min. of incubation at $30^{\circ}C$ revealed the release of 5 xylose oligomers plus xylose. The optimum pH for xylan hydrolysis by the crude enzyme was approximately 5.

Crude xylanase (840 ml) was applied to a 1.5 x 26 cm column of CM-Sephadex (C-25) equilibrated with 0.02 M Na acetate buffer (pH 5.0). The column was then washed with 150 ml buffer and 7.5 ml fractions collected;

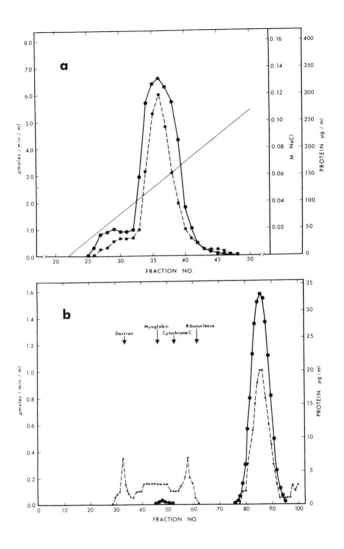

Figure 1.   Xylanase purification scheme: a) CM-Sephadex (C-25) ion
exchange column chromatography.   b) Elution pattern in Bio-gel
P-10 (1.8 x 85 cm) in the presence of 0.05 M Na acetate buffer
(pH 5.0) and 0.1 M NaCl.   The markers indicate the elution
peaks of the standard blue Dextran (mol. wt 2,000,000), myoglobin (mol. wt 17,500), cytochrome C (mol. wt 12,300), and
ribonuclease (mol. wt 13,700).

the buffer wash was followed by a NaCl gradient (250 ml buffer + 250 ml 0.15 M NaCl in buffer). The xylanase activity and conductivity of each fraction was determined (Figure 1a). The peak xylanase fractions (85 ml) contained 90 μg protein/ml and had a specific activity of 58.7. The peak fractions were dialyzed, lyophilized and taken up in 8 ml 0.05 M Na acetate buffer containing 0.1 M NaCl. Two ml portions of the reconstituted enzyme were applied to a 1.8 x 85 cm column of Bio-Gel P-10 equilibrated with 0.05 M Na acetate buffer (pH 5.0) containing 0.1 M NaCl and eluted with the same buffer-salt mixture. Two ml fractions were collected. The elution patterns of Blue Dextran 2000 (mol. wt 2,000,000), myoglobin (mol. wt 17,500), cytochrome C (mol. wt 12,300) and ribonuclease (mol. wt 13,700) were determined under the same conditions in the Bio-Gel column. Based on its elution volume in Bio-Gel P-10, xylanase appeared to have a very low molecular weight (Figure 1b). The peak fractions of xylanase were pooled; they exhibited a protein content of 40 μg/ml and a specific activity of 137.1. Thus, the two step purification procedure resulted in a 28-fold purification.

The purified xylanase exhibited a pH optimum of 5 and an isoelectric point (pI) of 9.7. When it was subjected to SDS gel electrophoresis only 1 protein band was detected and the estimated molecular weight of this protein was approximately 15,000 Daltons (Figure 2). Paper chromatography

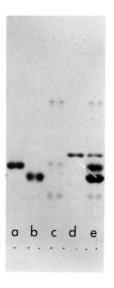

Figure 2. Electrophoresis of purified xylanase on 13% SDS slab gels. a) Ribonuclease (2 μg), mol. wt 13,700; b) Cytochrome C (2 μg), mol. wt 12,300; c) Chymotrypsinogen (2 μg), mol. wt 24,000 (note degradation products of about mol. wt 13,000 and mol. wt 11,000); d) Xylanase (6 μg), estimated mol. wt 15,000; d) Standards plus xylanase.

Bateman et al. - Cell-wall degradation

of reaction mixtures containing the purified enzyme revealed the release of xylose oligomers but not xylose from xylan. When araban, carboxymethyl-cellulose, β-1,4 galactan, Na polypectate or polygalacturonic acid were used as potential substrates for the purified enzyme no release of reducing groups was detectable even after 24 hr of incubation.

When the purified endo-β-1,4 xylanase was incubated with isolated corn and bean cell walls, carbohydrate was solubilized (Figure 3). The

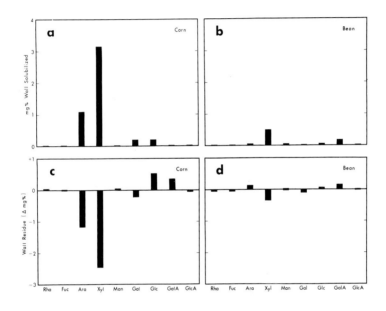

Figure 3. Solubilization of cell wall sugars with purified xylanase. Residues were spun down and both supernatants and cell wall residues analyzed. Results are expressed in mg %. a, b) Carbohydrates found in supernatants of xylanase treated corn and bean cell walls, respectively. c, d) Constituents of wall residues in xylanase treated corn and bean cell walls, respectively, compared to buffer treated controls. (Rha = rhamnose; Fuc = fucose; Ara = arabinose; Xyl = xylose; Man = mannose; Gal = galactose; Glc = glucose; GalA = galacturonic acid; GlcA = glucuronic acid.)

corn walls appeared to be more susceptible to this enzyme; approximately 4.5 mg % of the wall sample was solubilized, while only about 0.6 mg % of the bean wall was solubilized in comparable reaction mixtures. Xylose and arabinose were found in the xylanase soluble fraction of the corn walls; galactose and glucose were present in lesser amounts. Xylose appeared to be the only sugar in significant amounts in the xylanase soluble fraction from bean walls. Analysis of cell wall residues following the xylanase treatment revealed decreases in xylose and arabinose contents in the case of corn walls and a decrease only in xylose content of bean walls when compared to analysis of buffer treated controls (Figure 3).

Discussion

The crude xylanase system produced by T. pseudokoningii released xylose as well as oligomers of xylose from β-1,4 linked xylan; the purified xylanase released only oligomers of xylose. It thus appears that the xylanase complex of T. pseudokoningii consists of an endo-β-1,4 xylanase plus a β-xylosidase. In this regard the xylanase system of T. pseudo-koningii is similar to that of other microorganisms (King & Fuller, 1968; Strobel, 1963; Walker, 1967).

The endo-β-1,4 xylanase component produced by T. pseudokoningii was easily purified by a relatively simple two-step scheme (Figure 1). Purification was facilitated because of its high isoelectric point (pI 9.7) and its low molecular weight (ca. 15,000 Daltons). This enzyme is very stable and can be stored for at least 9 months in 0.05 M Na acetate buffer (pH 5.0) containing 0.1 M NaCl at -20^{0}C without significant loss of activity.

Our analysis of carbohydrates in isolated cell walls from corn and bean seedlings are similar to those reported for walls of monocots and dicots grown in cell suspension cultures (Burke et al., 1974; Wilder & Albersheim, 1973), although the percentage composition of a given sugar detected was somewhat lower in the present study. This was not unexpected since the cell walls from seedlings would be expected to have some secondary thickening and thus be more difficult to hydrolyze.

The purified endo-β-1,4 xylanase from T. pseudokoningii readily solubilized xylose and arabinose from corn cell walls (Figure 3). We assumed that the arabinose solubilized was covalently linked to xylose oligomers released from the arabinoxylan. The relatively small amount of

Bateman et al. - Cell-wall degradation

xylose released from bean cell walls was expected on the basis of their
composition (Wilder & Albersheim, 1973). The xylose released from the bean
cell walls probably represents the action of the enzyme on β-1,4 xylan
deposited in the secondary wall (Nevins et al., 1968). Our study demonstrates that xylan in cell walls is susceptible to hydrolysis by endo-β-1,4
xylanase without the prior action of a "wall-modifying enzyme" (Karr &
Albersheim, 1970). The endo-β-1,4 xylanase of T. pseudokoningii should
prove to be a valuable tool in helping to further elucidate plant cell wall
structure.

References

Barnett, N.M. (1974). Release of peroxidase from soybean hypocotyl cell
    walls by Sclerotium rolfsii culture filtrates. Canadian Journal of
    Botany 52, 265-271.
Bateman, D.F., Jones, T.M. & Yoder, O.C. (1973). Degradation of corn cell
    walls by extracellular enzymes produced by Helminthosporium maydis
    race T. Phytopathology 63, 1523-1529.
Bateman, D.F., Van Etten, H.D., English, P.D., Nevins, D.J. & Albersheim,
    P. (1969). Susceptibility to enzymatic degradation of cell walls
    from bean plants resistant and susceptible to Rhizoctonia solani
    Kuhn. Plant Physiology 44, 641-648.
Burke, D., Kaufman, P., McNeil, M. & Albersheim, P. (1974). The structure
    of plant cell walls. VI. A survey of the walls of suspension-cultured
    monocots. Plant Physiology 54, 109-115.
Jones, T.M. & Albersheim, P. (1972). A gas chromatographic method for
    determination of aldose and uronic acid constituents of plant cell
    wall polysaccharides. Plant Physiology 49, 926-936.
Karr, A.L. & Albersheim, P. (1970). Polysaccharide-degrading enzymes are
    unable to attack plant cell walls without prior action by a "wall-
    modifying enzyme". Plant Physiology 46, 69-80.
King, N.J. & Fuller, D.B. (1968). The xylanase system of Coniophora
    cerebella. Biochemical Journal 108, 571-576.
Klug, D.D. (1963). Hydroxyethyl ethers of cellulose and their analytical
    determination. pp. 315-317 in Methods in Carbohydrate Chemistry, ed.
    by R.L. Whistler. Vol. III, Academic Press, New York.

Lowry, O.H., Rosebrough, N.J., Farr, A.L. & Randall, R.J. (1951). Protein measurement with the Folin phenol reagent. Journal of Biological Chemistry 193, 265-275.

Maizel, V. Jr. (1971). Polyacrylamide gel electrophoresis of viral proteins. pp. 179-246 in Methods in Virology, Vol. 5.

Meier, H. (1958). Barium hydroxide as a selective precipitating agent for hemicelluloses. Acta Chemica Scandinavica 12, 144-146.

Mullen, J.M. & Bateman, D.F. (1975). Polysaccharide degrading enzymes produced by Fusarium roseum 'Avenaceum' in culture and during pathogenesis. Physiological Plant Pathology 6, 233-246.

Nelson, N. (1944). A photometric adaptation of the Somogyi method for the determination of glucose. Journal of Biological Chemistry 153, 375-380.

Nevins, D.J., English, P.D. & Albersheim, P. (1968). Changes in cell wall polysaccharides associated with growth. Plant Physiology 43, 914-922.

Strobel, G.A. (1963). A xylanase system produced by Diplodia viticola. Phytopathology 53, 592-596.

Walker, D.J. (1967). Some properties of xylanase and xylobiase from mixed rumen organisms. Australian Journal of Biological Sciences 20, 799-808.

Wilder, B.M. & Albersheim, P. (1973). The structure of plant cell walls. IV. A structural comparison of the wall hemicellulose of cell suspension cultures of sycamore (Acer pseudoplatanus) and of Red Kidney bean (Phaseolus vulgaris). Plant Physiology 51, 889-893.

DISCUSSION

Chairman: R.J.W. Byrde

Eriksson: When I was asked earlier this morning to comment upon procedures we use to separate polysaccharidases from each other which are difficult to separate, I forgot to mention one method which has proven to be very useful. In our attempts to separate xylanase and cellulase activities, we have mixed carboxymethylcellulose, a substrate for the cellulase, with the enzyme mixture. This mixture is then subjected to gel filtration in columns of either Biogel or Sephadex. The columns are equilibrated with buffer containing

Bateman et al. - Cell-wall degradation

carboxymethylcellulose with the pH adjusted to approximately 2 pH units off
the optimum for activity of the cellulase. The columns are run at a
temperature of 4 to 5°C. In this system you have a constant association
with the cellulase and its substrate while undergoing gel filtration. As
a result of this type of interaction we have been able to obtain some
separation of the two enzyme activities. Although you may not get complete
separation of the two enzymes, we feel this is a good procedure to demonstrate that you are dealing with two enzymatic entities as opposed to a
multi-enzyme particle.

Albersheim: How do you explain the fact that you got such a small
percentage of the xylose present in corn walls released by your endo-
xylanase?

Bateman: The amount of xylose released from corn cell walls by our
endoxylanase amounted to more than 3% of the total wall sample assayed.
Based on our assay procedures, this was equivalent to approximately 20%
of the total wall xylose, so I don't consider that the amount of xylose
released by our enzyme as being insignificant. There are, however, several
possibilities as to why more xylose was not released. Perhaps the xylose
polymers within the wall are masked to a degree by other cell wall polymers.
Also, xylans in corn cell walls are branched polymers and it is likely that
our purified endoxylanase may not be able to completely degrade such a
substrate. With respect to our studies on bean cell wall, which you did
not ask about, there was very little xylose released. Also, these walls
contain relatively little xylose. But because these walls were isolated
from seedlings, I would expect that some of the xylose present would be in
the form of xylan, as opposed to xyloglucan which is a known constituent
of the primary cell wall.

Albersheim: Xyloglucan would not be susceptible to your endoxylanase. I
would also suspect that xylan in secondary wall would probably be masked
by lignin. This may also be true for the corn cell walls. When you are
analyzing a plant cell wall for its constituents and get out only a small
percentage of a component, you have to ask the question whether this is due
to the fact that you have exposed some of the normally masked material
during the process of preparing the walls.

Bateman: I am perfectly aware of the point you are making. Also, I am not offering the data in this manuscript as being something absolute. We will be looking more thoroughly into the questions raised by using a battery of enzymes individually and in various combinations. It is quite likely that if some of the polymers in the corn or bean cell walls were solubilized then the xylan polymers would be more exposed and more readily degraded by our endo-xylanase. We will be carrying out such studies in the future.

Spencer: Could you clear up for me why you assume a covalent linkage between xylose and arabinose?

Bateman: The xylan in corn cell walls is supposedly an arabinoxylan. Since the purified endo-xylanase we were using was unable to cleave arabinosyl linkages, it was assumed that the arabinose solubilized by our enzyme was covalently linked to xylose moieties. I believe that our interpretation is compatible with the known composition of corn xylan.

Albersheim: Those of you interested in corn cell walls would be interested to know that Hall & Garbo have done a detailed study of corn cell wall structure. I don't think this work has been published, but they have a lot of information on the arabinoxylan and other polymers in corn cell walls. Also, I believe that they have an enzyme that releases large amounts of the arabinoxylan.

Jarvis: Can I draw attention away from the lignification for a minute? Xylans as you find them in plant cell walls are normally fairly heavily acetylated, at least in the monocots. The acetyl groups are generally removed when fractionating cell walls with alkali. I would suggest that the presence of acetyl groups may be one means by which xylans in intact walls could be protected against xylanase activity. I would also suggest that we ought to know a bit more about the degree of acetylation of other cell wall polysaccharides as well as xylan.

Bateman: That's a good point. Acetyl groups in xylan could serve as a possible protection mechanism against enzymatic decomposition. I am not sure about corn cell walls but in certain species you can have a high degree of acetylation.

Selvendran: Could you speculate on the possible role of uronic acid residues and acetyl groups in xylan and how these groups may influence

Bateman et al. - Cell-wall degradation

xylanase activity. Also, I was wondering whether or not you checked
arabinase activity on araban in your studies. If you have xylan as normally
occurs in monocots and treat it with xylanase, one tends to get disaccharides and trisaccharides. I'm a bit puzzled by your results.

Bateman: The uronic acid in acetyl groups attached to the xylan chain could
prevent xylan chains from binding to each other within the wall structure.
This would tend to make this wall fraction amorphous as opposed to crystalline. It is my view that these residues would also restrict the amount of
xylan degradation by a purified endo-xylanase as like the one used in our
studies. With regard to the second point, we have checked our enzyme
preparation for its ability to degrade araban. Our purified xylanase does
not contain any arabanase activity. You have to remember that after
treatment of cell walls or xylan with our purified enzyme, the fraction
solubilized was further hydrolysed prior to quantitation by gas-liquid
chromatography.

Selvendran: In my work I found that if you isolate xylan and purify it,
then treat it with xylanase from Trichoderma one obtains almost complete
breakdown, but if you treat bean fibers you can dissolve only about 25% of
the xylose in these fibers. The resistance of the xylan in plant tissue
to xylanase, I feel, is related to its close association with lignin.

Bateman: Yes, but you don't have lignification in all plant tissues.

Selvendran: No, but you are getting out the xylan. If you have xylan in
a tissue it must also be lignified.

Bateman: The point I wish to make is that the xylan preparation used in
our enzyme studies was highly purified. I do not believe that it contained
lignin. In fact, we could account for more than 90% of the material in our
preparation as carbohydrate by the analysis we employed. So, I am confident
that lignin was not a significant factor in the xylan preparation used in
our studies.

Selvendran: I don't like to labor on the point but, in our studies where
we purified xylan from bean fibers and subjected this material to hydrolysis
with xylanase, we could only solubilize about 50% of the total material we
started with. So I am pretty certain that there is degraded lignin

associated with the purified xylan preparation and I wouldn't be surprised if this were not the case with your material as well.

Bateman: I don't doubt your results at all. You were working with an entirely different type of plant tissue from that we employed in our studies. In my view, the role of lignin in cell walls of young seedlings is questionable. Perhaps someone else would like to comment on this matter.

Jarvis: If it is of any relevance, there is a small amount of 1,4 linked xylan in uninfected, unwounded potato tuber tissue. I wouldn't really like to say if such tubers contain lignin or not, but certainly there is not much lignin present.

Mussell: Did you use both the Barnett and Albersheim procedures to prepare your isolated cell walls? Have you made any comparisons to see if there are differences in what is removed before walls were extracted with organic solvents?

Bateman: We didn't make such comparisons on the preparations used in this study. We have in the past, compared the compositions of cell walls prepared by the two different procedures. As I recall, walls prepared by the Barnett procedure tend to have a bit more uronic acid but with respect to other carbohydrate constituents, walls prepared by the two procedures seem to be quite similar.

Mussell: Have you noted any differences in the susceptibilities of cell walls prepared by the two procedures to enzymatic degradation?

Bateman: Not really. I'm not sure that a critical comparison has been made.

CELL-WALL DEGRADATION IN SENESCENT FRUIT TISSUE IN RELATION TO PATHOGEN ATTACK

Michael Knee, East Malling Research Station, Maidstone, Kent, England

Fruits become more susceptible to attack by fungal pathogens as they ripen, possibly because degradation of cell walls during ripening aids hyphal penetration.

Apple fruit cell walls contain a relatively unbranched polyuronide, some of which becomes soluble on ripening, and a polyuronide with abundant sidechains of galactose and arabinose residues which remains insoluble (Knee, 1973; Knee, Fielding, Archer & Laborda, 1975).

The unbranched polyuronide is a component of the middle lamella and solubilization of this component is associated with cell separation. The stability of the middle lamella may depend on co-operative binding of polygalacturonate chains with calcium ions. Weakening of this structure during fruit ripening could occur due to methylation of free carboxyl groups. In some fruits polyuronide chains are broken during ripening but this does not occur in the apple.

Although cell walls become more susceptible to degradation by poly-galacturonase during apple ripening (Knee et al., 1975), autolytic degradation of the host tissue is thought to be a more important cause of increased susceptibility to fungal invasion.

As shown in Table 1 the ability to attack isolated cell walls can be related to the molecular weights of purified polygalacturonases and arabinosidases (Knee et al., 1975). A high molecular weight polygalacturo-nase from Phytophthora infestans (Cole, 1970) may not be able to permeate the gel structure of the middle lamella and wall matrix but it could be sufficiently active close to the surface of invading hyphae to aid pene-tration. By contrast a low molecular weight polygalacturonase from Sclerotinia fructigena can release all of the uronide residues from isolated

Table 1. Apple cell wall degrading ability of fungal cell walls degrading
          enzymes.

Enzyme	Source	M.W.	% total Monomer released
Polygalacturonase	S. fructigena	37,000	99
	S. fructigena	75,000	53
	P. infestans	ca 200,000	5
Arabinosidase	S. fructigena	40,000	47
	S. fructigena	220,000	3
	S. fructigena	350,000	5

In these assays equal activities of polygalacturonase enzymes against
polygalacturonic acid and arabinosidase isoenzymes against p-nitrophenyl
arabinoside were incubated with apple cell walls (Knee et al., 1975).
The MW's of polygalacturonases from S. fructigena are taken from Byrde
& Willets (1976) and the MW's of arabinosidases from Laborda, Fielding
& Byrde (1973).

walls. A polygalacturonase of intermediate molecular weight, also from
S. fructigena, releases about half of the total uronic acid residues and is
thought to degrade the middle lamella selectively. These low M.W. enzymes
would diffuse away from the hypha causing more extensive damage to plant
tissue.

Arabinosidase activity is unlikely to weaken the cell wall but may help
to satisfy the nutritional requirements of a pathogen. The role of
galactanase may be in nutrition or, possibly, structural attack.

References

Byrde, R.J.W. & Willets, H.J. (1976). The Brown rot fungi of fruit, their
     biology and control. Pergamon Press, Oxford.
Cole, A.L.J. (1970). Pectic enzyme activity from Phytophthora infestans.
     Phytochemistry 8, 337-340.
Knee, M. (1973). Polysaccharides and glycoproteins of apple fruit cell
     walls. Phytochemistry 12, 637-653.
Knee, M., Fielding, A.H., Archer, S.A. & Laborda, F. (1975). Enzymic

analysis of cell wall structure in apple fruit cortical tissue. Phytochemistry 14, 2213-2222.

Laborda, F., Fielding, A.H. & Byrde, R.J.W. (1973). Extra- and intra- cellular α-L-arabinofuranosidase of <u>Sclerotinia fructigena</u>. Journal of General Microbiology 79, 321-329.

DISCUSSION

Chairman: D.F. Bateman

Eriksson: I would like to comment on the relationship between the molecular weight of enzymes and their substrates. Ahlgren and I published a paper (Acta Chem. Scand. 21, 1193, 1967) on the molecular weight of different enzymes. We investigated some 20 to 30 enzymes from several different fungi and observed a relationship between enzyme size and the molecular weight of its substrate. The low molecular weight enzymes attacked high molecular weight substrates. The high molecular weight enzymes has low molecular weight substrates. For example, endo-β-1,4 glucanase had a lov molecular weight and it attacked polymeric cellulose, whereas the β-1,4 glucosidase had a high molecular weight and attacked only low molecular weight substrates. I haven't actually observed any exceptions to this rule.

Knee: I wish to point out that the polygalacturonase produced by <u>Phytophthora infestans</u> has a high molecular weight and it attacks high molecular weight substrates. There is heterogeneity within classes of enzymes. There may be a general rule but there are exceptions.

Eriksson: The fungi we studied were saprophytes, not pathogens.

Kauss: I have a comment on the observation of increased methylation of the pectins in ripening fruit. The methylation of poly uronides requires contribution of a methyl group from S-adenosylmethionine. We should not assume that such a nucleotide is available outside of the protoplast. The only way I can envisage how one would obtain increased methoxyl content of the pectic substances in the cell wall would involve a turnover of these materials. This would mean that there would have to be increased production

of the methylated pectin product and at the same time degradation of the
non-methylated pectin within the wall. Are there enzymes in the wall that
favor degradation of the non-methylated pectins?

Knee: I am aware of the problem you raised. In fact, it is unlikely that
you would have methylation from S-adenosylmethionine occurring within the
wall. As far as turnover of pectins is concerned, I think this is also
unlikely because of the absence of polyuronide degrading enzymes in apple.
Also, I've shown in strawberry ripening that when the berries are labelled
with $^{14}CO_2$ when they're very young, that the pectin which is solubilized,
is also highly methylated and is derived from presynthesized pectin. This
is not a turnover situation.

Bateman: It does appear that we have a bit of a biochemical dilemma here.
Dr. Tronsmo indicated in his paper that the pectin in young apple fruit,
for example, has a low methoxyl content and that the methoxyl content
increases with maturation. It would appear that the problem under discussion
has been confirmed by Dr. Tronsmo's observations.

Knee: Remembering Albersheim's work on methylation, I think he certainly
found evidence that turnover of methyl groups is certainly cytoplasmic.

BIOCHEMISTRY OF PATHOGENIC GROWTH OF <u>BOTRYTIS CINEREA</u> IN APPLE FRUIT

<u>Arne Tronsmo</u> and Anne Marte Tronsmo, Institute of Biology and Geology, University of Tromsø, N-9001 Tromsø, Norway

<u>Botrytis cinerea</u> is the cause of dry eye rot on apple, which is one of the most serious diseases of apples in Western Norway. The pathogen infects the sepals of apple flowers and lives in a latent condition inside the tissue until fruit maturation when it expands in the tissue near the sepals and causes a rot (Tronsmo & Raa, 1977).

Electron microscope studies have shown that <u>B. cinerea</u> is restricted to the middle lamellae. However, at a late stage of disease development the pathogen can extend through the cellulose layer of the host cell wall. This mode of pathogenic growth is consistent with growth characteristics and enzyme production in vitro. The pathogen can grow on low molecular weight sugars and on pectine, but not on Ca-polygalacturonate, cellulose, cellophane and gelatine. Pectinase measured by viscosimetry was produced on all the media which support growth, but cellulase was detected only when the medium contained both pectine and cellulose. This ability to produce some cellulase when exposed to cellulose in addition to pectine, may explain why the pathogen at a late stage of disease development can extend through the cellulose rich layer of the host cell walls.

The predominant role of pectinase in the pathogenesis of dry eye rot seems inconsistent with the observation that no pectinase activity could be detected in a water extract of diseased tissue (Table 2). However, pectinase produced by <u>B. cinerea</u> binds to cell walls with ionic bound proteins (Table 2). This may explain why no pectinase can be detected in diseased tissue.

Cell walls of dicotyledons contain proteins which inhibit endopolygalacturonase from plant pathogens. Albersheim & Anderson-Prouty (1975) have suggested that this inhibition may be a general mechanism of defence against plant pathogens. However, inhibition of pectinase and binding of

Table 1. Binding of pectinase to apple cell walls. Figures show the
         activity of pectinase in the culture filtrate of <u>Botrytis
         cinerea</u> which had grown in shake culture with apple pectine
         as the sole source of carbon (a), in the same culture filtrate
         after exposure to host cell walls which have ionic bound
         proteins (b) and after exposure to the host cells without
         ionic bound proteins (c). Enzyme activities are expressed as
         reduction of specific viscosity during 10 minutes. Figures
         in parentheses show the initial flow time in seconds and the
         flow time after 10 minutes. The pH in the incubation mixture
         was 4.5.

Material	Pectinase activity
a) Culture filtrate	1.4 (100.8 - 69.7)
b) Culture filtrate exposed to cell walls with ionic bound proteins	0.3 (100.8 - 95.3)
c) Culture filtrate exposed to cell walls without ionic bound proteins	1.2 (100.8 - 74.6)

it to the host cell walls may also be essential for the pathogenicity of
the fungus. Due to binding and inhibition the destruction of pectine may
be restricted to the very vicinity of the pathogen. In this way the
pathogen may grow without provoking a hypersensitive defence reaction of
host cells.

Growth experiments have shown that B. cinerea is unable to grow on
Ca-polygalacturonate. During the period of latency there is low degree of
methylation of the pectine, which then may be highly crosslinked by
divalent ions. During maturation when the fruit becomes susceptible to
B. cinerea the degree of methyl estrification increases. In the same
period there is a marked decrease in the content of Ca and Mg. Change
from resistance to susceptibility during fruit maturation may accordingly
be due to increased methylation of the pectine component of the middle
lamellae, which then becomes available as a carbon source.

In the first period of latency, which follows immediately after
infection, the degree of methylation is high. Latency in this period can
accordingly not be due to low methoxyl content of the pectine. We have no
explanation of latency in this period. It may be that the middle lamella

Table 2. Inhibition of pectinase produced by <u>Botrytis cinerea</u> by components extracted from apple cell walls. Figures show the activity of pectinase in the culture filtrate of <u>Botrytis cinerea</u> (a), in the culture filtrate after exposure to water extracts from apple cell walls with ionic proteins (b), after exposure to a 1M NaCl extract of the apple cell walls (c), after exposure to an extract in water of dry eye rot tissue (d), and the pectinase activity of an extract in water of dry eye rot tissue (e). For details see Table 1.

Material	Pectinase activity
a) Culture filtrate	1.5 (135.9 - 103.0)
b) Culture filtrate + a water extract of cell walls with ionic proteins	1.5 (135.9 - 103.4)
c) Culture filtrate + a 1M NaCl extract of cell walls with ionic proteins	1.0 (135.9 - 116.8)
d) Culture filtrate + a water extract of B. cinerea dry eye rot tissue	0.9 (135.9 - 116.8)
c) Pectinase activity of a water extract of dry eye rot tissue	0 (135.9 - 135.8)

in the immature fruit is so tightly crosslinked that the pectinase produced by B. cinerea cannot penetrate into the matrix (Knee, 1976).

References

Albersheim, R. & Anderson-Prouty, A.J. (1975). Carbohydrates, proteins, cell surfaces and the biochemistry of pathogenesis. Annual Review of Plant Physiology 26, 31-52.

Knee, M. (1976). Cell wall degradation in senescent fruit tissue in relation to pathogen attack. This symposium.

Tronsmo, A. & Raa, J. (1976). Life cycle of the dry eye rot pathogen Botrytis cinerea Pers. on apple. Phytopathologische Zeitschrift (in press).

DISCUSSION

<u>Chairman</u>: R.J.W. Byrde

<u>Goodman</u>: Do I dare inject some plant pathology like the control of disease in some of the discussion? I was not aware of <u>Botrytis</u> being such an early pathogen in apple as you indicate. This recalls to my mind the situation in grape, where infection occurs in extremely early time during the period of bloom, but we see no change in condition of the fruit until just before harvest. At that time we see what is known as bunch rot and the entire cluster of grapes collapses, it is entirely parasitized. Thus, it seems that the situation with grape infection and that described by you are quite parallel. It might also be of interest that fungicides applied early on, let's say at the time of blossoming or shortly thereafter, seems to be a fairly effective way of controlling the disease. In the past, people have tried to control such diseases by frequently spraying during the growing season. But as you know, and I know, the damage has already been done since the infection had occurred shortly after or during the blooming period.

<u>Albersheim</u>: I think the idea that a phytoalexin might keep the fungus in a latent state is a very fine idea and could be easily examined.

CELL WALL ALTERATIONS IN LOCALIZED PLANT VIRUS INFECTION

C. Hiruki, Department of Plant Science, University of Alberta, Edmonton, Alberta T6G 2E3, Canada

The development of localized necrotic lesions is generally assumed to involve "a hypersensitive reaction" of cells to virus and to be associated with profound metabolic alterations in the affected cells (Diener, 1963). Since virus replication takes place prior to the formation of local lesions, the rapidity and completeness of the development of some limiting factors in infected cells or in neighbouring cells would influence cell-to-cell virus movement and contribute to the restriction of virus within the limited area.

Potato virus M (PVM) lesions appeared 3 days after manual inoculation of Red Kidney bean (Phaseolus vulgaris L.). The number of lesions increased rapidly between the 3rd and 4th days. Lesion size reached a maximum 6 to 7 days after inoculation, and averaged 0.6 mm in diameter (Hiruki, 1970, 1973; Hiruki et al., 1974). The PVM lesion consisted of a dark brown necrotic centre spot (necrotic cells) and a chlorotic halo (sub-necrotic cells). The lesion was surrounded by normally green tissue (non-necrotic cells). The alteration of cell wall due to PVM infection was investigated using electron microscopy and fluorescent microscopy (Hiruki & Tu, 1972).

Main cell wall alterations in the leaf tissue containing PVM lesions were (i) callose (β-1,3-D-glucan) deposition in both sub-necrotic cells and the non-necrotic cells immediately outside the sub-necrotic cells (Hiruki & Tu, 1972), and (ii) thickening of the secondary walls in the sub-necrotic cells (Tu & Hiruki, 1971). The former is probably a general host reaction to cell injury; in this case, an injury caused by virus infection. The thickening initially occurred on the inner wall of non-necrotic cells immediately adjacent to sub-necrotic ones. The location of this thickening coincided with a fluorescent zone around sub-necrotic cells. The thickening was associated with the appearance of boundary formation (Esau et al.,

1966), and subsequent increase of multivesicular bodies in extraprotoplasmic space. The abnormal thickening of secondary cell walls in a few layers of sub-necrotic cells surrounding the necrotic cells was a striking feature of the PVM lesions. These cells still contained cellular organelles of variable morphological intactness, including nucleus, chloroplast, mito-chondrion, dictyosome and endoplasmic reticulum. Since virus aggregates were detectable in sub-necrotic cells developing secondary wall thickening, this thickening was considered a barrier that enclosed virus particles to restrict virus spread, not a defense barrier. It is probable that the deposition of callose in non-necrotic cells constitutes a stage preceding secondary wall thickening in sub-necrotic cells. The fact that the material deposited in the zone of sub-necrotic cells does not fluoresce suggests that (i) the chemical nature of callose may be modified, and/or (ii) callose may be embedded in other cell wall constituents in such a way that it does not fluoresce or is prevented from doing so. Our observations appeared to support the latter, but the former could not be excluded. It should also be noted that callose deposition was not found in potato that was systemi-cally infected with PVM (Tu & Hiruki, 1970).

The following two types of blocking mechanisms, presented as a hypothesis based on our investigations are not necessarily mutually exclusive and may be in operation at the time of intense virus-host interaction resulting in the formation of local lesions. The first mechanism involves blocking by callose that is deposited quickly in non-infected cells immediately adjacent to infected cells. Since the deposits of callose on cell walls were found at greater radial distances from the necrotic centre-zone than virus particles in the cytoplasm, callose may be capable of serving as an effective defense barrier. This virus barrier may become more effective by sealing the plasmodesmata between the two zones of cells, the sub-necrotic and non-necrotic cells. The second mechanism involves blocking of virus by more complex "secondary wall thickening" which occurs in the infected cells surrounding necrotic cells. Complete plugging and sealing of intercellular cytoplasmic connections are achieved with the accumulation of tubules and vesicles. Callose may be seen embedded in other cell wall deposits. Cytoplasm containing virus particles is thus enclosed in a cell having thickened walls. The ultra-structure of this type of thickening is more complex than mere deposition of callose substance.

To understand the process(es) involved in the host parasite interaction in localized virus infection, in particular cell wall alterations, it is essential to establish more firmly the cellular factors and the specific functional roles of each organelle involved.

References

Diener, T.O. (1963). Physiology of virus-infected plants. Annual Review of Phytopathology 1, 197-218.

Esau, K., Cheadle, V.I. & Gill, R.H. (1966). Cytology of differentiating tracheary elements. II. Structures associated with cell surfaces. American Journal of Botany 53, 765-771.

Hiruki, C. (1970). Red Kidney bean, a useful bioassay host for qualitative and quantitative work with potato virus M. Phytopathology 60, 739-740.

Hiruki, C. (1973). Detection of potato virus M in tubers, sprouts, and leaves of potato by the French bean test. Potato Research 16, 202-212.

Hiruki, C. & Tu, J.C. (1972). Light and electron microscopy of potato virus M lesions and marginal tissue in Red Kidney bean. Phytopathology 62, 77-85.

Hiruki, C., Pountney, E. & Saksena, K.N. (1974). Factors affecting bioassay of potato virus M in Red Kidney bean. Phytopathology 64, 807-811.

Tu, J.C. & Hiruki, C. (1970). Ultrastructure of potato infected with potato virus M. Virology 42, 238-242.

Tu, J.C. & Hiruki, C. (1971). Electron microscopy of cell wall thickening in local lesions of potato virus M-infected Red Kidney bean. Phytopathology 61, 862-868.

DISCUSSION

Chairman: D.F. Bateman

Goodman: Later in this symposium you will see that the vesicles that you have demonstrated are almost exactly the same as those we have observed. The ontogeny of the flat residues is as follows. The plasmalemma convolutes and subsequently vesiculates. These vesicles then migrate to the cell wall surface and discharge, we believe "wall substances". In our system these substances appear as wall fibers and form a region of wall fiber apposition. The empty vesicles flatten and fragment. The fragmented vesicular membrane passes through the wall and eventually emerges from the outer wall surface as small vesicles or granules 25-30 nm in diameter. These granules participate in the localization of HR-inducing bacteria (see Phytopathology 66, 754-764).

Bateman: Dr. Hiruki, do you agree with this interpretation?

Hiruki: Yes.

WOUND INDUCED RESISTANCE TO PLANT VIRUS INFECTIONS

W.C. Kimmins, Department of Biology, Dalhousie University, Halifax, N.S., Canada

During the past few days, we have discussed methods of resistance to plant pathogens, in particular the role of carbohydrate hydrolases and phytoalexins. I believe the available evidence indicates that neither of these are involved with hypersensitive resistance to plant viruses. Dr. Goodman referred to the phenomenon of immunity or induced resistance and from the work of Ross and his colleagues we do know that this is an effective defense reaction to plant viruses. Another defense reaction which is effective in limiting the spread of plant viruses is the hypersensitive response (field resistance). The mechanism for this expression of resistance is the subject of the following paper. Kimmins & Brown (1973) suggested that the injury accompanying the introduction of the virus causes a wound-induced redifferentiation of the cell wall. This may localize the virus if it occurs in advance of virus spread. We also proposed (Kimmins & Brown, 1975) that the intercellular spread of a non-localized virus infection may be associated with suppression of the wound-induced cell wall modification.

To determine the nature of the changes in cell wall composition, we undertook a histochemical study of the tissue bordering virus and mechanical lesions. Positive reactions were noted for lignin, suberin and callose in cells at the border of mechanical lesions and at later periods for virus lesions (Table 1). However, the staining reactions were difficult to interpret and to relate to any resistance mechanism. To obtain more reliable data for lignification in leaves with a localized virus infection, a lignin-containing fraction was isolated and its lignin content estimated by ionization difference spectra and gas-liquid chromatography of the nitrobenzene oxidation products.

The primary leaves of 11-day old bean plants were treated in one of the

following ways (a) inoculated with TNV at 350 μg/ml which produced an
average of 264 lesions per leaf, (b) abraded (virus omitted from the
inoculum), (c) left uninoculated (control plants). There were 360 plants
in each treatment. Primary leaf samples from 60 plants were harvested at
24 hourly intervals for 120 hours from the time of treatment. The leaves
were deribbed and weighed. By extraction of non-lignin constituents a fine
white powder of lignin-containing material was obtained (LCM).

Results

The ionization difference spectra of extracts from control and abraded
plants showed two major peaks, at approximately 250 nm and 300 nm, which are
characteristic of the phenolate moiety of simple substituted aromatic
hydroxyl compounds. In samples from the virus infected leaves at 72 hours,
a third maximum was detected between 350-400 nm. This was also present in
samples from control and abraded leaves by 96 hours, although to a lesser
extent, and is probably due to phenolic hydroxyl groups associated with
carbonyl groups, carbon-carbon double bonds of biphenyl groups. The higher
absorptivity differences between 350-400 nm in the virus treatments,
suggest that increased synthesis of a conjugated polymer such as lignin had
occurred earlier and to a greater extent than in leaves with abrason wounds.

The formation of lignothioglycolic acid derivatives is a good
indication for both the presence and the amount of lignin in herbaceous
plants (Table 2). Where wounding had occurred (abraded leaves), lignification was greater at 24 hours than in the control leaves, and at 120 hours
the yield was higher by a factor of 1.68 than that from non-abraded tissue.
In the virus-infected leaves at 48 hours (prior to lesion formation), the
yield had increased to 3.34%. Between 48 and 72 hours, when lesions had
developed, the lignin had more than doubled and increased further to 19.59%
at 120 hours. Similar results were obtained from the absorbance in N,
N-dimethylformamide which at 278 nm is proportional to the amount of
ligninthioglycolic derivative present.

The yield of phenolic aldehydes from the nitrobenzene oxidation
provided more conclusive evidence of lignification. In samples from
control plants all three aldehydes increased in amount during the experiment
(Table 3), but the increase was much larger in samples from the abraded and
virus extracts. In extracts from abraded leaves yields of the aldehydes
increased 24 hours after wounding and a similar increase was found in the

virus treated plants. Between 48 and 120 hours yields of aldehydes from the
virus-infected plants exceeded those from abraded leaves.

Discussion

The results from difference spectra, absorbance measurements in N,
N-dimethylformamide, lignothioglycolic acid isolation and estimation of
phenolic aldehydes formed by nitrobenzene oxidation show that lignin was
isolated from all three groups of plants. Greater amounts were detected in
abraded leaves than in leaves from untreated plants showing that lignin was
more rapidly synthesized in response to wounding. In virus-infected leaves,
the amount of lignin at 24 hours was similar to that in leaves of abraded
plants. All the techniques used for lignin estimation indicate that around
48 hours lignin levels in the virus-infected plants were too high to be due
to the wounding effect from inoculation alone. It is possible that at this
time, the disruption of the tissue by virus infection became significant
enough to cause additional wound or injury induced lignin.

In conclusion, the results indicate that lignification is associated
with injury from the inoculation procedure and also with virus-induced
necrosis. These observations support the view that virus localization
may be caused by the induction of secondary cell wall thickening which is
induced by the inoculation wound.

Table 1. Cell wall staining reactions for tobacco mosaic virus (TMV), tobacco necrosis virus (TNV) and mechanical lesions (M) on Phaseolus vulgaris.

Stain	TMV	TNV	M
Aniline Blue (V)	-VE	-VE	-VE
Aniline Blue (U.V.)	48-168[a]	48-120	24-168
Phloroglucinol	144-168	-VE	72-168
Chlorine-Sulfite	48-96	48-96	-VE
Maule	-VE	-VE	-VE
$KMnO_4$	144-168	-VE	-VE
KOH	48-168	48-96	-VE
Sudan IV	96-168	-VE	72-168
Gentian Violet	144-168	-VE	72-168
$1KI-H_2SO_4$	-VE	-VE	-VE
Zinc-Chloro-Iodine	-VE	-VE	-VE
$FeCl_3$	-VE	-VE	-VE

Table 2. Determination of lignin in extracts from control, abraded and TNV inoculated dwarf bean leaves.

| Harvest time (HRS) | Yield of lignothioglycolic acid as % of dry weight | | |
	Control	Abraded	TNV
0	0.75	0.78	-
24	0.86	1.54	1.30
48	1.83	2.47	3.34
72	2.13	3.03	8.58
96	2.66	4.20	14.18
120	4.69	7.90	19.59

Table 3. Yield of nitrobenzene oxidation products from control, abraded and TNV inoculated dwarf bean leaves.

Treatment and post-inoculation time (hours)	Yield (mgs/g dry weight)		
	p-hydroxybenzaldehyde	vanillin	syringaldehyde
Control			
0	0.101	0.540	0.171
24			
48	0.128	0.529	0.228
72	0.323	0.353	0.484
96	0.331	0.663	0.323
120	0.542	0.508	1.100
Abraded			
24	0.278	0.646	0.183
48	0.396	0.670	0.447
72	0.415	0.469	0.692
96	0.543	1.140	0.511
120	0.655	0.700	1.620
TNV			
24	0.205	0.704	0.243
48	0.416	0.864	0.295
72	0.623	0.923	0.538
96	1.215	1.298	0.585
120	1.323	1.174	2.211

Kimmins - Plant virus infection

References

Kimmins, W.C. & Brown, R.G. (1973). Hypersensitive resistance. The role
    of cell wall glycoprotein in virus localization. Canadian Journal of
    Botany 51, 1923-1926.
Kimmins, W.C. & Brown, R.G. (1975). Effect of a non-localized infection by
    SBMV on a cell wall glycoprotein from bean leaves. Phytopathology 65,
    1350-1351.

DISCUSSION

Chairman: D.F. Bateman

Delmer: When you isolated cell walls in your study, did you try to
isolate walls from localized regions around the wounds or from whole
leaves?

Kimmins: The leaves we worked with were pretty well saturated with
lesions. They were about twice the size of a two-cent box of matches
and contained an average of 264 lesions. The lesions occupied quite a
substantial portion of the total leaf area. Cell walls were isolated
from the total leaf.

LIGNIFICATION

J. Friend, Department of Plant Biology, University, Hull, HU6 7RX, England

Paper read by: M. Jarvis

Hijwegen (1963) found that lignification occurred after inoculation of resistant but not susceptible cucumber plants with Cladosporium cucumerinum. He proposed that lignification could be an active form of resistance in host-parasite relationships.

There is an increase of a lignin-like material in potato tuber discs following inoculation with Phytophthora infestans (Friend & Knee, 1969). In a cultivar containing the R_1 gene for resistance to P. infestans the rate of accumulation of the lignin-like material is faster than in a cultivar not containing major genes for resistance (Friend, Reynolds & Aveyard, 1973; Friend, 1973). The relative rates of increase in the lignin-like material in the two cultivars correlate with increases in activity of phenylalanine ammonia lyase (PAL) activity and also with caffeic acid O-methyl transferase (COMT) activity (Friend & Thornton, 1974).

Using a different race of the fungus it is possible to produce a susceptible reaction in a major gene containing resistant variety. A comparison of a compatible and an incompatible reaction on an R_1 and an R_3 cultivar shows that in the case of each variety there is always more lignification and a faster PAL increase in the incompatible reaction (Henderson, 1975).

Deposition of lignin-like material is also found in cell walls of potato leaves inoculated with an incompatible race of P. infestans or, after an initial wound, with Botrytis cinerea which is a non-pathogen of potato. B. cinerea also induces the deposition of the lignin-like material in other Solanaceous plants to which it is non-pathogenic, such as tomato,

tobacco and <u>Datura</u> (experiments in collaboration with Mrs. M. Ward &
Professor T. Swain). Similar results were obtained when wounded wheat
leaves were inoculated with non-pathogens (Ride, 1975).

The lignin-like material is measured by an ultra-violet difference
spectrum of material extracted from the cell wall by hot alkali; the
method which is based on that of Stafford (1960) gives little indication
of the structure of the material other than its phenolic content. More
recent experiments indicate that part of the phenolic material is esterified
cinnamic acids; an appreciable quantity of material was dissolved in alkali
by saponification for a few hours at room temperature under nitrogen and it
was shown both by 2-dimensional tlc and by glc after trimethylsilyation to
contain ferulic and p-coumaric acids. However, it is not clear whether
these acids are esterified to lignin, to protein, or to carbohydrate,
although some of the cell wall preparations do contain protein which is not
necessarily covalently bonded to the acids. There are indications also
that the acids are esterified to two different components; they are easily
released from one by saponification and from the second after a partial
breakdown of possibly a high molecular weight component.

In order to explain esterification of cell walls as a component of a
resistance reaction, it is postulated that the ferulic and p-coumaric acids
are esterified to either C_2 or C_3 of the sugar monomers of the major polymeric carbohydrates of the wall. These modified substrates would then be
sufficiently altered so that they would not be able to act as substrates
for the fungal polysaccharide hydrolases. In the case of potato cell wall,
an important polysaccharide component is galactan (β-1,4-linked galactose
units) which is hydrolysed by galactanases secreted by P. infestans; other
potato pathogens secrete polygalacturonases and pectate lyases. To resist
attack by these enzymes there would need to be ferulyl or p-coumaryl
esterification of both the galactan and the linear rhamnogalacturonan
polymers. This mechanism would be the equivalent of a chemical barrier
which is distinct from a physical barrier which prevents the substrate from
coming into direct contact with the pathogen and/or its enzymes, and which
would require greater chemical changes to take place.

References

Friend, J. (1973). Lignification. pp. 383-396 in Fungal Pathogenicity and the Plant's Response, ed. by R.J.W. Byrde & C.V. Cutting. Academic Press, London & New York.

Friend, J. & Knee, M. (1969). Cell-wall changes in potato tuber tissue infected with Phytophthora infestans (Mont.) de Bary. Journal of Experimental Botany 20, 763-775.

Friend, J., Reynolds, S.B. & Aveyard, M.A. (1973). Phenylalanine ammonia lyase, chlorogenic acid and lignin in potato tuber tissue inoculated with Phytophthora infestans. Physiological Plant Pathology 3, 495-507.

Friend, J. & Thornton, J.D. (1974). Caffeic acid-O-methyl transferase, phenolase and peroxidase in potato tuber tissue inoculated with Phytophthora infestans. Phytopathologische Zeitschrift 81, 56-64.

Henderson, S.J. (1975). Role of lignin in the defence reaction of Solanum tuberosum L. following attach by Phytophthora infestans (Mont.) de Bary. Ph.D. Thesis, University of Hull.

Hijwegen, T. (1963). Lignification, a possible mechanism of active resistance against pathogens. Tidskrift for Plantenziekten 69, 314-317.

Ride, J.P. (1975). Lignification in wounded wheat leaves in response to fungi and its possible role in resistance. Physiological Plant Pathology 5, 125-134.

Stafford, H.A. (1960). Difference between lignin-like polymers formed by peroxidation of eugenol and ferulic acid in leaf sections of Phleum. Plant Physiology 35, 108-118.

INTERACTIONS OF POTATO TUBER POLYPHENOLS AND PROTEINS WITH ERWINIA CAROTOVORA

R.K. Tripathi[1], M.N. Verma and R.P. Gupta, Department of Plant Pathology, G.B. Pant University of Agriculture and Technology, Pantnagar, Nainital, India

Kufri Dewa, an Indian potato variety, is resistant to bacterial soft rot (Erwinia carotovora var carotovora) and can be stored at room temperature, with very little loss, for up to 8 months. In contrast, another variety, C-1769, started rotting in 2 months and almost all the tubers rotted within 6 months. Studies on the mechanism of resistance in Kufri Dewa to bacterial soft rot revealed the importance of interaction of polyphenols with the growth and pectic enzyme activity of E. carotovora. A part of this study is reported here.

Throughout the storage period from March to October, both peel and pulp tissues of Kufri Dewa contained much higher phenolic content than did C-1769. Although the phenolic compounds decreased with time in both varieties, the rate of decrease in Kufri Dewa was much slower, with a 30% decrease in peels in 4 months as compared to 70% in C-1769 (Table 1; Tripathi & Verma, 1975). By 5 months peels of Kufri Dewa still contained more than 60% of the original phenol content and only 19% rotting had occurred. In C-1769, however, the peels contained less than 25% of the original phenols with about 80% rotting. Of the five phenolic compounds observed by paper chromatography of a 95% boiling ethanol extract (Kuć et al., 1956), chlorogenic acid and caffeic acid were the major compounds, the identity of which was confirmed by multiple solvent chromatography and ultra-violet absorption spectroscopy.

Both peels and pulps of Kufri Dewa tubers showed higher polyphenol

---

1) Present address: Institut für Phytopathologie, Justus Liebig Universität, D-6300 Giessen, Federal Republic of Germany

Table 1. Changes in total phenol content in tubers of Kufri Dewa and
C-1769 during storage.

Months of storage	Kufri Dewa		C-1769	
	Total phenols (mg/g fresh wt)	Percent decrease	Total phenols (mg/g fresh wt)	Percent decrease
	Peels			
0	6.39	0	4.66	0
2	5.62	11.9	3.28	33.9
4	4.49	29.6	1.51	69.5
	Pulp			
0	0.78	0	0.46	0
2	0.70	10.4	0.38	18.80
4	0.62	20.72	0.233	50.2

C.D. 0.068 and 0.092 at 5 and 1% probability, respectively.

oxidase and peroxidase activities assayed by the methods of Weaver & Hautala
(1970) and Sakai et al. (1964), respectively (Tripathi & Verma, 1975;
Tripathi et al., 1974). Starch gel electrophoresis (Scandalios, 1969) of
the peels of Kufri Dewa and C-1769 showed 8 and 5 peroxidase isoenzymes,
respectively (Tripathi et al., 1974). The pulp tissues of both varieties
showed only one isoenzyme.

Peeled tuber discs of Kufri Dewa remained resistant upon incubation
under aerobic conditions, but the resistance was lost upon incubation under
anaerobic conditions created by the method of Gottlieb & Tripathi (1968),
by immersion in water or in the presence of cycloheximide ($\geq$ 5 ug/ml).
Thus resistance was oxygen and protein synthesis dependent. The loss of
resistance in water was partially reversed and the resistance restored by
periodical bubbling of air (Table 2). Under aerobic conditions, the discs
synthesized phenylalanine ammonia lyase (PAL)(Gupta & Tripathi, 1976), as
shown by the assay of the enzyme by the method of Koukol & Conn (1961),
phenols and turned brown in about 12 hrs and under these conditions the
discs remained resistant to rotting (Table 2). The discs incorporated
phenylalanine-1-C[14] into protein (9500 dpm/mg protein) and into phenolic
compounds extracted in 95% ethanol (3600 dpm/mg phenols). The paper

Table 2. Synthesis of phenylalanine ammonia lyase, phenols, browning of peeled tuber discs of Kufri Dewa and resistance to rotting under different conditions of incubation.

Incubation conditions	Relative PAL activity*	Increase in phenol content over zero time (ug/g fresh weight)	Percentage of discs showing browning	Percent weight loss**
In air				
-cycloheximide	100	35.55	100	11.0
+cycloheximide, 20 µg/ml	0	0	0	89.7
Under nitrogen	3	0	0	86.6
Immersion in water	7	6.0	5	77.5
Immersion in water with air bubbled at 30 min intervals	68	21.7	39	32.5
Treated with 8-hydroxy-quinoline (50 µg/ml)	-	30.0	0	59.0
Treated with 2,4-dinitro-phenol (50 µg/ml)	-	13.0	0	79.7
Azaguanine, 50 µg/ml	-	15.0	96	16.0
Sodium malonate, 50 µg/ml	-	15.2	94	17.5
Sodium azide, 50 µg/ml	-	11.0	0	0

*The relative PAL activity is the percentage of activity (units per mg protein) taking the activity in discs incubated in air as 100.

**The discs were inoculated with 0.2 ml water suspension of a log phase bacterial culture (10^8 bacteria per ml) per disc after 14 hr of incubation under above conditions. The weight loss was measured 48 hrs after inoculation.

chromatography of phenolic compounds followed by radiochromatogram scanning showed the maximum radioactivity in chlorogenic acid followed by caffeic acid and an unidentified compound with an Rf of 0.22 (Figure 1). In the presence of cycloheximide no incorporation into protein or phenols was observed, the discs remained white and were susceptible to rotting. Under nitrogen and in the presence of cycloheximide, PAL activity was not present and no phenols were synthesized. Upon inoculation with a log phase bacterial suspension (10^8 cells/ml) these discs rotted, while those incubated under aerobic conditions and without the inhibitor remained resistant. In the discs immersed in water, the synthesis of PAL and phenols and browning

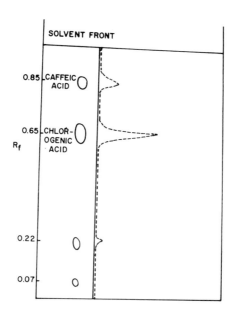

Figure 1. Incorporation of phenylalanine-1-C^{14} into phenolic compounds by
the peeled tuber discs of Kufri Dewa. The peeled discs were
incubated on a thin film of water containing 10 μc of the
tracer/ml. After a 12 hr incubation, the discs were washed
with sterile distilled water and were ground and extracted in
boiling 95% ethanol for 5 min. After concentration in a
rotary vacuum evaporator, the extract was chromatographed on
Whatman No 1 paper, developed in butanol-acetic acid water
(4:1:5, upper phase), spots located under ultraviolet lamp
followed by radiochromatogram scanning.

reactions occurred at a very much reduced rate (Table 2). The loss of
resistance was always associated with loss of PAL activity, phenol synthesis
and oxidation of phenols to result into browning.

The loss of resistance upon treatment with cycloheximide was probably
due to the inhibition of PAL and cinnamoyl-4-hydroxylase in the peeled
discs. When the discs first treated with cycloheximide for 16 hr were later
treated with 10^{-3}M cinnamic acid, the product of PAL, these remained white
and susceptible to rotting, However, upon treatment with caffeic acid, the
product of cinnamoyl hydroxylase, the white discs turned brown and the
resistance was restored to a large extent.

Tripathi et al. - Pectic enzymes and polyphenols

Table 3. Effects of extracts of Kufri Dewa tuber discs incubated under
different conditions and of chlorogenic acid on the activity of
pectin transeliminase.*

Source of extract/compound	Pectin transeliminase activity, 232/mg protein/hr
None	0.37
Freshly peeled discs	0.35
Discs incubated in air for 24 hrs	0.19
Discs incubated in air for 48 hrs	0.11
Discs incubated in air for 72 hrs	0.09
Discs incubated under nitrogen for 72 hrs	0.31
Discs incubated for 72 hrs immersed in water	0.24
Discs incubated for 72 hrs with cycloheximide, 20 µg/ml	0.35
Chlorogenic acid, 50 µg/ml	0.31
Chlorogenic acid, 50 µg/ml after incubation with polyphenol oxidase for 30 min	0.12
Polyphenol oxidase, 200 µg/ml protein	0.30
Freshly peeled discs + polyphenol oxidase	0.21

*The bacteria were grown in potato broth for 48 hrs. The culture filtrate
after dialysis and filtration through sterile Seitz filter was saturated
with 80% ammonium sulphate. The precipitate, dissolved in Tris buffer,
0.1 M, pH 8.0, was used as the enzyme source. The assay mixture was
incubated at 30°C.

A 95% alcohol extract of brown discs of tubers, upon concentration
under infra red lamp, when added to the assay mixture for pectin trans-
eliminase (Cronshaw & Wood, 1973), inhibited the enzyme activity (Table 3).
This extract at a concentration of higher than 50 µg/ml phenols was also
inhibitory to growth of E. carotovora. A similar extract of freshly peeled
discs had no effect. Inhibition of transeliminase activity was also
observed with chrologenic acid oxidized by polyphenol oxidase (Worthington
Enzymes), but not by chlorogenic acid or polyphenol oxidase separately.

A protein extract of 24 hr incubated peeled discs of Kufri Dewa in
Tris-HCl buffer (0.1 m, pH 8.0) on gel chromatography through Sephadex G-75
showed 13 bands, while from freshly peeled discs only 8 bands. Thus at
least 5 new buffer soluble proteins were synthesized during the 24 hr
incubation period. The first band fractions, which were brown in colour,
when concentrated under infra red lamp and added to transeliminase assay

tubes, inhibited the activity by 56%. This fraction seemed to be a complex
of protein and phenol or quinones. No other band was inhibitory to trans-
eliminase. However, when the fourth band was added to the first band and
the mixture was concentrated, the inhibitory activity of the first band
fractions increased to 78%. This fourth band gave a positive Molisch test
for sugars, Lowry's test for proteins and was probably a glycoprotein. The
details on the properties of these fractions and their interaction with each
other is not yet worked out.

Acknowledgements

The participation of the first author in the Symposium of Cell Wall
Biochemistry as Related to Host-parasite Interaction, held at Tromsø, Norway,
was supported by Alexander von Humboldt Foundation. This support and the
support at Institute of Phytopathology, Justus Liebig University, Giessen
is gratefully acknowledged. The research reported herein was supported by
the Experiment Station, G.B. Pant University of Agriculture and Technology,
Pantnagar, India and Indian Council of Agricultural Research by a junior
fellowship to the second author. Thanks are also due to Professor Eckart
Schlösser for going through the manuscript.

References

Cronshaw, D. & Wood, R.K.S. (1973). Annals of Botany 37, 463-471.
Gottlieb, D. & Tripathi, R.K. (1968). Mycologia 60, 571-590.
Gupta, R.P. & Tripathi, R.K. (1976). Indian Journal of Experimental
    Biology (in press).
Koukol, J. & Conn, E. (1961). Journal of Biological Chemistry 236,
    2692-2698.
Kuc, J., Henze, R.E., Ullstrup, A.J. & Quackenbusch, F.W. (1956).
    Journal of American Chemical Society 78, 3123.
Sakai, R., Romiyama, K. & Takemori, T. (1964). Annals of Phytopathologi-
    cal Society, Japan 29, 120.
Scandalios, J.G. (1969). Biochemical Genetics 3, 37.
Tripathi, R.K. & Verma, M.N. (1975). Indian Journal of Experimental
    Biology 13, 414-416.
Tripathi, R.K., Verma, M.N. & Gupta, V.K. (1974). Indian Journal of

Tripathi et al. - Pectic enzymes and polyphenols

Experimental Biology 12, 591-592.
Weaver, M.L. & Hautala, E. (1970). American Potato Journal 47, 457.

DISCUSSION

Chairman: J. Raa

Byrde: Did you try the effect of polyphenolase alone without any phenols on the enzymes?

Tripathi: Yes I did. There was no effect. Also we tried peroxidase alone and there was no effect.

LIGNIFICATION AND THE ONSET OF PREMUNITION IN MUSKMELON PLANTS

A. Touzé and M. Rossignol, Centre de Physiologie végétale - L.A. nº 241, 31077 Toulouse Cédex, France

Kuć, Shockley & Kearney (1975) reported the use of a living pathogen to provide systemic protection against the same pathogen. As the experimental system they used was familiar to us we tried to reproduce such a protection.

The inoculation of one cotyledon of a muskmelon seedling, with a virulent isolate of Colletotrichum lagenarium, followed by the spray of a low level of inoculum on the entire seedling effectively triggered a protection against infection by this strain: plants are less infected and recover health.

In order to understand the mechanisms of this acquired resistance, we began investigations at the cell wall level.

A frequent response of plant tissues to injury caused by fungal attack is the formation of lignin-like compounds. Lignification occurs in muskmelon following anthracnose as shown by an anatomical investigation and by estimation of the lignin content (Table 1).

Because of this result it seemed worthwhile to investigate the importance of this process with respect to premunition. Five days after infection, staining of hypocotyl sections showed that thickening and lignification of the walls were much more noticeable in protected plants.

Since lignin is presumed to be biosynthetically derived from phenylalanine, it was of interest to study the changes of phenylalanine-ammonia lyase (PAL) occurring after infection of both susceptible and protected muskmelons. There are marked increases of PAL activities (Table 2) in both cases but no significant difference between susceptible and protected tissues.

As this result did not allow any conclusion, an experiment was undertaken to compare the dynamics of lignin synthesis in infected and protected muskmelons. For this purpose, at 0, 2, 4, or 6 days after infection cuttings were placed in a solution of U-^{14}C L-phenylalanine, and illuminated for 24

TABLE 1: LIGNIN ESTIMATION

(ionization difference spectra)

	$\triangle$ D.O. 245 nm		$\triangle$ D.O. 300 nm		$\triangle$ D.O. 350 nm	
	H	I	H	I	H	I
1 mg of C.W.	0,125	0,500	0,025	0,100	0,003	0,060
1 stem C.W.	3,6	12,1	0,8	2,4	0,1	1,4

C.W. : CELL WALL   H : HEALTHY   I : INFECTED

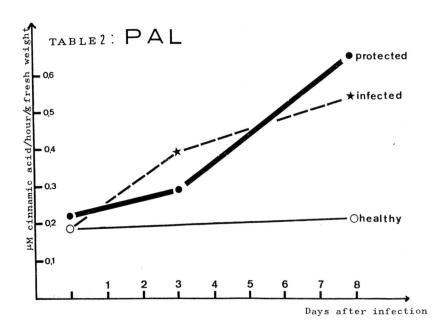

TABLE 2 : PAL

- protected
- ★ infected
- ○ healthy

µM cinnamic acid/hour/g fresh weight

Days after infection

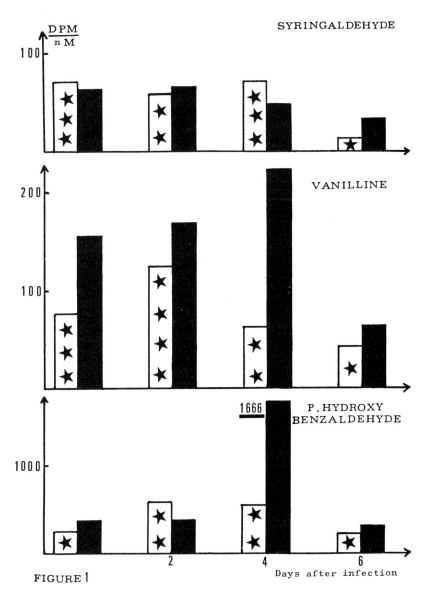

FIGURE 1

Specific radioactivities of syringaldehyde, vanilline
and p. hydroxybenzaldehyde in protected (■) and
unprotected (★) plants.

hours. Isolated stem cell walls, treated to remove phenolic acids in the
form of glycosides, were subjected to alkaline nitrobenzene oxydation. The
different aromatic aldehydes (p-hydroxybenzaldehyde, vanillin, syringaldehyde) obtained from lignin were separated on a polyclar column (Alibert &
Puech, 1976), measured at 280 nm and counted in a liquid scintillation
spectrometer. The evolution of the lignin synthesis potentialities
(expressed as specific activities, Figure 1) shows that, in protected
plants:

- the rate of synthesis of the syringyl lignin does not differ from that
  of unprotected plants,
- to the contrary, the rates of synthesis of the guaiacil lignin (as soon
  as infection starts) and of the p-coumaryl lignin (when the symptoms
  appear) are clearly increased.

Finally the possible role of lignification in the resistance of muskmelon to anthracnose, through the early protective inoculations described,
will be discussed.

References

Alibert, G. & Puech, J.L. (1976). Separation et dosage automatique des
    aldéhydes benzoïques et cinnamiques par chromatographie en phase
    liquide. Accepted for publication in Journal of Chromatography.
Kuč, J., Shockley, G. & Kearney, K. (1975). Protection of cucumber
    against Colletotrichum lagenarium by Colletotrichum lagenarium.
    Physiological Plant Pathology 7, 195-199.

DISCUSSION

Chairman: J. Raa

Jarvis: I would like to ask what you mean by your unidentified nonpolymeric
polymer which you referred to at the end?

Touzé: Those have been obtained after treatment of the cell wall. Perhaps
it's phenolic acids which bind to the carbohydrates in the cell wall.

Jarvis: These appear to be phenolic; they are u.v. absorbing but they are of low molecular weight. Is that right?

Touzé: They have low molecular weights, but we have been unable to identify them.

Jarvis: Do I understand rightly in saying that after you had extracted phenolic material for the Stafford procedure different spectrum method, and before you measured the different spectrum, you put it through a Sephadex G-25 column.

Touzé: Yes, because in this manner the small phenolics are separated.

Jarvis: Did you find any small phenolics?

Touzé: Yes, but in the method we have explained the phenolics were not identified.

Jarvis: That's interesting because we have the same sort of thing.

Paxton: Do you find that ethylene does anything to lignification?

Touzé: We have not measured ethylene.

Paxton: There is an obvious parallel there.

Touzé: You can see that the stem is larger and thicker and I have shown this and perhaps it can be an effect of ethylene. All these experiments are, however, in progress.

Paxton: Did you look for production of ethylene by the fungus?

Touzé: The fungus does not produce ethylene.

GENERAL DISCUSSION ON PLANT IMMUNIZATION

Chairman: J. Raa

Albersheim: There is just one point that I wanted to make about immunization that pertains to phytoalexins. Kuć et al. (Physiol. Plant Pathol. 7,
195-199, 1975) reported that phytoalexins were not found in trees in the
parts of the plants that were away from protection infection. So that it
is not due to increased levels of phytoalexins that the plant is being
protected.

Goodman: A lot of preimmunization experiments have been conducted over the
years. If one looks through the literature these types of experiments have
been done for viruses, bacteria and for fungi mainly using not virulent
strains but frequently avirulent or attenuated strains or isolates of the
pathogen. I think that we have been led to believe that the plant just
doesn't have an immune system. I think that the work of Kuć and now Prof.
Touzé (this symposium) and some of the others ought to be looked at a bit
more carefully.

Paxton: Immunization poses an interesting problem because it implies if
you carry it to its logical end that once a plant gets a disease or a spot
on the leaves then that's it for that disease; and if the plant can continue
to go on you don't see the progression of diseases that you actually do in
the field, that's hard to resolve. Why do you see a disease build up and
in many cases continue to infect new leaves as they are produced when in
fact theoretically they should be immunized after their first infection?

Goodman: We did some experiments with tobacco and also with apples in which
we infiltrated tissue with avirulent strains of in one case Erwinia amylovora
and in the other case Pseudomonas pisi. Although the experiments are not
complete at this time, the ones with P. pisi gave us the results as follows:
During a six hour period of time you could go back to the tissue and extract
something that was pelletable, that would agglutinate the inducing or the

challenge strain and there was an increase in titer over a six hour period.

Albersheim: I would like to try to answer Jack Paxton's statement. In these protection experiments protection is carried out deliberately with low levels of inoculum when the plants are particularly healthy and young and you have a slow rate of killing presumably. In the field it may be that you get this build up of inoculum so quickly that the leaves are killed before they can export whatever it is that is going to protect the other leaves in the plant and it could simply be a rate phenomenon you are talking about.

Mussell: If I could just amplify that a little bit if I understand what Kuč has been explaining to me about his system. The protection is of tissue that then grow, not of the tissues that were infected, so that it is not that you are arresting development of disease once it has started in one spot. You are preventing it from starting in new tissues that develop after that infection.

Delmer: There have been some studies on plants on stress in general which show that pre-exposure to one type of stress predisposes subsequent resistance to other kinds of stress. This is not speaking of infection but of water stress predisposing the plant to be more resistant to subsequent ion stress and so on. I think this is kind of a similar situation to the pathogenesis situation that Prof. Touzé is talking about. Invariably absisic acid (ABA) is produced under these situations, and as a matter of fact pre-exposure to ABA subsequently make plants more resistant to other stresses. I haven't heard ABA mentioned here once in a plant pathology symposium, and I am wondering if there might be some correlation between this type of resistance and pathogenesis and if anyone has any information about absisic acid and its role in infection.

Pegg: Wright & Hiron (Nature 224, 719, 1969) showed that plant tissue placed under water stress showed an almost immediate increase in abscisic acid. This has to be considered in the context of vascular wilt diseases where the inhibitor, growth substance metabolism of the leaves may be affected from indirect causes. Similarly, if experimental plants wilt from lack of water the ABA content remains high even after the plant has been restored to full turgor and for several days later. Such plants would be.

General discussion on plant immunization

experimentally quite different from ones maintained throughout on full turgor. The significance of this may be seen in comparative experiments on detached plant parts and whole plants where the former may suffer a temporary water deficit.

HYDROXYPROLINE-RICH GLYCOPROTEINS OF THE PLANT CELL WALL AS AN INHIBITORY
ENVIRONMENT FOR THE GROWTH OF A PATHOGEN IN ITS HOST

M.T. Esquerré-Tugayé and A. Toppan, Université Paul Sabatier, Centre de
Physiologie Végétale, L.A. CNRS 241, 118, route de Narbonne, 31077
Toulouse-Cédex, France

In the understanding of such phenomena as recognition and defence
mechanisms in plant-microorganism interactions, the study of proteins and
glycoconjugates of the outer layers of the cell might be helpful. The cell
wall of higher plants contains extensin, a hydroxyproline-rich glycoprotein,
which is characterized by the presence in the molecule (Lamport, 1973) of
the repeating pentapeptide

$$
\begin{array}{c}
\text{Gal} \\
| \\
-\text{Ser}-\text{Hyp}-\text{Hyp}-\text{Hyp}-\text{Hyp}- \\
| \\
(\text{ara})_{1 \text{ to } 4}
\end{array}
$$

where serine (ser) and hydroxyproline (hyp) may or may not be glycosilated
by galactose (1 unit) and arabinose (up to 4 units).

In Melon (Cucumis melo) seedlings infected with Colletotrichum lagenarium the cell wall becomes ten times richer in hydroxyproline (Esquerré-
Tugayé, 1973), and this accounts for an accumulation of extensin, as shown
by selective extraction and separation of the hyp-arabinosides (Esquerré-
Tugayé & Mazau, 1974) (Figure 1). This accumulation begins after an early
increase in the rate of C_2H_4 production (Figure 2). We have demonstrated
that the fact of inhibiting the formation of C_2H_4 with canaline or benzyl
isothiocyanate (BITC) also lowers the rate of extensin deposition in the
cell wall (Figure 3). Thus, ethylene seems to be involved in the described
extensin response. A similar behavior has not, to date, been reported for
diseased plants. In order to understand the biological significance of such
cell wall modifications in the case of Melon anthracnose, two types of
experiments were set up.

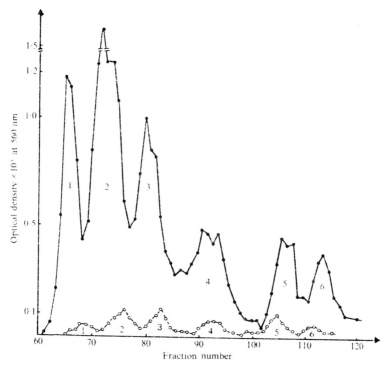

Figure 1. Fractionation on Aminex AF 50 x 8, H^+ form, of the G25 hydroxy-proline-rich compounds from He (500 mg -o-o-) and I (500 mg (-•-•-) cell walls, using a pH gradient (200 ml of H_2O in mixing chamber and 200 ml of 0.2M HCl in the reservoir); the fraction volume is 3 ml of which 1 ml is used for the hydroxyproline estimation.

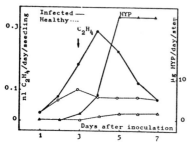

Figure 2. Rates of C_2H_4 production and cell-wall hyp enrichment during the infection of C. melo by C. lagenarium.

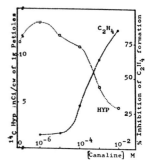

Figure 3. Inhibition of ethylene formation and repercussion on the cell wall hyp enrichment.

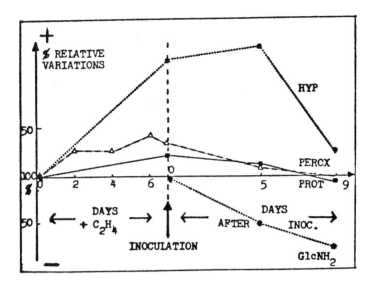

Figure 4. Relationships between the cell wall hyp level and the resistance of Melon seedlings to the infection by *C. lagenarium*. 1. Effect of a pretreatment by C_2H_4.

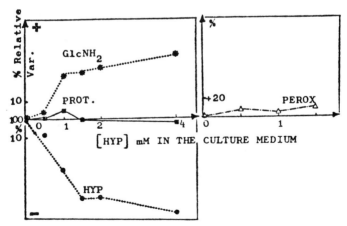

Figure 5. Relationship between the cell wall hyp level and the resistance of Melon seedlings to the infection by *C. lagenarium*. 2. Effect of a treatment by free hyp.

In the first one, the cell walls of healthy seedlings are enriched in hyp prior to infection by C. lagenarium (Figure 4), by treatment with ethylene (Ridge & Osborne, 1970). It appears that the resulting plants are much more resistant to infection than the control: infected seedlings not pretreated by ethylene.

In the second one, the accumulation of cell wall hyp which follows the inoculation, is lowered by adding free hyp to the culture medium of the seedlings (Figure 5); this amino acid is supposed to inhibit proline hydroxylation (Cleland, 1967). Infected seedlings, not treated with hyp, are used as the control. Throughout the two types of experiments, the determination of hyp in the cell wall and of the fungal mass in the seedlings (expressed as µg glycosamine, Toppan, Esquerré-Tugayé & Touzé, 1976), show a close correlation between the amount of extensin and the resistance of the seedlings to anthracnose.

The way in which an enrichment of extensin would bring about modifications of the cell surface that could slow the spread of a pathogen, will be discussed.

References

Cleland, R. (1967). Inhibition of formation of protein-bound hydroxyproline by free hydroxyproline in Avena coleoptiles. Plant Physiology 42, 1165-1170.

Esquerré-Tugayé, M.T. (1973). Influence d'une maladie parasitaire sur la teneur en hydroxyproline des parois cellulaires d'épicotyles et petioles de plantes de Melon. Comptes Rendus de L'Académie des Sciences de Paris, Série D, tome 276, 525-528.

Esquerré-Tugayé, M.T. & Mazau, D. (1974). Effect of a fungal disease on extensin, the plant cell wall glycoprotein. Journal of Experimental Botany 25, 509-513.

Lamport, D.T.A. (1973). The glycopeptide linkages of extensin: O-D-galactosyl serine and O-L-arabinosyl hydroxyproline. pp. 149-164 in Biogenesis of plant cell wall polysaccharides. Academic Press, Inc., New York & London.

Ridge, I. & Osborne, D.J. (1970). Hydroxyproline and peroxydases in cell walls of Pisum sativum: Regulation by ethylene. Journal of Experimental

Botany 21, 843-856.

Toppan, A., Esquerré-Tugayé, M.T. & Touzé, A. (1976). An improved approach for the accurate determination of fungal pathogens in diseased plants. Submitted to Physiological Plant Pathology.

DISCUSSION

Chairman: J. Raa

Delmer: I have two questions. The first is whether only extensin in the cell wall changes if that's the only component of the cell wall that you see changing in composition? The second question is whether you see any increases in cytoplasmic hyp-arabino galactan type molecules that is the kind of cytoplasmic hyp containing glycoproteins that Lamport has seen?

Esquerré-Tugayé: I have looked for protein in the cell wall and for enzymes and it seems that the oxidase and invertase are stimulated during the infection, but it seems that some other enzymes, like β-galactosidase, are not affected.

Delmer: What about the neutral sugar fraction of the cell wall? I mean other hemicellulosic, polysaccharides or cellulose components of the cell wall.

Esquerré-Tugayé: There is a slight decrease in the hemicellulose content and also in that of the pectic substances. But there is also an increase in hydroxyproline in the cytoplasm which I think is because extensin is synthesised while it's on the way to the cell wall.

Valent: Yes, but the arabino galactan containing hydroxyproline proteins might be expected to have galactan activity.

Esquerré-Tugayé: I have not checked that. I don't know that they are going to the cell wall. At least when we studied serine glycoside and hydroxyproline we always have a constant ratio between serine glycoside and hydroxyproline between infected plants.

Albersheim: I would like to suggest that perhaps these molecules work by some means already suggested earlier this week, interacting with glucan elicitors. It was suggested that they might bind them, but I wonder whether they might be receptors and actually the interactions lead to a positive interaction of turning on defense mechanisms in the plant where there isn't a connection between these; because there have been a number of papers now that have been alluded to where these hydroxyproline rich arabinose containing proteins bind to beta glucans. They haven't looked at the structure of beta glucans in detail.

GLUCANOHYDROLASES OF HIGHER PLANTS: A POSSIBLE DEFENCE MECHANISM AGAINST
PARASITIC FUNGI

George F. Pegg, University of London, Wye College, Wye near Ashford,
Kent TN2 5AH, England

The involvement of glucanohydrolases in host parasite relations as a
possible host defence mechanism has received relatively little attention.
Prior to 1970 nothing had been reported and since that time fewer than six
papers on the subject have appeared. Most have been concerned with infection per se but not necessarily with a host in which fungal hyphae were
undergoing lysis.

The classical picture of lysis of the parasite is seen with the
mycorrhizal endophytes of orchids. Orchids typically have no carbohydrate
reserve in the seed and the successful development of the seedling protocorm
depends on the establishment of a symbiotic relationship with a fungal
parasite in the surface soil or humus. The symbionts, usually strains of
the imperfect stage of Thanetephorus, Ceratobasidium, or Rhizoctonia and,
less commonly, Armillaria, are restricted to the cortical tissues of the
host by a process of lysis in which coils or peletons of hyphae in the
cells clump together and then undergo progressive lysis, from which is has
been assumed that the products are beneficially absorbed by the host
(Plate I). Burgess (1939) described distinct zones of lysis in Orchis
mascula infected with Rhizoctonia. Expressed sap from protocorm tissue,
when applied to plate culture of Rhizoctonia induced staling and lysis of
the mycelium. This report by Burgess was the first to indicate a possible
host lytic effect, although Bernard in 1909 had suggested a host mediated
defence mechanism in orchid parasitism.

Another host parasite relationship in which lysis occurs as a regular
feature is in fungus vascular wilt infection caused by Verticillium spp.
Dixon & Pegg (1969) showed a 70% reduction in tomato plants infected with

Plate I

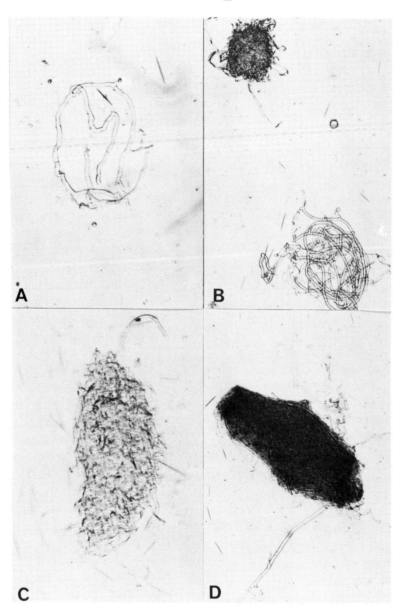

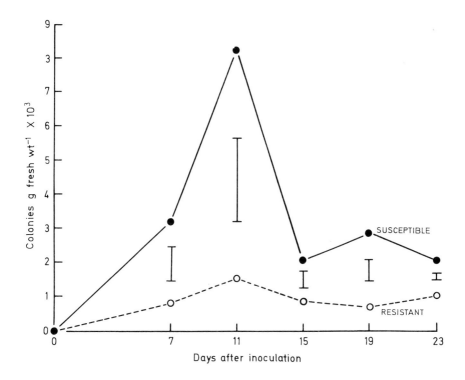

Figure 1. Colonisation and mycelial lysis in stems of tomato cv.
Craigella, susceptible and resistant to Verticillium albo-
atrum. Vertical bars = LSD for ten replicates (P = 0.05).

Plate I. Mycelial peletons of a symbiotic Rhizoctonia isolate in
various stages of lysis, from a macerate of protocorm cells
of the orchid Dactylorhiza purpurella grown as a tissue
culture on a cellulose minimal agar.
A. Initial infection stage prior to lysis.
B. Early peleton (bottom) and one in an advanced stage of
lysis (top).
C. & D. Terminal stages in peleton lysis.

a weakly compatible strain of V. albo-atrum. Lysis commenced after two or three weeks, depending on the host cultivar (Figure 1). A similar phenomenon has been reported for olive trees (Wilhelm & Taylor, 1965) and apricot (Taylor & Flentje, 1968) infected with V. dahliae. In both of these trees and in the orchid Galeola septentrionalis infected with A. mellea, lysis followed a seasonal pattern to a point where no pathogen could be isolated. Perennial infection in all these hosts is thought to be maintained by seasonal re-infection.

Hyphal wall composition

By far the majority of fungi, Chytridiomycetes, Ascomycetes, Basidiomycetes and Deuteromycetes have a cell wall chemical structure which is broadly similar and are included in Bartnicki-Garcia's (1968) Chitin-Glucan group V. Non-cellulosic glucans present are of two kinds, an alkali soluble or S-glucan, consisting mainly of β-1-6 linkages and an alkali insoluble yeast glucan. The most commonly occurring link is β-1,3, but β-1,3 and β-1,4 have been recorded. Claims for the presence of β-1,4 glucan in the wall of Fusarium have not been substantiated by the critical examination of x-ray powders.

The cell wall of Verticillium albo-atrum

According to Wang & Bartnicki-Garcia (1970) and our own researches, the wall of V. albo-atrum contains at least three different polysaccharides. These are: i) an alkali-soluble heteropolysaccharide of D-glucuronic acid, β-D glucose, D-mannose and D-galactose, ii) β-linked glucans, and iii) chitin. The alkali soluble heteropolysaccharide and protein probably represent surface and/or interfibrillar components. Following alkali hydrolysis the wall loses its granular appearance and a randomly interwoven network of glucan and chitin microfibrils is revealed. The sugar residues in the alkali-soluble heteropolysaccharide are all found, with the exception of D-galactose in the hemicellulose fraction of Angiosperm plant walls. It is not possible at this stage to make more precise comparisons of Verticillium wall components and tomato hemicellulose, nevertheless, the hydrolytic activity on the fungal heteropolysaccharide of glucosidase, mannosidase, glucuronidase and glucanase of host origin would not be unexpected.

The major component of V. albo-atrum, the glucan is known to contain 3
linkage types 1→3, 1→4 and 1→6. From enzymic digestion studies of the
alkali-insoluble wall fraction by Streptomyces QMB814 glucanase, and the
laminarin oligomers produced, 1→3 linkages would appear to be predominant.
β-1,4-linked cellobiose was also found in this hydrolysate. Extensive
specific tests for the presence of cellulose, however, were all negative
(Wang & Bartnicki-Garcia, 1970). The possibility of cellobiose arising by
transglucosidation during enzymolysis as with 1,4-β- glucosidase (Reese et
al., 1967) was discounted following the low specific activities of oligosaccharides obtained after digestion of the wall residue in the presence of
^{14}C glucose. The enzymes most involved in lysis of the glucan would be
endo and exo 1,3 or 1,6-β-glucanases and 1,4-β-glucosidase. The possibility
exists that 1,3-β-glucanase could break some 1,4-β-linkages, since the
specificity of the enzyme may be determined by the oligosaccharide rather
than the linkage per se. Thus endo 1,3-β-glucanase has an affinity for
laminaribiose and splits the adjacent linkage, whether it be 1→3 or 1→4
(Perlin & Reese, 1963). Similarly, 1,4-β-glucosidase is completely non-
specific with regards to the cleavage of diglucose molecules and attacks
all linkages to different extents (Table 1). In this way the enzyme may

Table 1. The non-specificity of 1,3-β-glucanase and 1,4-β-glucosidase
from different sources on different glucose dimers (relative
activity). (After Reese et al., 1967.)

Substrate	1,4-β-glucosidases		Exo-1,3-β-glucanases	
	Aspergillus niger	Almond emulsin	Basidiomycete	Sporotrichum pruinosum
Diglucose				
link 1 - 1	21	30	7 7	12
1 - 2	40	35	12	10
1 - 3	97	100	100	100
1 - 4	100	37	1	0
1 - 6	72	8	0	0
Laminarin	3	6	2,000	10,000

substitute for laminaribiose or suggest erroneously the activity of exo-
1,3-β-glucanase.

Chitin, a linear polymer of β-1,4-linked N acetylglucosamine, is a
major constituent of the microfibrillar network of the Verticillium cell
wall, comprising 7.6-10% of the total wall dry weight. From the limited
information available little can be said regarding the chemical and physical
similarity between Arthropod exo skeletons and fungal chitin. In the
latter, however, it is usually in association with small amounts of other
polysaccharides and protein substances which may protect against the independent action of chitinase. Enzymes which would be expected to be involved
in chitin degradation would be an endo chitinase, chitiobiase, active on the
dimer and trimer, lysozyme and N acetylglucosaminidase which removes terminal
non-reducing N acetylglucosamine residues.

Higher plant glucanohydrolases

Many of the enzymes that might be expected to be involved in fungal
wall lysis have been shown to be constitutive in the tissues of higher
plants and show properties similar to the enzymes concerned with microbial
autolysis. Some of the more important ones are listed below.

Endo-1,3-β-glucanase EC 3.2.1.39

This enzyme has been reported in a wide range of Angiosperm species
(Dillon & O'Colla, 1951; Peat et al., 1952; Manners, 1955; Eschrich, 1959
and Clarke & Stone, 1962). The latter authors reported the presence of
endo and exo glucanase (EC 3.2.1.58) in more than 20 species of plants,
with especially high activity in tobacco leaves and seeds of Soya and
germinating wheat. Moore & Stone (1968) in a pioneering paper demonstrated
increased glucanase in TMV-infected Nicotiana glutinosa. Increased enzyme
levels were later reported in Uromyces-infected Phaseolus vulgaris (Abeles
et al., 1971), tree species (Wargo, 1974), Colletotrichum-infected melon
(Rabenantoandro et al., 1976) and in Verticillium-infected tomato (Pegg,
1976).

Chitinase EC 3.2.1.14

Found originally as a contaminant of 1,4-β-glucosidase in almond emulsin (Grassmann et al., 1934) and later in bean and other seeds (Powning & Irzykiewicz, 1965) and Phaseolus vulgaris plants (Abeles & Forrence, 1970) and forest trees (Wargo, 1974), Pegg & Vessey (1972) described its involvement in Verticillium-wilt of tomato. No chitin, or free N acetylglucosamine is present in tomato, but acetylglucosamine is detected following alkaline hydrolysis of tissue (Ride & Drysdale, 1972).

N Acetylglucosaminidase EC 3.2.1

Now including chitobiase (3.2.1.28) responsible for the hydrolysis of terminal non-reducing 2-acetamido 2-deoxy β-D glucose residues in chitobiose and higher analogues.

Lysozyme EC 3.2.1.17

This muramidase with activity similar to EC 3.2.1.14 was first detected in Fig (Ficus carica) by Fleming (1922) and later by Meyer et al. (1946). Howard & Glazer (1969) and Glazer et al. (1969) have shown it to be present in large quantities in Papaya and Fig (Ficus carica) respectively. Pegg & Vessey (1972), however, could find no lysozyme activity in tomato.

1,4-β-Glucosidase EC 3.2.1.21

The main hydrolase of sweet almond emulsin has been associated with changed metabolism in many plant diseases. Its action is most commonly attributed to the hydrolysis of β-D-glucosides, but Reese et al. (1967) have shown the non-specificity of this enzyme with the ability to cleave all linkages of di-glucose. It is most probable, therefore, that it has a role in hydrolysis of wall glucosyl polymers partially degraded by endo-glucanases.

Purification of host enzymes

Most studies have been concerned with crude aqueous extracts of healthy and infected tissue but, depending on the specificity of the assay it is difficult to comment on the presice nature of the enzyme involved. Abeles et al. (1970) attempted a several stage purification of an endo-glucanase from Phaseolus vulgaris , a method which has been used subsequently by other workers. Water homogenates of ethylene-treated leaves

Table 2. Stages in the purification of a 1,3-β-glucanase from Phaseolus vulgaris (after Abeles et al., 1970).

Step	Purification	Glucanase units[1]	Units recovered	Protein recovered	Protein recovered	Specific activity
			%	mg	%	Units glucanase/ mg protein
1	Crude Homogenate	122,000	100	2,730	100	45
2	60 C for 10 min.	93,100	76	1,515	55	62
3	DEAE-cellulose	78,400	64	700	25	112
4	60% $(NH_4)_2SO_4$	76,000	62	427	16	178
5	Lyophilized	70,000	58	200	7.3	350
6	Hydroxyapatite	31,800	31	60	2.2	530
7	CM-Sephadex	15,200	12	22	0.8	695

1) One unit = 1 mg glucose equivalent/hr at $50^{o}C$.

were centrifuged at $10,000_g$, and the supernatant heated at $60^{o}C$ for 10 minutes to remove heat labile protein (see Rabenantroandro, 1976). The glucanase was eluted from DEAE cellulose and a 60% $(NH_4)_2SO_4$ fraction was partially puri- fied by ion-exchange chromatography on hydroxylapatite and gel filtration on CM Sephadex. In this separation a major peak of glucanase was separated from chitinase and peroxidase but was contaminated with RNase. The loss of total activity associated with increased specific activity in the different stages of chromatography is shown in Table 2.

Attempts to achieve reasonable purity with tomato glucanase using this technique showed that the glucanase peak was contaminated with chitinase, NAGase, glucosidase and peroxidase. Excellent first stage purification with the tomatu enzyme, however, was achieved by stepwise elution of a Tris HCl pH 8.0 buffer. Endo-glucanase, assayed by the Nelson-Somogyi determination of reducing groups from hydrolysed laminarin, eluted as a sharp peak with no other hydrolase present (Figure 2A). The peak fractions were concentrated through P10 Amicon membranes and subjected to gel-permeation chromatography on Bio-Gel-150 (Figure 2B) and the glucanase peak finally purified by iso- electric focusing (IEF). The conditions for IEF were a sucrose density gradient and ampholine carriers giving a gradient of pH from 4.0 to 6.0. The enzyme had a pH of 5.7. Preliminary results suggest a molecular weight

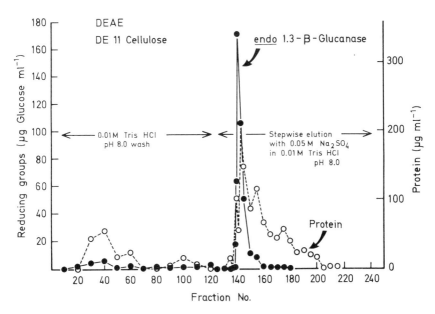

Figure 2A. Stages in the purification of an endo 1,3-β-glucanase from tomato leaves (Lycopersicon esculentum Mill.) cv. Craigella. Ion-exchange chromatography on DEAE (DE11) cellulose (30 x 3 cm column fractions = 10 ml).

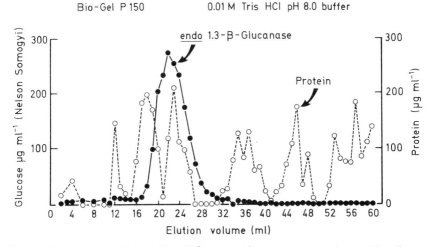

Figure 2B. Same as Figure 2A. Gel-permeation chromatography on Biogel P-150 polyacrylamide gel (30 x 1.5 cm column).

in the region of 30,000. Purification was associated with some loss of
activity but storage in bovine serum albumin partly protected against this.
Hyphal walls of Verticillium albo-atrum bathed in dialysed fractions of the
enzyme concentrated by IEF showed a reduction in dry weight associated with
increased reducing groups in the bathing solution. Digests were carried
out for periods up to 24 h in 0.05% NaN$_3$.

Enzyme induction by different pathogens

Studies by Moore & Stone (1968, 1972), Abeles et al. (1971), Pegg &
Vessey (1972), Wargo (1974) and others have shown that increased glucanohydrolase activity is associated with virus and fungal infection. Infection
of tomato cv. Craigella with 4 quite different pathogens was studied in
relation to the induction of chitinase, N acetylglucoseaminidase (NAG-ase),
1-3-β-glucanase and 1,4-β-glucosidase in stems and leaves. The pathogens
selected for different cell wall compositions were Verticillium albo-atrum
involving a predominantly chitin-glucan wall structure, Phytophthora
cryptogea with a non-chitin, cellulose-glucan wall, the wilt bacterial
pathogen, Pseudomonas solanacearum and tobacco mosaic virus. Plants were
assayed in the pre-symptom period, 48 h after inoculation for P. solanacearum and 7 days for the other pathogens. The results (Table 3) showed no
correlation between the wall structure of the non-viral pathogens and their
ability to induce glucan hydrolases. P. solanacearum infection induced
significant increases in all hydrolase activity in leaves but only in
glucanase and chitinase activity in stems, the site of the pathogen at
that time. Root inoculation by zoospores of Phytophthora cryptogea induced
a large increase in chitinase and NAG-ase in stems only, demonstrating that
the increase in enzyme was not related to specific pathogen wall inducers.
In these experiments no stimulation of glucanase was observed following TMV
inoculation. The enzyme was assayed using the glucose oxidase method
(Morley et al., 1968) which is specific for glucose. Moore & Stone (1972),
however, measuring reducing groups with the Nelson Somogyi method showed
large and consistent increases in glucanase activity with virus infection.
It is likely, therefore, that the endo-glucanase in tomato is incapable of
splitting small glucose polymers to glucose, a function which is most probably carried out by 1,4-β-glucosidase.

The results with all four pathogens suggest a non-specific, host-

Table 3. Glucanohydrolase activity in tomato leaves and stems following inoculation with pathogens of different cell wall chemical composition. (Means of ten replicate plant extracts.)

Pathogen			Nagase $(\mu g\ PNP\ ml^{-1}h^{-1})$	Chitinase $(\mu g\ NAGA\ ml^{-1}h^{-1})$	$1,4-\beta-$ Glucosidase $(\mu g\ PNP\ ml^{-1}h^{-1})$	$1,3-\beta-$ Glucanase $(\mu g\ glucose\ ml^{-1}h^{-1})$
Phytophthora cryptogea	Leaf	H	502	64	32	6
			ns	ns	ns	ns
		D	451	72	30	7
	Stem	H	174	21	12	1
			*	*	*	ns
		D	328	117	22	2
TMV	Leaf	H	197	52	15	-
			*	*	*	
		D	359	61	37	-
	Stem	H	155	18	5	-
			*	*	ns	
		D	235	35	4	-
Verticillium albo-atrum	Leaf	H	528	75	53	21
			*	*	ns	ns
		D	670	108	55	26
	Stem	H	128	9	2	5
			*	*	*	*
		D	254	42	6	10
Pseudomonas solanacearum	Leaf	H	393	44	29	3
			*	*	*	*
		D	508	56	40	9
	Stem	H	119	6	4	2
			ns	*	ns	*
		D	106	18	3	11

Nagase = N acetylglucosaminidase. PNP = Paranitrophenol. NAGA = N acetylglucosamine. * = Results statistically significant (P = 0.05).

mediated response, possibly as a metabolic stress reaction, which will be discussed later in relation to endogenous ethylene.

Chitinase activity in <u>Verticillium</u> infection of tomato

Pegg & Vessey (1972) demonstrated that chitinase activity in tomato cv Moscow was constitutive and an <u>endo</u> enzyme based on viscometric assays on colloidal chitin (Table 4). It was present in the xylem exudate and vacuum extracted xylem fluid of healthy stems increasing on inoculation with a compatible strain. Whereas lysis was most pronounced in incompatible combinations, activity was highest in compatible reactions, in direct proportion to the quantity of the invading mycelium (Table 5).

Table 4.   Constitutive chitinase activity in sterile grown <u>Lycopersicon</u> <u>esculentum</u> cv Moscow seedlings.

Seedling age (days)	Experiment	Chitinase activity (μg N-AGN)
17	1	3.80
	2	4.12
	3	3.90
28	1	4.31
	2	5.38
	3	5.23

N-AGN = <u>N</u> acetylglucosamine nitrogen.

Table 5.   Chitinase activity of tomato cv Potentate stems showing resistant and susceptible reactions to hop (<u>Humulus lupulus</u>) and tomato strains of <u>Verticillium albo-atrum</u>.

Days after inoculation	Chitinase activity (μg <u>N</u> acetylglucosamine N)		
	Control	Tomato strain (susceptible reaction)	Hop strain (resistant reaction)
21	1.0	2.2	1.5
28	1.1	2.3	1.6

A scheme for the partial purification of the tomato chitinase is presented in Figure 3 giving an increase of 3.7 in the specific activity of the enzyme. The K_m and V_{max} values for a chitin dispersion in pH 5.2 0.1 M-acetate buffer were 0.408 mg ml^{-1} and 3.7 μg NAG nitrogen h^{-1}ml^{-1} respec-

PARTIAL PURIFICATION OF TOMATO CHITINASE

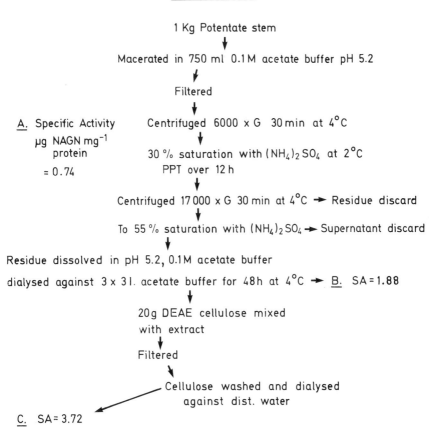

A. Specific Activity
μg NAGN mg^{-1}
protein
= 0.74

1 Kg Potentate stem

Macerated in 750 ml 0.1M acetate buffer pH 5.2

Filtered

Centrifuged 6000 x G 30 min at 4°C

30 % saturation with (NH$_4$)$_2$SO$_4$ at 2°C
PPT over 12 h

Centrifuged 17 000 x G 30 min at 4°C ➤ Residue discard

To 55 % saturation with (NH$_4$)$_2$SO$_4$ ➤ Supernatant discard

Residue dissolved in pH 5.2, 0.1M acetate buffer

dialysed against 3 x 3 l. acetate buffer for 48h at 4°C ➤ B. SA = 1.88

20g DEAE cellulose mixed
with extract

Filtered

Cellulose washed and dialysed
against dist. water

C. SA = 3.72

Figure 3.   Partial purification of a tomato chitinase (NAGN = N acetylglucosamine nitrogen).

Table 6.   Effect of enzyme inhibitors on the chitinases from microorganisms and tomato plants.

Inhibitor 0.5 M	Replicate	V. albo-atrum	Streptomyces antibioticus	Tomato
Control	1	0.7	3.2	2.1
	2	0.7	3.5	2.1
	3	0.6	3.8	1.9
NaN$_3$	1	0.5	3.6	1.9
	2	0.5	3.9	1.6
KCN	1	0	0	0
	2	0	0	0
HgCl$_2$	1	0	0	0
	2	0	0	0
CuSO$_4$	1	0	0.1	0
	2	0	0.1	0.1

tively. The response of chitinases from V. albo-atrum, Streptomyces anti-
bioticus and healthy tomato to a range of inhibitors (Table 6) were identical.
Concentrations of NaN$_3$ up to 0.5 M reduced only slightly the velocity of
chitin and laminarin hydrolysis and 0.05 M NaN$_3$ was used in prolonged
incubations as a bacteriocide. Activity was optimal at pH 5.4 with evidence
of a minor peak at 5.8 and at 40°C, declining rapidly with higher temperatures and pH values. Comparisons of the mycelial content of susceptible
infected tomatoes maintained at 21°C for 4 weeks with plants transferred to
32°C at after 7 and 14 days (Figure 4) showed a reduction with the increased
temperature.

Endo-1,3-β-glucanase activity in tomato

Pegg (1976) has shown that a 1,3-β-glucanase is similarly constitutive
in tomato in equal amounts in cv Craigella S and the single-gene resistant
isoline Craigella R, with higher levels in leaves than in stems. As with
chitinase, more enzyme was formed in infected susceptible plants than

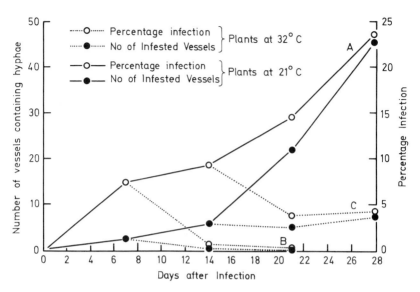

Figure 4. Effect of temperature colonisation of tomato stem xylem vessels
by Verticillium albo-atrum. A, plants grown continuously at
21°C, B and C, plants transferred to 32°C at 7 and 14 days,
respectively.

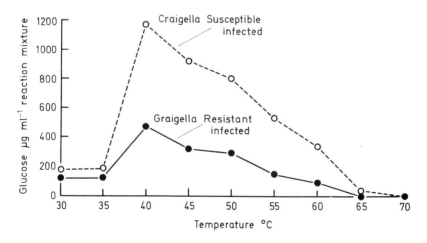

Figure 5. Effect of temperature on 1,3-β-glucanase from stems of
Verticillium-resistant and susceptible tomato plants.

resistant (Figure 5). The distribution of activity within the plant,
Table 7, was highest in the basal stem internodes where most mycelium was
concentrated.

Table 7. The distribution of 1,3-β-glucanase activity in healthy and
Verticillium-infected Craigella (susceptible) tomato plants.

Internode	Healthy	Infected
1 - 3	265^i	740
4 - 6	225	610
7 - 9	215	485

i) μg glucose ml^{-1} reaction mixture.

Glucanase and chitinase activities have also been studied in relation
to Verticillium dahliae-infection of potato (Solanum tuberosum), grown under
long and short photoperiods. In long days (15 h photoperiod - LD) plants
show resistance to symptom development and this is reflected among other
things in the total leaf area (Figure 6) and survival of mycelium in the
plant (Figure 7). Colonisation in LD & SD plants is identical 2 weeks after
inoculation but thereafter lysis occurs in LD plants reducing the mycelial
content 24-fold to near zero by 5 weeks. Conversely, in SD plants, mycelial
growth increased up to 4 weeks before declining. Chitinase activity (Figure
8), was higher in both LD and SD infected plants compared with controls but
was highest after 2 weeks in LD plants agreeing with the idea of enzyme-
induced hyphal lysis. This was also confirmed with the glucanase data
(Figure 9), where the partial lysis in SD plants between weeks 5 & 6 was
reflected in rising glucanase at the end of the experiment. These results
are different from those found in single gene resistant plants. Hydrolase
activity in plants resistant under extended photoperiods appears to be
inversely proportional to the level of pathogenic hyphae.

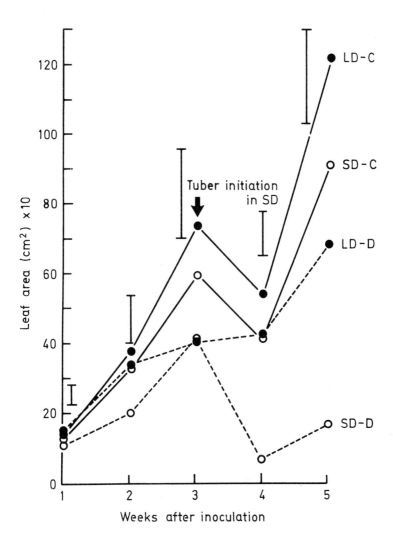

Figure 6. The effect of <u>Verticillium dahliae</u>-infection on the leaf area
of potato (<u>Solanum tuberosum</u>) cv. King Edward grown under
different photoperiods. LD and SD represent photoperiods of
16 and 8 hours respectively. C = controls, D = infected
plants. Vertical bars = LSD (P = 0.05).

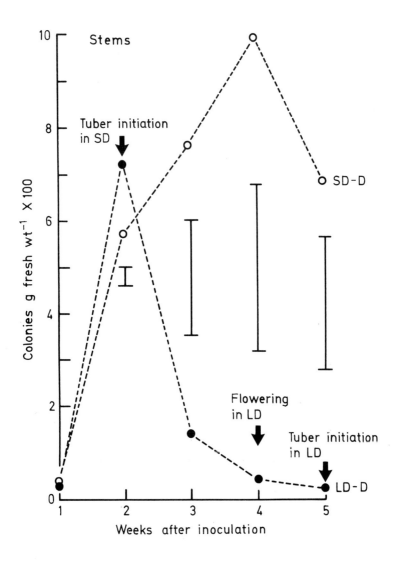

Figure 7.   Colonisation and mycelial lysis in potato stems infected
with Verticillium dahliae under long (LD) and short (SD)
photoperiods.

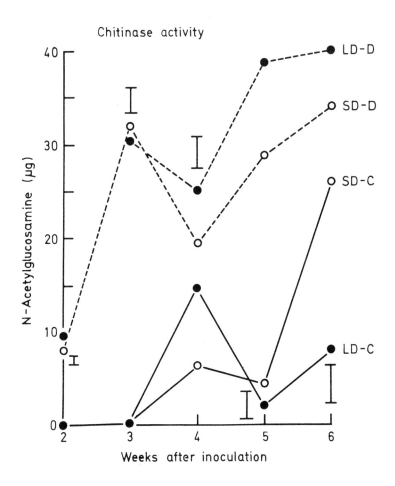

Figure 8.   Chitinase activity in potato stems from plants infected by
Verticillium dahliae under different photoperiods (legend
as in Figure 6).

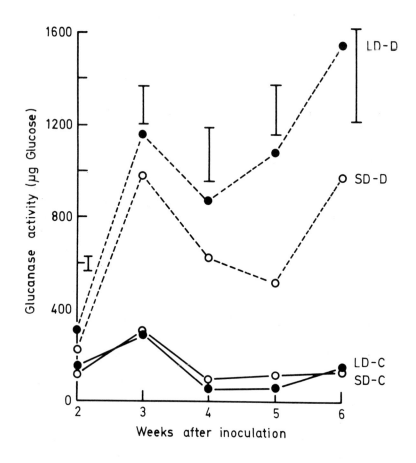

Figure 9.   1,3-β-glucanase activity of Verticillium dahliae-infected
potato stems.   Remainder of legend as for Figure 6.

Glucanohydrolase activity in forest trees

Wargo (1974) established the presence of glucanase and chitinase activity in phloem and cambium tissue of Quercus rubra, Q. velutina (red oak group), Q. alba (white oak) and Acer saccharinum (sugar maple) (Table 8) - a first report of chitinase in the tissues of a woody perennial. Interest was centered

Table 8. 1,3-β-glucanase and chitinase activities in phloem tissue of the stem and root of several forest trees (after Wargo, 1974).

Tree species	Tissue	Glucanase μm glucose equivalent	Chitinase μm N-acetyl-glucosamine
Red Oak group	Stem	0.95	0.54
	Root	0.80	0.11
White Oak	Stem	0.70	0.66
	Root	0.65	0.21
Sugar Maple	Stem	1.68	0.55
	Root	0.60	0.10

on the resistance of tree species to Armillaria mellea under certain conditions (Thomas, 1934 and Wargo & Houston, 1974), and the possibility of host-induced lysis of the pathogen as in the orchid Gastrodia elata (Kusano, 1911). Mycelium of A. mellea was incubated for 48 h at 38^{0}C in phloem extracts of sugar maple with predominant glucanase activity and in a chitinase preparation from Serratia marcescens containing no glucanase. Table 9 shows the dry weight loss, reducing sugar and N acetylglucosamine loss induced by each enzyme and the synergistic action of the combined extract. While this work may be criticised on the lack of sterility during prolonged incubation at high temperature and the possibility of contaminant bacterial enzyme activity, the findings nevertheless suggest a real role for host enzymes in hyphal lysis and agree with the results of host glucanase in vitro digests of Verticillium hyphal wall. The precise conditions during which A. mellea might be vulnerable to in vivo host lysis have

Table 9. Loss of sugars and cell wall dry weight in <u>Armillaria mellea</u>
after bathing in 1,3-β-glucanase and chitinase solutions,
singly and in combination (after Wargo, 1974).

Enzyme preparation	% dry weight loss	Reducing sugar % loss	N-acetylglucosamine % loss
Sugar Maple (predominantly 1,3-β-glucanase)	37	1.2	2.3
<u>S. marcescens</u> (chitinase)	30	0.6	4.4
Combined enzymes	40	3.7	8.3

yet to be determined, but preliminary observations would suggest that
drought-induced stress is one predisposing factor.

1,3-β-glucanase in melon anthracnose

A conflicting account of the role of 'host' glucanase was given by
Rabenantoandro et al. (1976) in <u>Colletotrichum lagenarium</u>-infection of
melon. These authors showed that still cultures of <u>C. lagenarium</u> produced
<u>exo</u> and <u>endo</u> glucanases. Enzymes were partially purified by gel-filtration
on Sephadex G 75 and by ion-exchange chromatography on SP Sephadex C25 and
DEAE cellulose. Healthy hypocotyl and petiole tissue was shown to contain
a constitutive glucanase (laminarinase) (Figure 10A) appearing as a single
peak on Sephadex G 75. Infection by <u>C. lagenarium</u> led to a large increase
in enzyme activity which resolved into two peaks (Figure 10B), the second
of which Rabenantoandro et. al. (1976) claimed, corresponded to the fungal
one based on its ion-exchange characteristics. The results, however, are
not entirely conclusive, since the temperature stability of the suspected
fungal enzymes from hypocotyl tissue at 50°C did not correspond precisely
to either of the two culture filtrate enzymes. These authors further
claimed that host glucanase from healthy and infected melon plants failed
to hydrolyse <u>C. lagenarium</u> cell walls when incubated for periods of up to
72 h. Skujins et al. (1965), however, have demonstrated the inability of
1,3-β-glucanase and chitinase of microbial origin singly, to lyse hyphal
walls of <u>Aspergillus</u> and <u>Fusarium</u>. In combination, however, they were

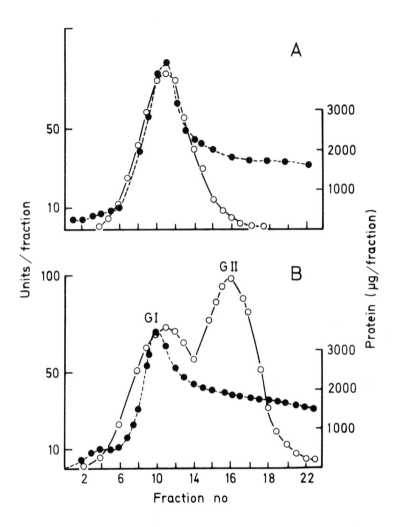

Figure 10.   Elution profiles of 1,3-β-glucanases o—o and proteins ●---●
from crude extracts of melon hypocotyls. A) healthy and B)
infected with <u>Colletotrichum lagenarium</u> run on Sephadex G 75.
GI and GII represent peaks coinciding with host and
<u>C. lagenarium</u> glucanases respectively (after Rabenantoandro
<u>et al.</u>, 1976).

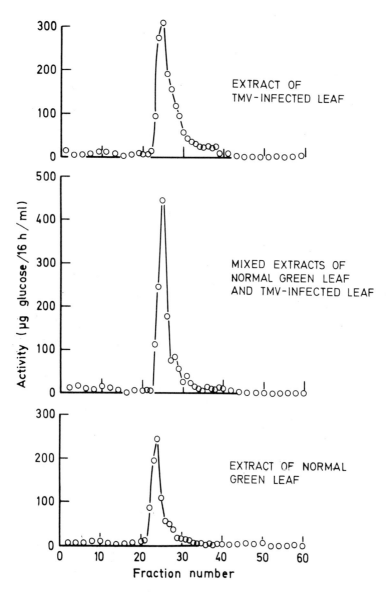

Figure 11. Electrophoretograms of 1,3-β-glucanase activity from healthy, TMV-infected <u>Nicotiana glutinosa</u> leaves and healthy and infected extracts combined (after Moore & Stone, 1972).

effective in dissolving the walls. This result and that of Potgieter &
Alexander (1965) suggest that negative lysis experiments involving single
enzymes are by no means conclusive. Moore & Stone (1972) presented electrophoretograms of healthy and TMV-infected Nicotiana glutinosa glucanase
in which the increase in enzyme in diseased tissue was identical to that in
healthy and appeared to be chemically identical (Figure 11).

Effect of ethylene on glucanohydrolase induction

Abeles & Forrence (1970) first implicated ethylene in glucanohydrolase
induction. Plants were subjected to ethylene at 10 μl l^{-1} and removed for
periods of up to three days. Abeles claimed that up to 4% of the soluble
leaf protein in C_2H_4-treated leaves was chitinase. Tissue exposed to ethylene
for three days showed a 47-fold increase in both chitinase and endo 1,3-β-
glucanase (Table 10) based on the action of crude tissue, water homogenates
on colloidal chitin and laminarin respectively.

Table 10. Effect of duration of exposure of 10 ppm ethylene on enzyme
activity in Phaseolus vulgaris (after Abeles et al., 1970).

Days of ethylene treatment	Protein mg/ml	Glucanase mg glucose ml^{-1}hr^{-1}	Chitinase μg NAG ml^{-1}hr^{-1}	Peroxidase 470 nm (ml 2 min)$^{-1}$
0	3.0	0.89	0.21	1.4
1	2.6	28.0	4.5	1.6
2	2.1	34.0	7.5	1.8
3	1.0	42.0	10.0	2.0

Ethylene production is a key feature of Verticillium infection of tomato
and other hosts (Pegg & Cronshaw, 1976) where it is produced as a peak in
stems and leaves (maximum rate of 24 nl g^{-1} dry wt h^{-1}) prior to, or
coinciding with, symptom induction and declining as symptoms become severe.
An interesting effect is seen in Resistant tissue where the single gene
resistance can be temporarily broken by decapitating the main root and
allowing shoots to flood with 10^7 spores ml^{-1}. Such plants showed a temporary rise in ethylene production coinciding with limited symptom appearance.
Spore germ tubes were rapidly lysed in resistant but not in susceptible stems

and such mycelium as developed was degraded, until after 14 days no fungus remained.

To study the effect of ethylene as an inducer of host resistance, whole plants were gassed with varying concentrations of ethylene prior to coinciding with inoculation. Symptoms in C_2H_4-treated plants were dramatically reduced compared with the controls (Plate II), particularly with the absence

Plate II. Ethylene-induced resistance to <u>Verticillium albo-atrum</u> in tomato.
A. Plant exposed to 5 ppm ethylene for 48 h prior to inoculation.
B. Ungassed leaf.

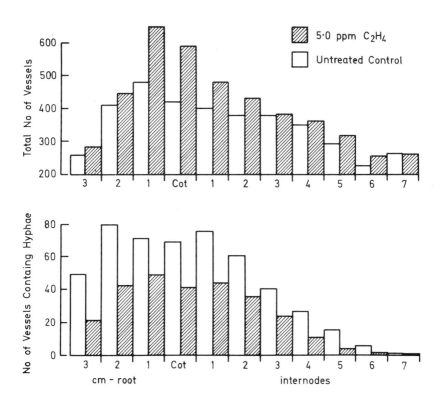

Figure 12.   The effect of exogenous ethylene treatment of tomato plants
(cv. Craigella susceptible) on xylem vessel differentiation
(upper) and reduction in Verticillium albo-atrum colonisation
(lower) of root and stem.

of overall wilting and leaf area reduction and necrosis. Figure 12 shows
the effect of gassing on the development and survival of mycelium in roots
and stem internodes. Mycelium lysis or failure to develop, was greatest in
the root and low stem internodes, as was the greater differentiation of new
xylem vessels in control plants. Further gassing experiments using a range
of C_2H_4 air mixtures from 0.1-100 ppm all showed the stimulating effect on
plant growth and the reduction in the total content of mycelium. Endo

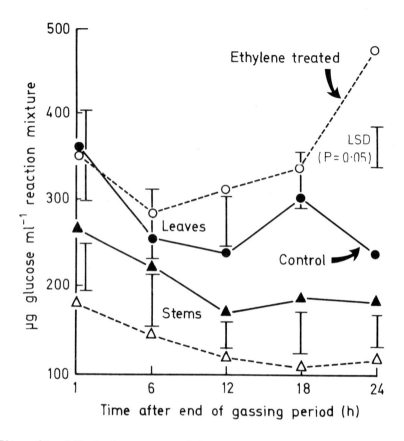

Figure 13. Effect of exogenous ethylene on 1,3-β-glucanase activity in leaves and stems of Craigella susceptible tomato plants. Whole plants exposed to 5.0 ppm ethylene for 48 h.

1,3-β-glucanase activity in leaves, but not in stems, was stimulated after gassing whole plants in ethylene (Figure 13). Ethylene had no effect, or inhibited chitinase activity in leaves and stems 48 hours post-gassing, a small increase in chitinase activity was observed in leaves only 144 hours after gassing. 1,4-β-glucosidase was inhibited in both leaf and stem tissue at all levels. These results, very different from those of Abeles & Forrence (1970) with Phaseolus vulgaris, clearly show the difficulty of generalising with regard to a single host genus. Since some experiments on P. vulgaris had been conducted on detached pulvini (Abeles & Forrence, 1970), stem slices of tomato were exposed to 1 ppm C_2H_4 for 48 h.   To investigate the possibility of tissue wounding interacting with ethylene, 2 g of stem as one segment was compared with 3 x 0.67 g sections and 6 x 0.34 g sections.   Chitinase activity although stimulated somewhat by wounding was significantly reduced below the control (ungassed) 2.0 g segments both at 48 and 144 hours after gassing.   Glucanase activity was unaffected by wounding and was increased by approximately 10% 2 and 6 days after gassing, unlike the stem result with gassed whole plants (Pegg, 1976).

On the evidence available to date, host glucanohydrolases offer the only explanation for the reduction in mycelium in C_2H_4-treated plants.   Extensive investigation into the possible induction of antifungal compounds by ethylene (Chalutz et al., 1969) and of rishitin especially have proved negative.   No antifungal substance in healthy ethylene-treated tomato tissue could be detected by bioassay or GLC.

The deposition of callose as a defence reaction

The thickening of cell walls and the partial or complete encapsulation of invasive fungal hyphae is well known.   Such deposits were described by De Bary in 1863 and were termed callosities by Young (1926) and lignitubers by Fellows (1928).   It is known that in many plants at least these deposits are identical to the callose of phloem sieve tubes and consist of homogeneous 1,3-β-glucan.

Of particular interest is the association of host callose deposition with the lysis of parasitic hyphae.   In the orchid species Gastrodia elata, G. callosa and Galeola hydra, the common endophyte Armillaria mellea is frequently invested by a sheath of callose along the whole length of the cell (Kusano, 1911; Burgeff, 1932), as a prelude to lysis.   The formation of callosities must involve the activity of host 1,3-β-glucanase, presumably in a very localised manner.   The stimulus for callose formation and the relationship between 1,3-β-glucanase as an inducer of a mechanical barrier to penetration compared with its potential role in wall lysis remains something of a paradox.

The suggestion has been made that callose formation is entirely non-specific and, as Ito (1949) has shown in sweet potato, may be induced by mechanical damage such as needle pricks.   Griffiths (1971), however, described

the initiation of lignitubers in pea and tomato roots following infection by
Verticillium dahliae in cortical cells adjacent to parasitised cells but
which had not been penetrated by the pathogen. Callosities formed from an
accumulation of paramural vesicles which coalesced after rupture and depletion of their contents. These vesicles which had been elaborated from the
host Golgi apparatus migrated through the host plasmalemma. It would be
tempting to assume that the vesicles were lysosomal in character, particularly with reference to 1,3-β-glucanase. After the establishment of lignitubers, hyphae from adjacent cells penetrated the innermost tangential wall
apparently by chemical dissolution of the callose. No hyphae were seen to
emerge from the lignitubers and 48 hours later the cytoplasm at the tip of
the penetrating hypha had broken down and lysis of the hyphae external to
the parasitised cell occurred. No breach in the host plasmalemma occurred
at any stage. In the case of a compatible host pathogen combination such
as V. albo-atrum and tomato, lignitubers may form in the root cortex and
stelar parenchyma but these are breached and no obvious lysis of the
pathogen occurs (Selman & Buckley, 1959; Griffiths & Isaac, 1966). In the
xylem vessels, however, a non-living system both chitinase and glucanase
are present at the site of lysis (Pegg & Vessey, 1972).

In the light of Griffith's work it is clear that recognition of the
pathogen must occur by chemical rather than physical processes. A striking
parallel between host-pathogen incompatibility and lignituber formation is
shown by the growth of the pollen tube into the stigma of angiosperms - a
process which could be regarded as parasitism. Heslop-Harrison and co-
workers (Heslop-Harrison & Dickinson, 1969; Heslop-Harrison et al., 1973a
and b; Heslop-Harrison et al., 1974; Heslop-Harrison et al., 1975; Heslop-
Harrison, 1975), have demonstrated the significance of gamete protein
(sporopollenin) and the genetic compatibility of the pollen grain and the
stigmatic surface. The parallel between the fungal hypha and the invasive
pollen tube is particularly apposite with regard to callose deposition. In
compatible systems, proteins from the pollen grain wall, notably the intine
are transferred onto a receptive surface of the stigma which can be destroyed
by protease treatment (Heslop-Harrison, 1975). Recognition is followed by
cutinolytic activity at the site of pollen tube contact (Heslop-Harrison et
al., 1974), and the pollen tube grows down within the cuticle or through the
pecto-cellulosic wall of the papilla. Passage through the style is through
the middle lamella with dissolution of pectin substrates by exo-enzymes in a

manner exactly analogous to fungal infection.

When pollen of an incompatible species or of self-incompatible species as illustrated by the Cruciferae, alights on a stigma, growth of the germ tube is arrested within a very short time by an intense deposition of callose. The effectiveness of this barrier can be seen by the inability of compatible pollen tubes to penetrate a stigma challenged by incompatible pollen. Evidence for the chemical induction of callose, confirming the work of Griffiths & Isaac (1966) is suggested by the fact that callose formation in the papilla may be induced prior to penetration and by the application of an agarose gel containing exine diffusates from incompatible pollen.

Table 11. Induction of 1,3-β-glucanase in Nicotiana glutinosa by tobacco mosaic and tomato spotted wilt viruses (after Moore & Stone, 1972).

| | Mg reducing sugar as glucose per 30 min. | | | | | |
| | TSWV-infected leaves | | TMV-infected leaves | | Control | |
	g fresh wt	g dry wt	g fresh wt	g dry wt	g fresh wt	g dry wt
Lesions	29	135	59	135	-	-
0.5 cm area around lesions	19	170	24	250	-	-
remainder of leaf	11	104	9	94	1	13

De Bokx (1967) reported that callose is formed in response to potato leaf roll virus infection and Wu et al. (1969) and Wu & Dimitman (1970) described callose formation in TMV-infected tobacco leaf. Moore & Stone (1972) showed 200-fold increase in endo 1,3-β-glucanase in TMV-infected Nicotiana glutinosa leaves on a fresh wt basis and a 40-fold increase on a dry wt basis (Table 11). The changes were greater in young than old leaves and were accompanied by only minor changes in 1,4-β-glucanase, acid phosphatase, chlorophyll and protein levels. Tomato spotted wilt virus (TSWV) and broad bean wilt virus (BBWV) induced similar large increases. Systemic infection with BBWV resulted in a rapid increase in enzyme up to a maximum at symptom development followed by a decline to levels below those in control leaves. The suggestion has been made (Esau, 1967), that virus spread could be localised by the blocking of plasmodesmata by callose. Wu et al. (1969) have shown that resistance of Phaseolus vulgaris to the spread of TMV is directly related to the deposition of callose around the lesions.

Since the cellular levels of 1,3-β-glucanase are so high in virus infected plants, for such a mechanism to work the hydrolase would have to be compartmentalised within the cell independently of the callose substrate. A further possibility is that since 1,3-β-glucanase can depolymerise callose the rapid spread in systemically-infected plants may be facilitated by the parallel rise in the enzyme.

Control of host enzyme activity

Since it has been shown that growth of a fungal pathogen may occur in the presence of enhanced levels of enzymes which have a propensity for hyphal lysis, it becomes necessary to postulate a control mechanism whereby enzymes are inactivated or walls rendered resistant. A classic example of this is seen with orchid endophytes such as Rhizoctonia in the protocorm of Dactylorhiza purpurella. In this tissue, cells containing peletons of the fungus in an advanced stage of lysis may be re-infected by apparently vigo- rously growing hyphae. This phenomenon would be difficult to explain in terms of the synthesis of anti-fungal chemicals.

A possible mechanism for such a control process has been described by Albersheim & Valent (1974). These authors showed that proteins from 8-day shake cultures of Colletotrichum lindemuthianum were capable of the rapid inhibition of Phaseolus vulgaris endo 1,3-β-glucanase hydrolysis of laminarin. Partial purification of the active fraction by ion-exchange chromatography on DEAE-cellulose using a 200-500 mM K acetate, pH 5.0 gradient and by gel-permeation on Bio-Gel P-20, suggested the existence of two proteins, with molecular weights in the region of 16,000 and 18,000. Glucanases from two bean cultivars were inhibited identically. A time course inhibition of glucanase activity (Figure 14) showed that the inhibi- tion of laminarin hydrolysis was instantaneous. This finding by Albersheim & Valent may provide an explanation for a subtle control mechanism of a host defensive system with the possibility of enzyme repressor activity being controlled by the nutritional status of the fungus.

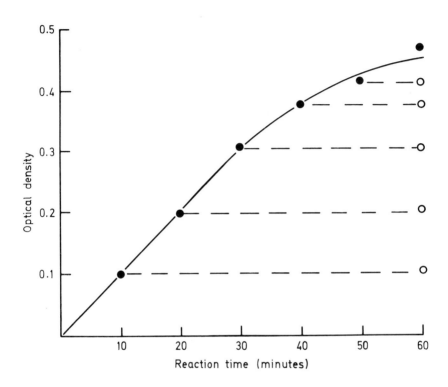

Figure 14. Time course of endo 1,3-β-glucanase hydrolysis of laminarin (solid line). Pecked line represents addition of protein inhibitor from Colletotrichum lindemuthianum cultures to the reaction mixture (after Albersheim & Valent, 1974).

Conclusion

From the limited evidence available to date, it would seem that higher plants possess a range of glucanohydrolases with the propensity to attack certain wall polymers of fungal parasites. Notwithstanding the current emphasis on phytoalexin-type resistance mechanisms in plants, the existence of in vivo lysis suggests that alternative means of host defence to invasive pathogens are available. This is particularly so where lysis cannot be attributed to microbial autolysis in the presence of fungistatic

- 337 -

levels of anti-fungal substances. Preliminary studies indicate an interaction between the defence mechanisms of host and parasite and the possibility of another basis for pathogen determined host specificity.

There is a need for more studies on the purification of a wide range
of host glucanohydrolases combined with experiments on the lysis of
pathogenic hyphae using sequential digests with specific enzymes. The
orchid protocorm system would seem to provide an excellent subject for
this purpose. Finally, the question of host specificity in relation to
the production by the pathogen of lytic enzyme inhibitors would be worthy
of a detailed further investigation.

References

Abeles, F.B., Bosshart, R.P., Forrence, L.E. & Habig, Q.H. (1970).
Preparation and purification of glucanase and chitinase from bean
leaves. Plant Physiology 47, 129-134.

Abeles, F.B. & Forrence, L.E. (1970). Temporal and hormonal control of
β-1,3 glucanase in Phaseolus vulgaris. Plant Physiology 45, 395-400.

Abeles, F.B., Leather, G.R., Forrence, L.E. & Craker, L.E. (1971).
Abcission: Regulation of senescence, protein synthesis and enzyme
secretion by ethylene. Journal of Horticultural Science 6, 19-24.

Albersheim, P. & Valent, B.S. (1974). Host-Pathogen interactions. VII.
Plant pathogens secrete proteins which inhibit enzymes of the host
capable of attacking the pathogen. Plant Physiology 53, 684-687.

Bartnicki-Garcia, S. (1968). Cell wall chemistry, morphogenesis, and
taxonomy of fungi. Annual Review of Microbiology 22, 87-108.

Bernard, N. (1909). L'Evolution dans la symbiose des Orchidees et leurs
champignons commensaux. Annales des Sciences Naturelles IX, 1.

Burgeff, H. (1932). Saprophytismus und symbiose. Studien an tropischen
Orchideen. Fischer, Jena.

Burgess, A. (1939). The defensive mechanism in orchid mycorrhiza. New
Phytologist 38, 273-283.

Chalutz, E., De Vay, J.E. & Maxie, E.C. (1969). Ethylene-induced
isocoumarin formation in carrot root tissue. Plant Physiology 44,
235-241.

Clarke, A.E. & Stone, B.A. (1962). β-1,3-Glucan hydrolases from the grape vine (Vitis vinifera) and other plants. Phytochemistry 1, 175-188.

De Bary, A. (1863). Recherches sur le development de quelques champignons parasites. Annales des Sciences Naturelles Botanique et Biologie Végétale 20, 5-148.

De Bokx, J.A. (1967). The callose test for the detection of leaf roll virus in potato tubers. European potato Journal 10, 221-232.

Dillon, T. & O'Colla, P. (1951). Enzymic hydrolysis of 1,3-linked polyglucosans. Chemistry and Industry III.

Dixon, G.R. & Pegg, G.F. (1969). Hyphal lysis and tylose formation in tomato cultivars infected by Verticillium albo-atrum. Transactions of the British mycological Society 53, 109-118.

Esau, K. (1967). Anatomy of plant virus infections. Annual Review of Phytopathology 5, 45-76.

Eschrich, W. (1959). The enzymic decomposition of callose with papain. Naturwissenschaften 46, 327-328.

Fellows, H. (1928). Some chemical & morphological phenomena attending infection of the wheat plant by Ophiobolus graminis. Journal of Agricultural Research 37, 647-661.

Fleming, A. (1922). On a remarkable bacteriolytic element found in tissues and secretions. Proceeding of the Royal Society (B) 93, 306-317.

Glazer, A.N., Barel, A.O., Howard, J.B. & Brown, D.M. (1969). Isolation and characterisation of fig lysozyme. Journal of Biological Chemistry 244, 3583-3589.

Grassmann, W., Zechmeister, L., Bender, R. & Toth, G. (1934). Über die Chitinspaltung durch Emulsin-Präparate (III Mitteil über enzymatische Spaltung von Polysacchariden). Berichte der Deutschen Chemischen Gesellschaft 67, 1-5.

Griffiths, D.A. (1971). The development of lignitubers in roots after infection by Verticillium dahliae Kleb. Canadian Journal of Microbiology 17, 441-444.

Griffiths, D.A. & Isaac, I. (1966). Host parasite relationships between tomato and pathogenic isolates of Verticillium. Annals of applied Biology 58, 259-272.

Heslop-Harrison, J. (1975). The physiology of the pollen grain surface. Proceedings of the Royal Society London B 190, 275-299.

Heslop-Harrison, J. & Dickinson, H.G. (1969). Time relationships of sporopollenin synthesis associated with tapetum and microspores in

Lilium. Planta 84, 199-214.

Heslop-Harrison, J., Heslop-Harrison, Y. & Barber, J. (1975). The stigma surface in incompatibility responses. Proceedings of the Royal Society London B 188, 287-297.

Heslop-Harrison, J., Heslop-Harrison, Y. & Knox, R.B. (1973a). The callose rejection reaction - a new bioassay for incompatibility in Cruciferae & Compositae. Incompatibility Newsletter 3, 75-76.

Heslop-Harrison, J., Heslop-Harrison, Y., Knox, R.B. & Howlett, B. (1973b). Pollen-wall proteins: 'gametophytic' and 'sporophytic' fractions in the pollen wall of the Malvaceae. Annals of Botany (London) 37, 403-412.

Heslop-Harrison, J., Knox, R.B. & Heslop-Harrison, Y. (1974). Pollen-wall proteins: exine held fractions associated with the incompatibility response in Cruciferae. Theoretical and applied Genetics 44, 133-137.

Howard, J.B. & Glazer, A.N. (1969). Papaya lysozyme. Terminal sequences and enzymatic properties. Journal of Biological Chemistry 244, 1399-1409.

Ito, K. (1949). Studies on "Murasaki-Mompa" disease caused by Helicobasidium mompa. Tanaka, Bulletin of the Government Forest Experiment Station, Tokyo, Japan 43, 1-126.

Kusano, S. (1911). Gastrodia elata and its symbiotic association with Armillaria mellea. Journal of the College of Agriculture IV, 1-66.

Macleod, A.M. (1960). Barley carbohydrate metabolism in relation to malting. Waterstein Laboratories Communications 23, 87-98.

Manners, D.J. (1955). Observations on Barley β-glucosidases. Proceedings of the Biochemical Society 343rd meeting. The Biochemical Journal 61, xiii.

Meyer, K., Hahnel, F. & Steinberg, A. (1946). Lysozyme of plant origin. Journal of Biological Chemistry 163, 733-740.

Moore, A.E. & Stone, B.A. (5) (1968). The occurrence of a β-1,3-glucan hydrolase in plants of Nicotiana glutinosa in normal and pathological states. Proceedings of the Federations of European Biochemical Societies, 182.

Moore, A.E. & Stone, B.A. (1972). Effect of infection with TMV and other viruses on the level of a β-1,3-glucan hydrolase in leaves of Nicotiana glutinosa. Virology 50, 791-798.

Morley, G., Dawson, A. & Marks, V. (1968). Manual and autoanalyser methods

for measuring blood glucose using Guiacum and Glucose Oxidase.
Proceedings of the Association of clinical Biochemists 5, 42-45.

Peat, S., Thomas, G.J. & Whelan, W.J. (1952). The enzymic synthesis and degradation of starch. XVII. Z-enzyme. Journal of the Chemical Society (London) 2, 722-733.

Pegg, G.F. (1976). The occurrence of 1,3-β-glucanase in healthy and Verticillium albo-atrum-infected susceptible and resistant tomato plants. Journal of Experimental Botany (in press).

Pegg, G.F. & Cronshaw, D.K. (1976). Ethylene production in tomato plants infected with Verticillium albo-atrum. Physiological Plant Pathology 8, 279-295.

Pegg, G.F. & Vessey, J.C. (1972). Chitinase activity in Lycopersicon esculentum and its relationship to the in vivo lysis of Verticillium albo-atrum mycelium. Physiological Plant Pathology 3, 207-222.

Perlin, A.S. & Reese, E.T. (1963). Dimensions of the substrate site involved in the enzymolysis of a polysaccharide. Canadian Journal of Biochemistry and Physiology 41, 1842-1846.

Potgieter, H.J. & Alexander, M. (1965). Polysaccharide components of Neurospora crassa hyphal walls. Canadian Journal of Microbiology 11, 122.

Powning, R.F. & Irzykiewicz, H. (1965). Studies on the chitinase systems in bean and other seeds. Comparative Biochemistry & Physiology 14, 127-133.

Rabenantoandro, Y., Auriol, P. & Touzé, A. (1976). Implication of β-(1-3) glucanase in melon anthracnose. Physiological Plant Pathology 8, 313-324.

Reese, E.T., Maguire, A.H. & Parrish, F.W. (1967). Glucosidases and exo-glucanases. Canadian Journal of Biochemistry 46, 25-34.

Ride, J.P. & Drysdale, R.B. (1972). A rapid method for the chemical esti-mation of filamentous fungi in plant tissue. Physiological Plant Pathology 2, 7-15.

Selman, I.W. & Buckley, R. (1959). Factors affecting the invasion of tomato roots by Verticillium albo-atrum. Transactions of the British mycological Society 42, 227-234.

Skujins, J.J., Potgieter, H.J. & Alexander, M. (1965). Dissolution of fungal cell walls by a Streptomycete chinitase and β-(1-3) glucanase. Archives of Biochemistry and Biophysics III, 358-364.

Taylor, J.B. & Flentje, N.T. (1968). Infection, recovery from infection and resistance of apricot trees to Verticillium albo-atrum. New Zealand Journal of Botany 6, 417-426.

Thomas, H.E. (1934). Studies on Armillaria mellea (Vahl) infection, parasitism and host resistance. Journal of Agricultural Research 48, 187-218.

Wang, M.C. & Bartnicki-Garcia, S. Structure and composition of walls of the yeast form of Verticillium albo-atrum. Journal of General Microbiology (1970) 64, 41-54.

Wargo, P.M. (1974). Lysis of the cell wall of Armillaria mellea by enzymes from forest trees. Physiological Plant Pathology 5, 99-105.

Wargo, P.M. & Houston, D.R. (1974). Infection of defoliated sugar maple by Armillaria mellea. Phytopathology 64, 817-822.

Wilhelm, S. & Taylor, J.B. (1965). Control of Verticillium-wilt of olive through natural recovery and resistance. Phytopathology 55, 311-316.

Wu, J.H., Blakely, L.M. & Dimitman, J.E. (1969). Inactivation of a host resistance mechanism as an explanation for heat inactivation of TMV-infected bean leaves. Virology 37, 656-666.

Wu, J.H. & Dimitman, J.E. (1970). Leaf structure and callose formation as determinant of TMV movement in bean leaves as revealed by UV irradiation studies. Virology 40, 820-827.

Young, P.A. (1926). Facultative parasitism and host ranges of fungi. Americal Journal of Botany 13, 502-520.

DISCUSSION

Chairman: J. Raa

Paxton: I would like to make a comment about some work (Howlett, B.M., Knox, R.B. & Heslop-Harrison, J., J. Cell Sci. 13, 603-619, 1973) which I think is quite exciting and which parallels what you were saying. I am referring to the ability to extract material from compatible pollen, which appears to be protein but which has by no means been proved to be protein, which is capable of inducing the susceptible or compatible reaction in what would otherwise be an incompatible reaction. In other words, the recognition here appears to be at the surface and once the compatible recognition takes place, callus is not formed and the pollen tube is allowed to grow into the plant even though theoretically the same materials are there that are capable of producing the callus. It is therefore possible to invert the system to make an accepting reaction that will not produce this callus.

Pegg: I didn't mention that. The technique is now being used experimentally in plant breeding to try to get crosses with hitherto incompatible reactions.

Mussell: I am curious, as to the location of these enzymes that you are talking about, within the plant, both in the healthy situation and when you get the increase in the infected situation. Do you know anything about where they are, such as on the wall?

Pegg: We do not know at this stage where they are. We suspect that some of them may well be on the wall, but we cannot answer that at present. Could I turn the question back to you and ask if you think it is surprising that they are in the xylem fluid?

Mussell: No, not at all!

Pegg: The bleeding sap is very dilute and the quantity of fluid one obtains is a function of the amount of water at the roots and yet the enzymes are present in relatively large amounts.

Kijne: This is only an affirmative question, you told us that stress could suddenly change the resistant plant into a susceptible one, what kind of stress did this author mean?

Pegg: What I had in mind was drought stress. It is well known that trees and woody shrubs under water stress show increased susceptibility to Armillaria mellea. The evidence suggests that roots may be surrounded by large quantities of the pathogen but penetration and death in certain species does not occur until the roots become water stressed. It may not be coincidental that under these conditions the activities of glucanase and chitinase fall to a very low level.

Kimmins: I wasn't quite sure from your reference to Moore & Stone (Virology 50, 791-798, 1972) about the increase of 1,3-β-glucanases, to what you would attribute the significance of the increase in this enzyme in the virus infected plant.

Pegg: I cannot speak from personal experience with viruses, but Wu et al. (Virology 37, 656-666, 1969), De Bokx (Eur. potato J. 10, 221-232, 1967) and Esau (Ann. Rev. Phytopathol. 5, 45-76, 1967) have speculated that callose plugs block plasmodesmata thereby preventing the migration of virus nucleo-proteins or free RNA through the wall. Moore & Stone have also claimed that in the case of a systemic virus infection, glucanases in some way modify the wall structure mediating the rapid movement of the virus around the plant. This, however, presupposes that there are 1→3 linked glucans in the wall unless the enzymes are non-specific. It ultimately depends upon the specificity of the enzyme. We know, for example, that there is no chitin per se in tomato plants. The enzyme from tomato we designate as chitinase is so named based on its action on arthropod chitin, but it is clearly a non-specific enzyme and probably acts on muco-polysaccharides in the tomato.

Cooper: You are not suggesting are you, that the increase in 1,3-β-glucanase as reported by Moore & Stone is associated with increased synthesis of callose?

Pegg: I am suggesting that it might be involved in the increased mobility of callose. I did not suggest that it is a callose synthetase, although this is quite possible.

Cooper: You see, I don't understand how an increase in the activity of a 1,3-β-glucanase can be related to the resistance of the plant to that virus. Some people have suggested that the 1,3-β-glucanase may very well participate in callose synthesis. I think if you look at the literature there is no evidence to support it whatsoever. At the same time, why should a plant synthesise 1,3-β-glucanase to mobilize callose which is probably there to inhibit the virus infection. And if you are going to mobilize callose all you will be doing presumably is aiding the virus infection. My own feeling about that work is that to analyse 1,3-β-glucanases in that system is very difficult and I think it would have to be repeated in other systems before one could really feel very confident about it.

Pegg: Yes, I don't feel so much that the analysis of the enzyme presents difficulty but rather the interpretation of what the enzyme is doing. I think that there is no doubt that there is a substantial increase in the endo glucanase but you know, the interpretation of how this may be affecting virus multiplication or transmission is still in the realm of speculation.

Callose is very much involved with the limitation of the orchid symbiont
Armillaria mellea and the reason the orchid is called Galiosa callosa is
because a tube of callose runs right across the cell as the hypha goes
through, such that the hypha is encased in glucan. There is clearly a role
for the enzyme there, but I agree with you in the virus work that one must
convincingly demonstrate a localization of virus based on callose.

Ballou: Just one comment, this idea of the involvement of synthetic
enzymes and degradative enzymes in reorganizing cell wall structure I think
is well illustrated in yeast cell division and could well apply here. That
is, if you are going to loosen up a wall structure and then resynthesise it
in a different morphology a degradative enzyme is very useful.

Pegg: I think that many microbial endo chitinases function as chitin
synthetases. Growth may be measured in terms of acetylglucosamine and
chitinase activity both of which are frequently greatest in log phase
cultures.

THE POSSIBLE PHYSIOLOGICAL ROLE OF LECTINS

Heinrich Kauss, Universität Kaiserlautern Biologi, Pfaffenbergstrasse, 6750 Kaiserlautern, Germany

Since the end of the last century it is known that certain proteins of plant origin are able to agglutinate red blood cells. For a while the agglutinating plant proteins have been referred to as "phytohemagglutinins", but the more general term "lectin" is now widely in use as similar agglutinins have also been isolated from animals.

Some of the basic properties of lectins will be explained shortly with the example of concanavalin A (ConA), the best explored lectin. The protein is typically made up from 4 subunits of 25,500 mol. wt, each of which appears to exhibit a saccharide binding site and binding sites for calcium and manganese. The number of subunits per molecule ConA is subject of some variability and is strongly dependent on the pH value and temperature. The saccharide binding site of ConA was thought to be identical with the site at which o-iodo-phenyl-β-D-glucopyranoside will bind (see Lis & Sharon, 1973). There is growing evidence now (Becker et al., 1976) that the latter glycoside is bound non-specifically via its hydrophobic aglycone in a hydrophobic pocket and the sugar binding site causing the lectin property is rather close to the ion binding sites. This close association would allow to understand the marked dependence of sugar binding activity on the presence of metal ions. Monovalent subunits of ConA are not able to agglutinate erythrocytes but still bind to the cell surface receptors, saccharides, Ca^{2+} and Mn^{2+} (Thomasson & Doyle, 1975). As far as known divalent metal ions appear to be of general importance for the configuration and related binding properties of lectins. Studies on lectins from the lentil (Paulova et al., 1971) and from wax beans, lima beans and soybeans (Galbraith & Goldstein, 1970) show that once the metal ions were removed, the lectins can either not or only partly be reactivated. The

participation of divalent ions with the lectin complex formation was one
of the reasons to include complexing agents during our own attempts to
solubilize lectins from plant organelles, as will be discussed later.

Most useful techniques to elucidate the sugar specificity of lectins
are inhibition studies with soluble sugar derivatives, often referred to as
"haptens" in analogy to the immunological terminology. In the case of
ConA the formation of the complex is inhibited by glycosidic derivatives
exhibiting α-mannose or α-glucose residues. Several other lectins are
inhibited also by monosaccharides or simple sugar derivatives. On the
other hand, lectins are known for which simple sugar inhibitors are not
known. This is understandable from the fact that for effective binding
on the erythrocyte surface a rather complex carbohydrate structure of the
receptors appears to be required. The affinity is partly dependent on
the sugars which are present but is also dependent on the anomeric forms,
spatial arrangement, distance of the monomeric units and nature of the
peptide chain by which the carbohydrate groups are bound to the cell
surface. Thus classification of a certain plant protein as a "lectin" is
much dependent on the availability of the appropriate carbohydrate groups
on the animal cells used by chance in the assay system. Therefore it
appears very likely that many plant proteins exhibiting carbohydrate
binding properties are not yet recognized as lectins due to the lack of
universal assay methods.

The list of known lectins has steadily increased in the last years,
their general properties are described in recent reviews (Sharon & Lis,
1972; Lis & Sharon, 1973; Boyd, 1970). The chemical composition of the
lectins so far purified is quite different, their molecular weight ranges
from about 25,000 to more than 400,000 and the sugar content may range
from zero to about 20 percent. Therefore lectins are not a natural family
but an artificial group of proteins; the main feature in common is the
ability to agglutinate animal cells.

The use of the term "lectin" requires some more remarks. Primarily
"lectin" designates a protein or glycoprotein which, as assayed with the
agglutination assay, has a special recognition site or binding site in
order to form a complex with the carbohydrate groups present on the surface
of red blood cells or other single animal cells. Monovalent subunits
derived from defined lectins can still bind to the same receptors but are
not able to agglutinate the cells. It appears feasibly to call them

lectins as well. However, some confusion may arise if plant proteins for which a carbohydrate binding property has been shown using quite different techniques (e.g. Jermyn & Yeow, 1975) are also named lectins. Caution in this direction should also be exercised with polysaccharides able to agglutinate erythrocytes, such as the "wheat flour agglutinating polysaccharide" which is composed of D-xylose and L-arabinose and is counteracted by tryptophan (Minetti et al., 1976).

For a long time the agglutinating activity of plant proteins has been regarded as a curiosity. More recently lectins were increasingly used as an investigative tool in cell biology and cancer research. The striking success of this work has drawn the attention of plant biochemists and plant pathologists to lectins and I will try to shortly point out which physiological role lectins may play in plants. More detailed reviews with a similar aim are avilable (Callow, 1975; Kauss, 1976).

Membrane glycolipids and glycoproteins direct their carbohydrate tails out of the membrane plan thus establishing a region of oligosaccharide nature adjacent to the membranes and around the cells. Lectins can bind to both types of compounds (Lis & Sharon, 1973; Young & Leon, 1974; Surolia et al., 1975). This ability is the basis of numerous recent studies with animal cell cultures on membrane and cell surface structure (Sharon & Lis, 1975).

Binding requires a sufficient density or mobility of receptors and thus involves clustering of individual receptors leading to patches of glycoproteins which in the membranes of unbound cells are normally in random distribution. Binding of lectin to cells does not only result in perturbation of the outer surface of the plasma membrane (plasmalemma) but also in correlated alterations at the inner surface (Ji & Nicolson, 1974; Guérin et al., 1974). This and other evidence has led to the postulation of the "fluid membrane model", regarding the membrane organization no longer as fully fixed but allowing at least certain protein molecules or complexes to move laterally (Nicolson, 1976). One of the most thrilling observations which heavily stimulated recent lectin research was that differences in membrane fluidity can be detected as one of the early events of cell transformation by oncogenic viruses (Barnett et al., 1974). This observation may help to understand many effects found earlier in relation to lectin interaction with malignant cells (see Sharon & Lis, 1972).

Binding of lectins to animal cell surface receptors may also result in complex responses manifested by various physiological phenomena. There is an increase in synthetic activities after the induction of mitosis in resting lymphocytes (Lis & Sharon, 1973) and an initiation of membrane transport processes (Inoue et al., 1975). This implies that the stimulus set by binding of lectins at surface receptors has to be transmitted into the cell. At the other hand, the presence of endogenous lectin-like binding proteins at the mammalian cell surface is indicated indirectly by the striking phenomenon of contact inhibition observed in animal cell cultures. Directly it has been shown by the recent isolation of lectins from animal cell membranes (Bowles & Kauss, 1976b; Nowak et al., 1976). Therefore, we feel that addition of lectins _in vitro_ to animal cells may mimic the action of carbohydrate binding proteins endogenously present at the surface of adjacent cells. The many implications and speculations resulting from lectin work on animal cells which may help for a better understanding of cell and tissue development cannot be discussed here in more detail (see Edelman, 1976).

There is much evidence now that in animal cells mutual recognition is mediated by lectin-like carbohydrate binding proteins and their corresponding receptors. In plants only in two systems a similar statement appears possible, and in both of them the respective binding proteins have not been shown to agglutinate red blood or other animal cells, that means they are not proper lectins.

Certain species of Chlamydomonas release membrane vesicles (so-called "gamone") from the flagellar tips into the medium (McLean et al., 1974). When added back to gametes of the opposite mating type, the membrane material will cause the gametes to adhere to each other with the flagellar tips, thus mimicing a true mating reaction although this is not followed by cell fusion. The membranes appear to carry glycoproteins which are thought to be responsible for the contact between receptors at the flagellar tips. As the activity of glycosyltransferases is increased after mixing "gamone" from different mating types, it has been speculated that the proteins responsible for contact are the glycosyltransferases themselves (McLean & Bosmann, 1975).

The second example is represented by the mating factors of the yeast Hansenula wingei. Cell fusion between opposite mating types is initiated by a sexual agglutination reaction brought about by complementary glycoproteins (Crandall et al., 1974).

Only in two plant systems sufficient experimental data connect lectins proper with clear-cut physiological events. The first one is the adhesion of symbiontic rhizobia to legume roots. I will not discuss this as several specialists will do so in the course of this symposium.

The second one is the aggregation of cellular slime molds. Following starvation the single-celled amoebae aggregate and differentiate to form a multicellular organism exhibiting special morphological features. It was shown by Rosen et al. (1973) that a lectin named discoidin can be isolated from cohesive cells of Dictyostelium discoideum but not from amoebae. The protein has been purified by affinity chromatography and was shown to consist of two distinct proteins, discoidin I and II. The carbohydrate binding specificity of both lectins is similar but not identical. The two proteins can clearly be distinguished as two biosynthetically different proteins by the distinct timing of their appearance during differentiation and by several other properties. A lectin different from discoidin I and II in mol weight and sugar inhibition specificity has been isolated from another cellular slime mold, Polysphondylium pallidum and was named pallidin (Frazier et al., 1975). Again this lectin was absent from vegetative cells, but appeared in food-deprived cells in parallel to differentiation (Rosen et al., 1974). Both discoidin and pallidin have been implicated in the adhesion and aggregation process of the amoebae and it has been shown that the lectins are also detectable on the surface of the cells (Chang et al., 1975).

Recently Reitherman et al. (1975) assayed the agglutinability of glutaraldehyde-fixed cells of Dictyostelium by purified discoidin and pallidin. Agglutinability increased in parallel to the appearance of discoidin and to the cohesiveness of unfixed cells. Formation of the lectins therefore seems to be accompanied by the appearance of the respective receptors on the cell surface. The association constant (K_a) of discoidin I and II for fixed differentiated cells was in the range of $10^9 M^{-1}$, more than twentyfold higher than the K_a of these lectins for fixed vegetative cells. In addition, the K_a of these lectins for fixed differentiated cells of Polysphondylium was an order of magnitude lower than for fixed differentiated Dictyostelium cells. Thus the demonstration of species-specific receptors on the surface of cohesive cells supports the earlier suggestion that lectins provide a basis for species-specific cell cohesion in cellular slime molds.

Our own studies on the possible physiological role of lectins in higher plants were based on two considerations. Most of the lectins isolated in substantial amounts for biochemical studies are derived from seeds where they appear to occur in a more or less soluble state. Whenever their function in vivo is related to their sugar binding capacity they should occur at least partly in the form of complexes. The complexes formed between lectins and their endogenous receptors are quite stable, for example association constants in the range of $10^9 M^{-1}$ have been observed at the cell surface of Dictyostelium (Reitherman et al., 1975). If such strong interactions also occur between lectins in plant cells and their endogenous receptors, extraction with sodium chloride as usually done is not likely to extract those lectin molecules which are "in function". Therefore conditions for extraction had to be chosen which presumably would weaken the binding intensity. In addition, it appeared possible that studies on the subcellular localisation of lectins may help to understand their physiological role in plant cells.

Therefore, for lectin extraction we used various solutions of chaotropic salts in combination with complexing agents, detergents and acidic pH values. It was recognized after a while that we had to walk on a rather narrow ridge. When the conditions are set in favour of an effective decomposition of the lectin complex they are always at the risk of fractionating the lectins into non-agglutinating subunits, e.g. with urea or guanidinium salts. In addition we know in the meanwhile that at least in some cases we are solubilizing the potential receptor molecules also, to which recombination occurs during dialysis or extract concentration.

The first success we had was the solubilization of lectin with 0.5 M phosphate buffer and with a solution containing EDTA plus Triton from mung bean hypocotyl cell wall preparations. Comparison of the specific activities of the lectins from total tissue with those of cytoplasm and cell walls indicated that the lectins were definitely associated with the wall and were not cytoplasmic contaminants (Kauss & Glaser, 1974). Appreciable binding activity could only be detected when the walls were treated with boiled cytoplasmic supernatant prior to extraction: Mn^{2+} is only partly effective in a similar way. Hydrolase activity in the extracts is strongly diminished by sequential washing of the walls with sodium dodecylsulphate; extensive washings, however, result in an inactivation of the hemagglutination activity. The carbohydrate binding activity of both extracts is strongly diminished towards acidic pH values and is inhibited by D-galactose and γ-D-galactonolactone. The lectin can also be solubilized from the wall preparation using slightly acidic buffers (Kauss & Bowles, 1976). The occurrence of the lectin in the wall suggests an involvement in the extension growth process. The extracts contain appreciable amounts of hydroxyproline. A lectin rich in hydroxyproline, arabinose and galactose has been isolated by Allen & Neuberger (1973) from potato tubers, and we thought, that the wall lectin could be identical with the predicted "extensin" (Lamport, 1973). More recent work (Haass & Kauss, unpublished), however, shows that most of the hydroxyproline can be separated from the lectin. Although we still feel that the lectin may play a role in wall extension growth, we can only speculate how this might be the case. Three functions appear to be possible: Firstly, new polysaccharide molecules could have a lectin handle during their transport and introduction into the wall. Secondly, lectins in the wall could allow a sliding of wall polysaccharides relative to each other, regulated by protons and/or divalent ions. Thirdly, lectins could reside at the wall surface and establish wall-membrane contacts.

During a search for possible cellular precursors of the wall lectin we could demonstrate that membrane fractions enriched in plasma membrane, Golgi apparatus and endoplasmic reticulum release proteins with lectin activity on extraction with appropriate agents (Bowles & Kauss, 1975, 1976a). The carbohydrate specificity of the lectin fractions differs for each membrane type. We also found that total membrane preparations from a variety of plant and tissue types contain lectins as well. These results do not allow a conclusion in regard to the function of the wall lectin. But they are the first indication for a possible association of lectins with membranes and show possibly a function in membrane fusion or in maintaining the membrane structure.

In the above studies on hypocotyl membrane lectins one possible explanation for the occurrence of lectins associated with membranes would be an artificial binding during homogenization of soluble cytoplasmic lectins to glyco-units of the membranes. Thus, in order to confirm that lectins are indeed membrane components, mitochondria were isolated from the endosperm of Ricinus communis. They were divided in outer and inner

membrane and the extracts of these samples were checked for lectins.
Activity was found to be by far highest in extracts of the inner membranes,
a membrane type for which artificial binding of cytoplasmic lectins is
most unlikely (Bowles et al., 1976). It is interesting that part of the
lectin, which was found to be identical with R. communis agglutinin of mol.
wt 120,000, can also be liberated by addition of lactose, the hapten. This
is strong evidence, that at least part of the lectin was held within the
membrane via its carbohydrate binding site.

The demonstration of lectins as parts of plant membranes poses again
the question of their function. As discussed before, work with animal and
slime mold cells indicate that lectins at plasma membranes most likely play
a role in cell recognition and cell aggregation processes. Although
experimental evidence for an occurrence of complementary receptors for the
plant membrane lectins is missing, one could envisage on the basis of a
working hypothesis a similar role for the lectins on or in the various
internal membranes. Thus membrane contact and fusion during membrane flow
could be enabled by lectins. Alternatively, lectins could participate in
the maintenance of structural integrity of functionally related membrane
protein complexes.

The examples discussed up to now clearly indicate that the carbohydrate
binding proteins artificially classified as lectins due to the assay system
most likely play a multifunctional role in plant cells. I will add some
more observations indicating even more functions.

As some lectins are toxic to animals (Sharon & Lis, 1972; Olsnes et
al., 1974) it has been assumed that they might protect seeds and young
embryos against infection by parasites. No reports are available, however,
showing clear-cut growth inhibition of bacteria by lectins. On the other
hand, fungal growth is inhibited by wheat germ agglutinin (Lotan et al.,
1975).

Another possible function of lectins may be an involvement in the
complex processes of pollen recognition and incompatibility reactions. It
was found by Southworth (1975) that concanavalin A and Red Kidney bean
agglutinine (PHA), which are both mitogenic in animal systems (Lis &
Sharon, 1973) will stimulate pollen germination in vitro by reducing the
lag-period before the emergence of the pollen tube. In analogy to animal
cell systems one is tempted to speculate that the effects normally set by
the protein pellicle or proteinous exudates of the stigma are minimized by

added lectins *in vitro*.

Still another unknown role must be aligned to lectin activity demonstrated in sieve-tube sap of Robinia pseudoacacia (Kauss & Ziegler, 1974). The most potent sugars found to inhibit the complex-formation of the soluble protein were N-acetyl-D-galactosamine and glycosides containing galactose. As galactosides are not transported in the sieve tubes of Robinia, most of the lectin activity appears to be not directly involved in sugar transport mechanisms.

References

Albersheim, P. & Anderson-Prouty, A.J. (1975). Carbohydrate, proteins, cell surfaces, and the biochemistry of pathogenesis. Annual Review of Plant Physiology 26, 31-52.

Allen, A.K. & Neuberger, A. (1973). The purification and properties of the lectin from potato tubers, a hydroxyproline-containing glycoprotein. Biochemical Journal 135, 307-314.

Barnett, R.E., Furcht, L.T. & Scott, R.E. (1974). Differences in membrane fluidity and structure in contact-inhibited and transformed cells. Proceedings of the National Academy of Science US 71, 1992-1994.

Becker, J.W., Reeke, G.N. Jr., Cunningham, B.A. & Edelman, G.M. (1976). New evidence on the location of the saccharide-binding site of concanavalin A. Nature 259, 406-409.

Bowles, D.J. & Kauss, H. (1975). Carbohydrate-binding proteins from cellular membranes of plant tissue. Plant Science Letters 4, 411-418.

Bowles, D.J. & Kauss, H. (1976a). Characterization, enzymatic and lectin properties of isolated membranes from Phaseolus aureus. Biochimica Biophysica Acta (in press).

Bowles, D.J. & Kauss, H. (1976b). Isolation of a lectin from liver plasmamembrane and its binding to cellular membrane receptors *in vitro*. FEBS-Letters (in press).

Bowles, D.J., Schnarrenberger, C. & Kauss, H. (1976). Lectins as membrane components of mitochondria from Ricinus communis. Biochemical Journal (in press).

Boyd, W.C. (1970). Lectins. Annals N.Y. Academy of Science 169, 168-190.

Callow, J.A. (1975). Plant lectins. Current Advances in Plant Science 7, 181-193.

Chang, C.M., Reitherman, R.W., Rosen, S.D. & Barondes, S.H. (1975). Cell surface location of discoidin, a developmentally regulated carbohydrate-binding protein from Dictyostelium discoideum. Experimental Cell Research 95, 136-142.

Crandall, M., Lawrence, L.M. & Saunders, R.M. (1974). Molecular complementarity of yeast glycoprotein mating factors. Proceedings of the National Academy of Science US 71, 26-29.

Edelman, G.M. (1976). Surface modulation in cell recognition and cell growth. Science 192, 218-226.

Frazier, W.A.; Rosen, S.D., Reitherman, R.W. & Barondes, S.H. (1975). Purification and comparison of two developmentally regulated lectins from Dictyostelium discoideum. Journal of Biological Chemistry 250, 7714-7721.

Galbraith, W. & Goldstein, I.J. (1970). Phytohemagglutinins: A new class of metalloproteins. Isolation, purification, and some properties of the lectin from Phaseolus lunatus. FEBS-Letters 9, 197-201.

Guérin, C., Zachowski, A., Prigent, B., Paraf, A., Dunia, I., Diawara, M.-A. & Benedetti, E.L. (1974). Correlation between the mobility of inner plasma membrane structure and agglutination by concanavalin A in two lines of MOPC 173 plasmocytoma cells. Proceedings of the National Academy of Science US 71, 114-117.

Inoue, M., Utsumi, K. & Seno, S. (1975). Effect of concanavalin A and its derivative on the potassium compartmentation of Ehrlich ascites tumour cells. Nature 255, 556-557.

Jermyn, M.A. & Yeow, Y.M. (1975). A class of lectins present in the tissue of seed plants. Australian Journal of Plant Physiology 2, 501-531.

Ji, T.H. & Nicholson, G.L. (1974). Lectin binding and perturbation of the outer surface of the cell membrane induces a transmembrane organizational alteration at the inner surface. Proceedings of the National Academy of Science US 71, 2212-2216.

Kauss, H. (1976). Lectins (Phytohemagglutinins). In Progress in Botany 38 (in press). Springer-Verlag, Berlin-Heidelberg-N.Y.

Kauss, H. & Glaser, C. (1974). Carbohydrate-binding proteins from plant cell walls and their possible involvement in extension growth.

FEBS-Letters 45, 304-307.

Kauss, H. & Ziegler, H. (1974). Carbohydrate-binding proteins from the sieve-tube sap of Robinia pseudoacacia L. Planta 121, 197-200.

Kauss, H. & Bowles, D.J. (1976). Some properties of carbohydrate-binding proteins (lectins) solubilized from cell walls of Phaseolus aureus. Planta 130, 169-174.

Lamport, D.T.A. (1973). The glycopeptide linkages of extensin: 0-D-galactosyl serine and 0-L-arabinosyl hydroxyproline. pp. 149-165 in Biogenesis of plant cell wall polysaccharides, ed. by F.A. Loewus. Academic Press, London.

Lis, H. & Sharon, N. (1973). The biochemistry of plant lectins (phytohemagglutinins). Annual Review of Biochemistry 42, 541-574.

Lotan, R., Galun, E., Sharon, N. & Mirelman, D. (1975). Interaction of wheat germ agglutinin with microorganisms. Abstract 1001 of the 10th FEBS-Meeting, Paris.

McLean, R.J. & Bosmann, H.B. (1975). Cell-cell interactions: Enhancement of glycosyl transferase ectoenzyme systems during Chlamydomonas gametic contact. Proceedings of the National Academy of Science US 72, 310-313.

McLean, R.J., Laurendi, C.J. & Brown, R.M. Jr. (1974). The relationship of gamone to the mating reaction in Chlamydomonas moewusii. Proceedings of the National Academy of Science US 71, 2610-2613.

Minetti, M., Aducci, P. & Teichner, A. (1976). A new agglutinating activity from wheat flour inhibited by tryptophan. Biochimica Biophysica Acta (in press).

Nicolson, G.L. (1976). Transmembrane control of the receptors on normal and tumor cells. I. Cytoplasmic influence over cell surface components. Biochimica Biophysica Acta 457, 57-108.

Nowak, T.P., Haywood, P.L. & Barondes, S.H. (1976). Developmentally regulated lectin in embryonic chick muscle and a myogenic cell line. Biochemical Biophysical Research Communications 68, 650-657.

Olsnes, S., Refsnes, K. & Pihl, A. (1974). Mechanism of action of the toxic lectins abrin and ricin. Nature 249, 627-631.

Paulova, M., Ticha, M., Entlicher, G., Kostir, J.V. & Kocourek, J. (1971). Studies on phytohemagglutinins. IX. Metal content and activity of the hemagglutinin from the lentil (Lens esculenta Moench). Biochimica Biophysica Acta 252, 388-395.

Reitherman, R.W., Rosen, S.D., Frazier, W.A. & Barondes, S.H. (1975). Cell surface species-specific high affinity receptors for discoidin: Developmental regulation in Dictyostelium discoideum. Proceedings of the National Academy of Science US 72, 3541-3545.

Rosen, S.D., Kafka, J.A., Simpson, D.L. & Barondes, S.H. (1973). Developmentally regulated, carbohydrate-binding protein in Dictyostelium discoideum. Proceedings of the National Academy of Science US 70, 2554-2537.

Rosen, S.D., Simpson, D.L., Rose, J.E. & Barondes, S.H. (1974). Carbohydrate-binding protein from Polysphondylium pallidum implicated in intercellular adhesion. Nature 252, 128-151.

Sharon, N. & Lis, H. (1972). Lectins: Cell-agglutinating and sugar-specific proteins. Science 177, 949-959.

Sharon, N. & Lis, H. (1975). Use of lectins for the study of membranes. pp. 147-200 in Methods in Membrane Biology 3, ed. by E.D. Korn. Plenum Press, New York.

Southworth, D. (1975). Lectins stimulate pollen germination. Nature 258, 600-602.

Surolia, A., Bachhawat, B.K. & Podder, S.K. (1975). Interaction between lectin from Ricinus communis and liposomes containing gangliosides. Nature 257, 802-804.

Thomasson, D.L. & Doyle, R.J. (1975). Monovalent concanavalin A. Biochemical Biophysical Research Communications 67, 1545-1552.

Young, N.M. & Leon, M.A. (1974). The affinity of concanavalin A and Lens culinaris hemagglutinin for glycopeptides. Biochimica Biophysica Acta 365, 418-424.

DISCUSSION

Chairman: B. Solheim

Ballou: Some of the properties and distribution of these other lectins you described in the membrane sound a bit like glycosyl-transferase. Have you tested any of the preparations for that property?

Kauss: No! We know of course there are transferases present but we have

not purified them. In Chlamydomonas people think the binding factor in
the mating reaction is the glycosyl transferase itself but I think even in
Chlamydomonas or in animal tissue it's not clearly shown that it is really
the case.

Delmer: Do you really want to stick with the hemagglutination assay as a
definition of lectins? Don't you think this is really quite unofficial
when you are thinking about plants and the role of these things?

Kauss: If you use a term for some thing you are associating special
properties to it. However, lectin properties are very diverse, and I
think one should be careful using the term lectin. Agglutination is not
an official assay for lectin, however, the term has to have some very
definite significance. An extract may have carbohydrate-binding properties.
Before you can say you are dealing with a lectin you have to purify the
compound. For impure preparations I will use the term carbohydrate-
binding properties.

Albersheim: I have to take the position that agrees with, perhaps, what
was being implied by Clint Ballou and Debbie Delmer. I think that any
enzyme or any protein that binds carbohydrates can become a lectin. The
lectin definition is a very, very poor one. Lysozyme, for example, which
is a hydrolyase has been shown to be a lectin if you make a dimer of this
molecule. Lysozyme has a molecular weight of about 14,600. This is a
small protein, but it has a large binding site, it binds six sugars in a
row. And if you take two lysozyme molecules and put them together the
dimer will bind your blood cells together, now you have the definition of
a lection. Are we going to call lysozyme a lectin? Any glycosyl transferase will be a lectin because it will have a carbohydrate substrate that
it binds to the receptor, and if you only give it one of the substrates and
not the other, the enzyme is going to bind things together as long as it
has two active sites. So! I think the term lectin is one that maybe should
be lost or just call them hemagglutinins or agglutinins or phytoagglutinins,
if you want to talk about animal cells or what have you. In any case, the
fact that a lectin will bind to a carbohydrate, and that any enzyme which
binds to carbohydrates can be called a lectin, pertains to the talk I want
to give this morning.

THE EFFECT OF LECTINS ON CELL DIVISION IN TISSUE CULTURES OF SOYBEAN AND TOBACCO

Indra K. Vasil and David H. Hubbell, Department of Botany and Department of Soil Science, University of Florida, Gainesville, Florida, USA

Certain proteins that possess the ability to agglutinate plant and animal cells are widely distributed in nature, particularly in the seeds of legumes and in some invertebrates (Boyd, 1970; Sharon & Lis, 1972). These are known by the generic name of lectins, and those isolated from plants are commonly termed phytohemagglutinins (PHA). Lectins are known to bind specifically to the cell membrane and are being increasingly used in studies of the structure and function of surface membranes.

There have been only a few studies of the effects of lectins on plants. PHA has been reported to induce parthenocarpic fruit set in a male-sterile mutant of tomato, and in the Bartlett pear (Bangerth et al., 1972). Nagl (1972a, b) found a temporary stimulation of growth and mitotic activity in Phaseolus coccineus and Allium cepa. Concanavalin A (Con A) agglutinates plant protoplasts in the same manner as animal cells (Glimelius et al., 1974), while Con A as well as PHA stimulate germination of lily pollen (Southworth, 1975). A weak and insignificant effect of PHA on tomato callus cells has also been reported (Levenko & Kiforak, 1975).

Legume lectins interact with Rhizobium (Bohlool & Schmidt, 1974; Dazzo & Hubbell, 1975a, b; Hamblin & Kent, 1973), and recent evidence indicates that the lectins participate in the determination of host specificity in the Rhizobium-legume symbiosis by binding specifically to surface poly-saccharides of the appropriate infective Rhizobium cells and allowing specific adsorption of these cells to the legume root surface (Bohlool & Schmidt, 1974; Dazzo & Hubbell, 1975b). Hubbell (personal communication) has recently speculated that the reported mitogenic activity of lectins (Robbins, 1964; Douglas et al., 1969; Nagl, 1972a) may play a key role in the nodulation of legume roots, and the lectin may actually enter legume root cells by virtue of its adsorption to the bacteria. The present report describes the effect of lectins on the growth of plant tissues in vitro.

Materials and Methods

Seeds of soybean (Glycine max) were sterilized by a 2 min immersion in ethanol, followed by 10 min in 10% chlorox. The seeds were then rinsed with sterile distilled water, placed in hydrogen peroxide for 3-4 minutes, and rinsed finally with six changes of sterile distilled water. The sterilized seeds were placed on moistened filter paper in Petri dishes at $25^{o}C$ in the dark for 3 days, or until the roots were about 3 cm long. The terminal 0.3 cm of each root was discarded, and 2-3 segments, each 3 mm thick, were isolated from the cut end of each root. Ten root segments were placed in each Petri dish, and there were a total of 20 segments for each treatment.

Tobacco (Nicotiana tabacum Wis. 38) stem pieces were sterilized for 15 min in 10% chlorox and rinsed several times with sterile distilled water. The stem pith was then extracted with a sterile canula (3 mm diam) and cut into 5 mm long segments. Ten pith segments were placed in each Petri dish, and there were 20 segments for each treatment.

Soybean root segments were cultured on B-5 medium (Gamborg et al., 1968) containing 1 mg/l 2,4-dichlorophenoxyacetic acid (2,4-D), $10^{-6}M$ kinetin, 2% sucrose, and 1% agar. Tobacco pith segments were cultured on Murashige & Skoog's (1962) medium containing 0.04 mg/l kinetin, 4 mg/l indole-3-acetic acid (IAA), 100 mg/l inositol, 2% sucrose, and 1% agar. All cultures were incubated at $25^{o}C$ in the dark. All experiments were repeated twice.

Lectins were dissolved in 0.01 M sodium phosphate buffer (pH 7.4), sterilized through a 0.45 μm Millipore filter, and then added to sterilized and warm agar medium. Soybean agglutinin (Miles Laboratories, Kankakee, Illinois) was used for soybean root segments, and phytohemagglutinin-M (Calbiochem, La Jolla, Calif.) prepared from Phaseolus vulgaris was used for tobacco stem pith cultures.

Growth of the cultured root or stem pith segments took place by cell proliferation and callus formation, and was determined by revoming two pieces from each treatment, blotting on filter paper, and determining the fresh weight.

Vasil & Hubbell - Lectins and cell division

Results and Discussion

Soybean. Fresh weight of soybean root explants taken from various media
after 2, 4, 8, 16, and 32 days of growth clearly shows that growth by cell
division and callus formation occurred only in the presence of 2,4-D and
kinetin (Table 1). There is no growth in the absence of the plant growth
substances, with or without the addition of various concentrations of
soybean agglutinin.

Table 1. Growth (fresh weight in mg) of soybean root segments on B-5
medium. Average weight of starting explants was 23.3 mg.

Days in culture	No SBA					SBA (µg/ml)				
	0H	+H	0.1 0H	0.25 0H	0.5 0H	1.0 0H	0.1 +H	0.25 +H	0.5 +H	0.7 +H
2	28	38	30	31	36	31	33	38	34	42
4	33	45	29	31	31	30	35	38	44	39
8	30	63	28	37	35	35	49	50	72	67
16	54	164	34	43	54	38	124	112	240	154
32	24	313	29	34	32	38	244	148	167	174

SBA = soybean agglutinin
0H = no auxin and no cytokinin
+H = 1 mg/l 2,4-D and 10^{-6}M kinetin

Table 2. Growth (fresh weight in mg) of tobacco stem pith explants
on Murashige & Skoog's medium. Average weight of starting
explants was 67.5 mg.

Days in culture	No PHA			µg/ml PHA				
	0H	+H	50 0H	100 0H	10 +H	50 +H	100 +H
4	92	103	97	88	93	101	78
6	101	198	120	114	168	156	129
16	153	394	146	159	594	495	376
32	208	1120	244	135	1267	1152	968

PHA = phytohemagglutinin-M
0H = no auxin and no cytokinin
+H = 4 mg/l indole-3-acetic acid and 0.04 mg/l kinetin

<u>Tobacco</u>. The growth of tobacco pith explants in media with or without PHA is shown in Table 2. Here again, as with soybean root segments, growth by cell proliferation and callus formation took place only in the presence of IAA and kinetin, with or without the added PHA.

Data on cell counts from soybean as well as tobacco cultures agreed with the fresh weight determinations.

The evidence from the above experiments clearly shows that soybean agglutinin and PHA do not increase or induce cell division in cultured soybean root and tobacco pith segments. The slight and transient increase in mitotic activity reported by Nagl (1972a, b) with PHA might be due to the presence of contaminants - possibly plant growth substances - in the relatively impure lectin preparations used. He observed stimulation of germination and early seedling growth in <u>Phaseolus coccineus</u> and enhancement of root growth in <u>Allium cepa</u>. As the growth enhancement takes place only during the early phase of growth and lasts for a relatively short period of time, it might be caused by an enhanced uptake of water and nutrients due to the changed nature of the plasma membrane resulting from the binding of the lectin to the latter. Southworth (1975) has reported stimulation of lily pollen germination by Con A and PHA which is also related to the shortening of the lag period before germination.

References

Bangert, F., Gotz, G. & Buchloh, G. (1972). Effects of phytohemagglutinin upon parthenogenetic fruit set of the Bartlett pear (William's) and a male sterile mutant of the tomato. Zeitschrift für Pflanzenphysiologie 66, 357-377.

Bohlool, B.B. & Schmidt, E.L. (1974). Lectins: a possible basis for specificity in the <u>Rhizobium</u>-legume root nodule symbiosis. Science 185, 269-271.

Boyd, W.C. (1970). Lectins. Annals of the New York Academy of Sciences 169, 168-190.

Dazzo, F.B. & Hubbell, D.H. (1975a). Concanavalin A: lack of correlation between binding to <u>Rhizobium</u> and specificity in the <u>Rhizobium</u>-legume symbiosis. Plant and Soil 43, 713-717.

Dazzo, F.B. & Hubbell, D.H. (1975b). Antigenic differences between

infective and non-infective strains of <u>Rhizobium trifolii</u>. Applied
Microbiology 30, 172-177.

Douglas, S.D., Kamin, R.M. & Fudenberg, H.H. (1969). Human lymphocyte
response to phytomitogens <u>in vitro</u>: normal, agammaglobulinemic and
paraproteinemic individuals. Journal of Immunology 103, 1185-1195.

Gamborg, O.L., Miller, R.A. & Ojima, K. (1968). Nutrient requirements of
suspension cultures of soybean root cells. Experimental Cell Research
50, 151-158.

Glimelius, K., Wallin, A. & Eriksson, T. (1974). Agglutinating effects of
Concanavalin A on isolated protoplasts of <u>Daucus carota</u>. Physiologia
Plantarum 31, 225-230.

Hamblin, J. & Kent, S.P. (1973). Possible role of phytohemagglutinin in
<u>Phaseolus vulgaris</u>. Nature, New Biology 245, 28-30.

Levenko, B.A. & Kiforak, O.V. (1975). Effect of phytohemagglutinin on the
mitotic activity and ploidy of tomato callus cells. Genetica 11,
167-170.

Murashige, T. & Skoog, F. (1962). A revised medium for rapid growth and
bioassays with tobacco tissue cultures. Physiologia Plantarum 15,
473-497.

Nagl, W. (1972a). Phytohemagglutinin: transitory enhancement of growth
in <u>Phaseolus</u> and <u>Allium</u>. Planta, Berlin 106, 269-272.

Nagl, W. (1972b). Phytohämagglutinin: Temporäre Erhöhung der mitotischen
Aktivität bei <u>Allium</u>, und partieller Antagonismus gegenüber Colchicin.
Experimental Cell Research 74, 599-602.

Robbins, J.H. (1964). Tissue culture studies of the human lymphocyte.
Science 146, 1648-1654.

Sharon, N. & Lis, H. (1972). Lectins: cell-agglutinating and sugar-specific
proteins. Science 177, 949-959.

Southworth, D. (1975). Lectins stimulate pollen germination. Nature 258,
600-602.

DISCUSSION

<u>Chairman</u>: J. Raa

<u>Ballou</u>: I would like to point out that there are many different kinds of

lectins and there are many different ways to study their effect on cells. Cell growth and division is not the only response that is observed. For example, one of the responses that people measure is DNA production which is classically done by looking at tritiated thymidine incorporation. This I think would still be worth looking at and it might even be worth testing some unlikely lectins with your plant tissue. I am not prejudicing the example by picking what seems to be the most reasonable lectin to try.

Hubbell: As a matter of fact, the tritiated thymidine is one experiment that we hope to do. With regard to trying other lectins it's a very appealing idea but the limitation of funds may possibly cut us short before we get to that. I agree with you that the range of lectins that appear available and is appearing daily makes it very risky to make too much of a generalization of this type of thing.

Kijne: I recommend you to do an ulstrastructural study on this system, because cell division does not always occur when you have conditions like partial dedifferentiation and you won't see it in your callus culture. I think auxin and kinetin together is most ideal for root explant; anything else would be superfluous, so omit the kinetin and add the lectin and see what happens.

Hubbell: That's an excellent suggestion. As far as the ultrastructure goes we do have that planned also, we didn't do that on these first experiments but in subsequent experiments we do hope to try some lectin-coded and non-coded cells of Rhizobium just to see if the bacteria actually will carry the lectin in, and we hope to do some ultrastructure studies to see what is going on.

Kauss: The good mitogenic effect of lectins is seen in the resting cell. I guess that your system is not really effective to show any mitogenic effect, because it has just stopped growing. That's the case for lymphocytes at least.

Bauer: May I add two small comments. First is that if you are going to work with soybean lectin you have to be very careful that it is not adsorbed to surfaces that you have present, glass, plastic or whatever. You were using very low concentrations and you may very well have no lectin

Vasil & Hubbell - Lectins and cell division

by the time you start your experiment. The second thing is that we have
found a substantial amount of soybean lectin produced in callus culture,
so that just adding a very small amount more you would not expect to see
any effect even if it did have one.

Hubbell: With regard to the adsorption of lectins to glassware this is
something that I wasn't aware of until it was mentioned here the other day.
My mind flashed back to this data and I asked myself the same question if
indeed there was any lectin in the system. But I believe in this case Vasil
did add the lectin to the agar medium so hopefully there was enough there.

CELL WALL STRUCTURE AND RECOGNITION IN YEAST

Clinton E. Ballou, Department of Biochemistry, University of California,
Berkeley, California 94720, USA

Specific recognition between yeast cells occurs during mating. The
restrictions on such recognition are that the cells must be in the same
mating group, each must be homozygous at the mating type locus, and they
must be of opposite mating type. If these conditions are met, intermixed
Hansenula wingei 5- and 21-cells undergo an immediate and massive agglutination, owing to the interaction of constitutive factors in the cell walls,
followed by cell fusion. Crandall & Brock (1968) have demonstrated that the
recognition specificity and agglutinative ability reside in mannoproteins.
A model for the 5-agglutinin was first reported by Taylor & Orton (1971)
and was later refined by Yen & Ballou (1974).
    In contrast to the constitutive agglutination in Hansenula wingei,
intermixed Saccharomyces cerevisiae haploid cells must "differentiate" before
they develop the sexual agglutinative property. This change is initiated by
the mutual exchange of diffusible pheromones (Duntze, MacKay & Manney, 1970).
Lipke & Ballou (in press) have demonstrated that S. cerevisiae a-cells
respond to the pheromone released by α-cells with the formation of a fuzzy
surface coat at the tip of the copulatory extension. This material may be
related to the development of the agglutinative state.
    Biochemical analyses of the changes in the cell wall of a-cells after
treatment with the pheromone indicate extensive reorganization of the glucan
and mannoprotein components. In particular, there is a reduction in the
amount of the longer mannan sidechains that are normally terminated by
α-1,3-linkages. This suggested that the newly synthesized cell wall should
react poorly with α-1,3-specific fluorescent antibodies in comparison with
the parent cell wall. In fact, the opposite result was observed with the
new cell extension becoming more strongly labelled than the original part of
the cell wall.

These results suggest caution in the interpretation of antibody labelling experiments. Thus, Bohlool & Schmidt (1976) have observed polar labelling of Rhizobium japonicum cells with immunofluorescent soybean lectin, and they suggest that this may be the site for the specific recognition involved in establishing the Rhizobium-legume symbiosis. In fact, it may only reflect an enhanced exposure of surface antigens owing to a disorganization of wall structure in the area of the rapidly growing cell extension.

References

Bohlool, B. & Schmidt, E. (1976). Immunofluorescent polar tips of
    Rhizobium japonicum: Possible site of attachment or lectin binding.
    Journal of Bacteriology 125, 1188-1194.
Crandall, M. & Brock, T. (1968). Molecular basis of mating in Hansenula
    wingei. Bacteriological Reviews 32, 139-163.
Duntze, W., MacKay, V. & Manney, T. (1970). Saccharomyces cerevisiae: A
    diffusible sex factor. Science 168, 1472-1473.
Lipke, P. & Ballou, C. (1976). Morphogenetic changes in Saccharomyces
    cerevisiae a-cells in response to α-factor. Journal of Bacteriology,
    in press.
Taylor, N. & Orton, W. (1971). Cooperation among the active binding sites
    in the sex-specific agglutinin from the yeast, Hansenula wingei.
    Biochemistry 10, 2043-2049.
Yen, P. & Ballou, C. (1974). Partial characterization of the sexual
    agglutination factor from Hansenula wingei Y-2340 type 5 cells.
    Biochemistry 13, 2428-2437.

DISCUSSION

Chairman: D.F. Bateman

Albersheim: I'd just like to mention an aside because Dr. Ballou indicated it might have pertinence to plant pathology. We have been looking at mannan-containing glycoproteins of Phytophthora megasperma sojae. The extracellular enzymes secreted by this pathogen carry the same mannans as the cell surface mannans we have found. Dr. Ernest Zigler in my laboratory has shown that the investases from three races of Phytophthora megasperma sojae have different mannans, differing in ways that probably reflect similar structural differences that Dr. Ballou has shown for different yeast species.

Kijne: Dr. Ballou, what do you think about horizontal transport of receptors in a yeast cell wall?

Ballou: My understanding of a yeast cell wall is that it is a fairly fixed structure. Probably there is no movement of receptors in the wall. One would like to know if there is a movement of receptors on the plasma membrane. One could probably look at this by studying protoplasts and I have attempted to do this. Now, protoplasts do bind very well specific antiserum made against the cell surface. This suggests that they have mannoprotein on them. These mannoproteins may be in the process of secretion, however. I attempted to see if they could be capped with antibody. But it turns out that a protoplast is just a little too small to visualize well with the usual photoflorescent equipment that is available. I think this is still an interesting question.

MOLECULAR DETERMINANTS OF SYMBIONT-HOST SELECTIVITY BETWEEN NITROGEN-
FIXING BACTERIA AND PLANTS

Peter Albersheim and Jack Wolpert, Department of Chemistry, University
of Colorado, Boulder, Colorado 80309, USA

The lectins of legumes and the O-antigen-containing lipopolysaccharides
of the rhizobia are interacting constituents of these organisms (Wolpert &
Albersheim, 1976). The specific interaction between legume-lectins and
the rhizobial-lipopolysaccharides was established as follows: four lectins
have been isolated and studied. These are Concanavalin A from Jack beans,
pea lectin, phytohemagglutinin from Red Kidney beans and soybean
agglutinin. These lectins were brought to a high degree of purity by
affinity chromatography as had been described in the literature for each
lectin. The purified lectins were then covalently attached to Agarose,
an inert matrix. The O-antigen-containing surface lipopolysaccharides were
isolated from the symbionts of the four legumes: R. japonicum for soybeans,
R. leguminosarum for peas, R. phaseoli for Red Kidney beans, and R. species
for Jack beans. These lipopolysaccharides were passed through the four
lectin columns. In each case, the lipopolysaccharide from a Rhizobium sp.
interacts with the lectin column of its symbiont but not with the other
lectin columns. This observation established that the O-antigen-containing
lipopolysaccharides of the rhizobia are involved in determining the specificity of symbiosis in legumes. These findings are the first identification
of a rhizobial constituent involved in this process, and also confirm
evidence from other laboratories that suggests that lectins are involved
in the specificity of symbiosis (Bohlool & Schmidt, 1974; Hamblin & Kent,
1973).
    The binding of the rhizobial lipopolysaccharides to the legume lectins
turned out to be incomplete, that is, not all of the lipopolysaccharide
bound to the lectin-containing columns. Also, the kinetics of the binding

Albersheim & Wolpert - Symbiont-host selectivity

are complex. These observations perplexed us for a while, but they led to
what is probably our most exciting discovery in this work. We believe we
have determined that the lectins of legumes are enzymes. Not only do the
rhizobial lipopolysaccharides interact with the lectin of their symbiont
hosts, but the lipopolysaccharides appear to be degraded by the lectins.
We have evidence that pea lectin degrades the lipopolysaccharide from the
R. leguminosarum and that the phytohemagglutinin from Red Kidney beans
degrades the lipopolysaccharide from R. phaseoli. These lectins do not
degrade the lipopolysaccharides from the non-symbiont rhizobia that have
been tested. Although Concanavalin A and soybean agglutinin have not been
examined directly, indirect evidence has been obtained that suggests that
these lectins are also enzymes.

The likelihood that lectins are enzymes which degrade the lipopolysaccharides of the symbiont rhizobia is in agreement with a previously
unrelated observation. Van Brussell, in a thesis published in Holland in
1973, reported on his studies of the cell walls of the bacterioids of
R. leguminosarum. The major conclusion of van Brussell's work was that
the cell walls of the bacterioids contain much less lipopolysaccharide
than the cell walls of intact bacteria. The enzymic activity of the lectins
of the host plants could lead to an explanation of the lower lipopolysaccharide levels of the bacterioids.

A possibly related observation comes from our studies on elicitors of
phytoalexin production in plants. Elicitors are common constituents of
the cell walls of fungi and probably, too, of bacteria. Preliminary
evidence suggests that the O-antigen-containing lipopolysaccharides of
bacteria are elicitors. The lipopolysaccharide of E. coli stimulates the
production of phytoalexin in soybean cotyledons, and living as well as dead
E. coli cells can also elicit phytoalexin production on soybean cotyledons.
It is interesting to speculate that the lipopolysaccharides of rhizobial
symbionts do not elicit phytoalexin production in their symbiont hosts
because a lectin in the host initiates the degradation of the lipopolysaccharide of the rhizobia. This suggests that the establishment of
nitrogen-fixing symbiont pairs requires that the host has enzymes capable
of degrading the lipopolysaccharides of its symbiont bacteria.

This research was supported by NSF grant PCM75-13897 and by The
Rockefeller Foundation RFGAAS 7510.

- 374 -

References

Bohlool, B.B. & Schmidt, E.L. (1974). Lectins: a possible basis for
    specificity in the Rhizobium-legume root nodule symbiosis. Science
    185, 269-271.
Hamblin, J. & Kent, S.P. (1973). Possible role of phytohaemagglutinin in
    Phaseolus vulgaris L. Nature New Biology 245, 28-30.
Wolpert, J. & Albersheim, P. (1976). Host-symbiont interactions. I.
    The lectins of legumes interact with the O-antigen-containing
    lipopolysaccharides of their symbiont Rhizobia. Biochemical and
    Biophysical Research Communications 70, 729-737.

DISCUSSION

Chairman: B. Solheim

Kauss: It is easy to speculate that lectins could be transferase or
transferase could be lectins. There are several points which I feel
indicate, at least for several lectins, that they have their own properties.
However, at least five lectins in the literature pertains to your talk of
lectins. They had hydrolase activity associated with them for several
years and finally after three or more papers the hydrolytic and lectin
activities were separated.

Albersheim: Well, hydrolases are the least carbohydrate-binding enzymes
to be called lectins because hydrolases are very likely to have their
second substrate, water, available to release the enzyme from the carbohydrate. Glycosyl transferases undoubtedly have physiological roles and
are therefore very important. I should point to one example in the
literature which is well documented, in fact a double example. Neuramini-
dases from both bacteria and animal viruses have been demonstrated to be
lectins. A coat protein of an animal virus has been demonstrated to be a
hemagglutinin without any question; the same protein is an enzyme,
neuraminidase, and a "lectin". Just as clearly, a bacterial enzyme has
been demonstrated to be a lectin. Although the lectin-enzymes do not come
from plants (yet), by your definition they are lectins. I don't think

that the two examples can be disputed.

THE INVOLVEMENT OF SOYBEAN LECTIN IN HOST PLANT RECOGNITION OF THE NITROGEN-FIXING SYMBIONT RHIZOBIUM JAPONICUM

Wolfgang D. Bauer, T.V. Bhuvaneswari and Steven G. Pueppke, C.F. Kettering Research Laboratory, 150 East South College St., Yellow Spring, Ohio 45387, USA

The hypothesis that plant lectins are involved in the recognition of symbiotic and pathogenic microorganisms has recently received considerable attention. If the hypothesis is correct, it would constitute a major breakthrough in our understanding of plant-microorganism interactions and may have important practical applications in agriculture. Careful testing of this hypothesis has become essential. The published studies (Hamblin & Kent, 1973; Bohlool & Schmidt, 1974; Dazzo & Hubbell, 1975a, b; Wolpert & Albersheim, 1976) are suggestive, but fragmentary, inconclusive, and to some extent inconsistant. We have selected the symbiotic association between soybean (Glycine max) and Rhizobium japonicum as a model system suitable for a thorough and critical examination of the lectin recognition hypothesis.

The work of Bohlool & Schmidt (1974) with soybean was the first demonstration of a good biological correlation between lectin binding and the host specificity of rhizobial symbionts. These authors reported that fluorescent-labelled soybean lectin, obtained as a relatively crude preparation from soybean seeds, bound to 22 of the 25 tested strains of the symbiont, R. japonicum. In contrast, they detected no binding of the labelled lectin to any strains of the non-symbiotic rhizobial species that they tested.

In attempting to confirm this correlation, we decided that it would be of importance to use highly purified soybean lectin, and to examine the binding of labelled lectin to living bacterial cells (rather than the heat-fixed cell smears used by Bohlool & Schmidt). We purified soybean lectin to apparent homogeneity by affinity chromatography, treated it to incorporate either a fluorescent or radioactive label, and then repurified the labelled lectin by affinity chromatography.

We have found that the labelled lectin does bind to live cells of several strains of the symbiont, R. japonicum, and that it does not bind to any of the four non-symbiotic rhizobial species tested. The affinity constant for lectin binding to R. japonicum strain USDA 138 was found to be about $4 \times 10^7 M^{-1}$, with about 5×10^5 binding sites per cell. Moreover, all of this binding was found to be both reversible and competitively inhibitable by monosaccharide haptens, thus showing that binding is not due to non-specific adsorption of the labelled lectin to the bacteria. These results are in accord with the original findings of Bohlool & Schmidt, and with the hypothesis that soybean lectin enables the host plant to distinguish symbiotic bacteria from other microorganisms.

However, we have also found a substantial number of strains of R. japonicum to which soybean lectin does not bind at all. These "negative" strains include the three negative strains recorded by Bohlool & Schmidt, several strains reported by these authors as giving a minimal lectin binding reaction (i.e. binding to less than 1% of the cells), and several strains not previously tested. These results do not conform to the hypothesis that the host lectin is required for recognition of the symbiont, and the existence of such negative strains requires explanation if the hypothesis is to remain viable. We are currently seeking a satisfactory explanation for the behavior of these negative strains.

The hypothesis that lectins function in the recognition of rhizobial symbionts also requires that the symbiont-binding lectin be present where the symbiont can interact with it in vivo, i.e. in, or on, the roots of the host. Bohlool & Schmidt (1974) reported that they were able to detect hemagglutinating activity in soybean root extracts. However, they did not determine that the hemagglutinating activity was due to the lectin (from seeds) that showed selective binding to symbiotic rhizobia. Therefore, we decided to quantitatively measure the amounts of the R. japonicum-binding seed lectin in soybean root extracts. Radiolabelled seed lectin of known specific activity was added to 3-15 day old seedling roots, and root extracts prepared. The lectin in the extracts was purified by affinity chromatography, and the amount of endogenous lectin determined by isotopic dilution. The results of these experiments demonstrated quite conclusively that soybean roots do contain significant and increasing amounts of the same R. japonicum-binding lectin that is present in seeds. During a 15 day seedling growth period, 15 to 60 μg of the lectin was detected per root system. Soybean callus and suspension-cultured cells also contained significant amounts of the same lectin.

Assuming that soybean lectin is involved in the recognition of the symbiont, R. japonicum, one would like to know the mechanism by which lectin binding leads to recognition. Dazzo & Hubbell (1975a) have suggested that lectin binding in the white clover (Trifolium repens) - R. trifolii association provides a physical linkage between the bacterial cells and the root hair surfaces. We suspect, however, that simple physical attachment of this sort may not be an important aspect of lectin function in the soybean-R. japonicum association. Rhizobia are rather sticky bugs in their own right, and we have found no correlation between the lectin binding properties of various rhizobia and their adherence to soybean root hairs. Instead, we tentatively favor the possibility that the binding of host lectin to characteristic receptors on the surface of the symbiont acts as a trigger for subsequent responses by the host and/or the symbiont which lead to successful infection. This contention is supported by our recent discovery that the binding of soybean lectin to living cells of the symbiont causes a rapid and apparently localized swelling of the bacteria. The swollen region of a cell is irregularly spherical in shape and has a diameter several times that of an untreated cell. When bacteria are treated with fluorescent-labelled lectin, all of the fluorescence is associated with the swollen regions of the cells - the normal, unswollen ends of the cells remain non-fluorescent. When exposed to soybean lectin, cells of the symbiont also show a fairly rapid and substantial decrease in respiration rate. Significantly, both the swelling phenomenon and the decrease in respiration rate can be quickly and completely reversed by the addition of 1mM N-acetylgalactosamine, an effective monosaccharide hapten of soybean lectin binding. Non-symbiotic species of rhizobia and "negative" strains of R. japonicum neither swell nor show a change in respiratory rate when treated with the lectin. We are currently examining the ultrastructure of the swollen cells by electron microscopy and studying the importance - if any - of the lectin-induced swelling in the infection process.

Bauer et al. - Lectin and host plant recognition

References

Bohlool, B.B. & Schmidt, E.L. (1974). Lectins: a possible basis for specificity in the Rhizobium-legume root nodule symbiosis. Science 185, 269-271.

Dazzo, F.B. & Hubbell, D.H. (1975a). Cross-reactive antigens and lectin as determinants of symbiotic specificity in the Rhizobium-clover association. Applied Microbiology 30, 1017-1033.

Dazzo, F.B. & Hubbell, D.H. (1975b). Concanavalin A: lack of correlation between binding to Rhizobium and specificity in the Rhizobium-legume symbiosis. Plant Soil 43, 717-722.

Hamblin, J. & Kent, S.P. (1973). Possible role of phytohemagglutinin in Phaseolus vulgaris L. Nature New Biology 245, 28-30.

Wolpert, J.S. & Albersheim, P. (1976). Host-symbiont interactions. I. The lectins of legumes interact with the O-antigen-containing lipopolysaccharides of their symbiont Rhizobia. Biochemical and Biophysical Research Communications 70, 729-737.

THE ROLE OF LECTIN IN SOYBEAN - RHIZOBIUM JAPONICUM INTERACTIONS

T.S. Brethauer and J.D. Paxton, Department of Plant Pathology, University
of Illinois, Urbana, Illinois 61801, USA

We became interested several years ago in the phenomena which govern
"recognition" between microorganisms and plants. After discovering an
inducer, produced by the fungal pathogen Phytophthora megasperma var.
sojae, which triggers phytoalexin production by the soybean plant (Frank &
Paxton, 1971) we turned our attention to a somewhat similar interaction,
nodulation of soybeans.

Our first attempts at examining this system used the soybean lectin
which is soluble in 40% saturated ammonium sulfate solutions but insoluble
in 60% saturated solutions (Liener & Pallansch, 1952). This preparation
reacted with material in the supernatant of 14-day-old liquid cultures of
4 of the 29 strains of Rhizobium japonicum and cowpea rhizobia tested,
forming precipitin bands in Ouchterlony double diffusion plates. The
precipitin band could be dissolved or the reaction blocked by incorporating
.005 M D-galactose in the agar. D-Glucose, L-Arabinose or D-Mannose at
0.1 M concentrations or lower did not affect the reaction. The precipitin
reaction exhibits the same specificity for D-Galactose as the Soybean
Agglutinin SBA (Lis et al., 1970).

To follow this reaction further, the soybean lectin preparation was
labelled with fluorescein isothiocyanate and the binding of labelled lectin
to Rhizobium japonicum as described by Bohlool & Schmidt (1974) was studied.
This proved to be difficult to repeat and impossible to quantitate.
Therefore we labelled the lectin with ^{125}I using chloramine T and attempted
to study binding to the bacterium by collecting the lectin-treated cells
on 0.2 μ pore filters. This gave no indication of specific binding to the
bacteria, and filtration was an unsatisfactory, time-consuming procedure.

We then constructed a SBA affinity column using CH Sepharose 4B and

D-galactosamine (Allen & Neuberger, 1975).

The affinity absorbent bound only 15% of the ammonium sulfate preparation. SDS gel electrophoresis indicated 1 major band of about 30,000 daltons and 1 higher and 2 lower MW, minor bands in this preparation. Lactoperoxidase was then used as a more gentle procedure for labelling this lectin with ^{125}I.

Bacteria were then incubated with 0.1 mg of this lectin for 30 minutes in either 0.1 M glucose or 0.1 M galactose in phosphate buffered saline (PBS). After incubation the bacteria were washed 2X in PBS and the 3,000 Xg pellet counted on a gamma counter with about 60% counting efficiency.

The results of binding experiments with 29 isolates of Rhizobium are presented in Table 1. These results indicate that there is no correlation between the ability of this labelled lectin to bind a Rhizobium isolate and the isolate's ability to nodulate soybeans.

Table 1.

ISOLATES BINDING IODINATED SOYBEAN AGGLUTININ

Strains nodulating soybean		Strains nodulating cowpea but not soybean	
Rudy-Patrick 5063	1403/97*	Nitragin Crotalaria 32Z1	46,311/2126
Rudy-Patrick 5104	48,557/3539	Rudy-Patrick Cowpea 6411e	839/44
Nitragin 61A24	1045/35	USDA Peanut 3G4b4	56,548/11,194
Nitragin 61A89	27,529/1159		
USDA 3I1b110	56,475/25,647		
Illinois 126	47,108/18,486		
Illinois 10140	56,185/16,458		

ISOLATES NOT BINDING IODINATED SOYBEAN AGGLUTININ

Strains nodulating soybean		Strains nodulating cowpea but not soybean	
Illinois 71	30/11*	Nitragin Peanut 8A11	31/17
USDA 3I1b83	31/9	Nitragin Mungbean 127D3	30/22
USDA 3I1b85	22/14	Nitragin Glycine wightii 61B14	16/14
Nitragin 61A72	36/13	Nitragin Glycine wightii 61B11	10/15
Nitragin 61A92	60/50	Nitragin Cowpea 176A22	62/41
Nitragin 61A93	34/34	Nitragin Lespedeza 93K1	40/25
		Nitragin Lima bean 127E12	30/22
		Nitragin Cowpea 32H1	26/25
		Nitragin Jackbean 22A5	14/31
		USDA Crotalaria 3c1a1	42/13
		USDA Cowpea 3I6n9	45/55
		USDA Jackbean CB756	36/38
		Rudy-Patrick Lupine 5425	18/32

*(counts per minute of 0.1 M glucose treatment)/(counts per minute of 0.1 M galactose treatment)

References

Allen, A.K. & Neuberger, A. (1975). A simple method for the preparation of an affinity absorbent for soybean agglutinin using galactosamine and CH-Sepharose. FEBS Letters 50, 362-364.

Bohlool, B.B. & Schmidt, E.L. (1974). Lectins: A possible basis for specificity in <u>Rhizobium</u>-legume root nodule symbiosis. Science 185, 269-271.

Frank, J.A. & Paxton, J.D. (1971). An inducer of soybean phytoalexin production and its role in the resistance of soybeans to Phytophthora rot. Phytopathology 61, 954-958.

Liener, I.E. & Pallansch, M.J. (1952). Purification of a toxic substance from defatted soybean flour. Journal of Biological Chemistry 197, 29-36.

Lis, H., Sela, B.A., Sachs, L. & Sharon, N. (1970). Specific inhibition by N-acetyl-D-galactosamine of the interaction between soybean agglutinin and animal cell surfaces. Biochimica et Biophysica Acta 211, 582-585.

DISCUSSION

Chairman: B. Solheim

Albersheim: There are two points I would like to make; one is that the inability to bind lectins to nodulating strains, the three times that Bohlool & Schmidt could not do this and the fact that Bauer (this symposium) can't do this half of the time, does not in itself destroy the hypothesis because some other component could mask the binding sites in some strains. The masking component might be the exopolysaccharide. In addition there may be more than one lectin involved. I call it lectin with Dr. Kauss here for the lectin that has been studied is one that agglutinates animal cells or red blood cells, but there can be other lectins in the plant that do not agglutinate animal cells. So! I don't think negative results in that a lectin fails to bind destroys the hypothesis. It's only when you get the wrong positive that one must worry, and then you have to know why you are getting the wrong positives. The other point I want to make is that I disagree very strongly with some of Dr. Bauer's statements here. I think his work is excellent and I am very impressed with it, but I have the right to disagree with his predictions. First of all, I don't feel that the interactions of the lectins and the lipopolysaccharides that we are trying to study has to be on the surface. The lipopolysaccharide is obviously on the surface of the bacteria, we know where that is, but I don't think the lectin has to be on the surface of the root. It could well be inside the root. Jan Kijne (this symposium) showed some pictures of the fact that the bacteria goes down the infection thread looking like a bacteria and it is a long time before bacteria become bacteroids. This happens deep in the root. Therefore alterations of the lipopolysaccharide could occur deep in the root and not on the root surface. We did not know the work of the Leyden group when we were doing our work or when we came up with the idea or realized that these lectins were probably enzymes working on the lipo-polysaccharides. Van Brusell several years ago in his Ph.D. thesis had shown that the lipopolysaccharide of bacteroids were altered. That's exactly what we are predicting now that these lectin enzymes will do. The lipopolysaccharides are altered such that they are lost during isolation. I don't know if there is just one enzyme involved with the alteration. We have no assay for the other enzymes. There could be enzymes such as esterases working in the lipid A. Any number of enzymes in the plant could be effecting the lipopolysaccharides and changing them. Although Bauer's pictures of the swelling bacteria of course are suggestive, I don't think you have to place all of the responsibility on the lectin. I disagree with Bauer's last prediction, too, because I don't think it's been demonstrated that the lectin-lipopolysaccharide surface interactions have to be involved in the initiation of the infection process. There could be some other interaction that we don't yet recognize. There has to be recognition, I agree, but the lectin-bacterial lipopolysaccharide interaction does not have to be responsible.

Paxton: I would agree with Bauer that the lectin or recognition has to occur on or close to the surface of the plant since you don't see infection threads in non-hosts. In addition we do have these false positives of cowpea nodulating isolates that bind the soybean lectin without being able to nodulate soybeans.

Albersheim: Why do you say that it has to be the lectins that are doing this? Why can't it be something else that you haven't recognized?

Paxton: It could be lots of things but the argument has been presented in the literature that the lectins are the compounds that are responsible for the recognition. It obviously, I think, has to be a large molecule because it has to carry a fair amount of information to select between strains and species of rhizobia. My guess is you would have to look for a large molecule that is carrying enough information to be selective between various isolates.

Bauer: My feeling is that the biology that we really have to keep in mind is that the plant cannot afford to make a mistake. The plant is really committing itself when it accepts the microorganism as a symbiont. In this case it's really allowing the symbiont or microorganism to get inside of the plant. That's why I feel the recognition has to occur at the start of the infection process and not when the organism is well inside of it. It does not have to be the lectin, it's just a prediction of this hypothesis.

Paxton: We looked at two isolines of soybeans (Harasoy), one nodulating and one non-nodulating, and we could not find any difference in the lectins from the two isolines. But nodulation is a multistep process and it obviously means that if lectins are involved there is some other step that is being interferred with.

Solheim: Can you make it clear what the controversy is between Bohlool & Schmidt's paper (Science 185, 269-271, 1974), Albersheim's results, Bauer's results and your own results?

Paxton: The basis of our controversy revolves somewhat around the initial hypothesis put out by Bohlool & Schmidt that a lectin is responsible for specificity of the binding between the host and its symbiont. In this case of soybean and Rhizobium japonicum, Bohlool & Schmidt's data showed that the lectin from soybean bound to 22 out of 25 Rhizobium japonicum strains tested and not to non-symbiont rhizobia. Both Bauer and I have found that if you take crude soybean lectin and try to bind it to the bacteria, you see no specificity whatsoever. It binds to both non-nodulating and nodulating Rhizobium japonicum and to other species of Rhizobium. So there appeared to be no specificity whatsoever. When we use purified lectin we do see specific binding to nodulating strains of R. japonicum, but again we do not see any specificity of binding with R. japonicum as opposed to other Rhizobium species, and therefore the lectins may not be involved in the specificity or the recognition reaction between the legume host and the Rhizobium bacterium symbiont.

Albersheim: I have no argument with that at all! The only thing is that I never assumed that Bohlool & Schmidt's work was a definitive explanation for infectivity. My goal is to look to see where there are specific interactions. There is nothing to demonstrate, for example, that Bohlool & Schmidt weren't seeing some binding to the exopolysaccharide that was bound to the surface of some of these cells, or, indeed, it is not sure that a specific lectin was involved for they had impure lectin preparations. When you are working with whole bacteria I don't think you can completely get rid of the exopolysaccharide.

Lippincott: I would like to agree with Albersheim. I think you are looking at this attachment as too specific a single mechanism. There may be several keys and locks to unlock as the bacterium progresses into the hair cells and even from that hair cell into the next cell. The initial attachment may in fact not have to be that specific.

Paxton: Something specifically lets it into the root hair cell, and that has to be somewhere in that cell wall or close to the surface.

Lippincott: It could be even at the membrane level for example, and the actual attachment to the surface of the hair cell would not require that recognition factor but some other recognition factor.

Kijne: One of the conflicting points which is raised now is if these lectins are binding to the extracellular polysaccharides or not. Albersheim says that they are not binding to extracellular polysaccharides, but you got some agglutination with your immuno-diffusion technique. I don't know if your extracellular polysaccharides could be contaminated with lipopolysaccharides.

Paxton: In the immuno-diffusion experiments we simply used the culture filtrates and it very well could have LPS in it. I might mention that we have also tried to co-chromatograph radio labelled soybean lectins with exopolysaccharides from several strains of Rhizobium japonicum and have never found any evidence for a binding interaction between the two.

CROSS REACTIVE ANTIGENS AND LECTIN AS DETERMINANTS OF HOST SPECIFICITY IN THE RHIZOBIUM-CLOVER SYMBIOSIS

Frank B. Dazzo, Department of Bacteriology, Center for Studies of Nitrogen Fixation, University of Wisconsin, Madison, Wisconsin 53706, USA

The purpose of this research was to test whether host specificity is based on a recognition system involving complementary macromolecules present on both the plant and microbial symbionts. Host specificity in Rhizobium-legume symbioses is expressed prior to the formation of root hair infection threads (Li & Hubbell, 1969). Therefore, proof of complementary macromolecules which play a role in determining host specificity requires that they interact before the invagination of the root hair cell wall which initiates the infection process (Napoli & Hubbell, 1975). All subsequent events, e.g. formation of nodular tissue containing nitrogen fixing bacteria, occur after the compatible host recognition has been expressed. The Rhizobium-clover system was selected for study for two reasons. First, the clover has small seedling roots which are ideal for microscopic studies of the infection process. Second, infective and related non-infective mutant strain combinations of R. trifolii whose phenotypes differ in the ability to initiate infection threads in clover root hairs are available.

The first clue that suggested the requirement of macromolecules to express the infective phenotype came from the results of immunodiffusion and immunoelectrophoretic studies of various R. trifolii strains using hyperimmune antisera (Dazzo & Hubbell, 1975a). Slow diffusing antigens which were shared among all the infective strains were uniformly absent from the non-infective mutants.

The Fahraeus slide technique (Fahraeus, 1957) was modified to develop a quantitative assay to examine the adsorption of Rhizobium into clover root hairs (Dazzo, Napoli & Hubbell, 1976). Within a 12 hr incubation period, 4-5 times more cells of infective R. trifolii (including infective

revertants) than non-infective R. trifolii or R. meliloti were firmly
adsorbed to clover root hairs. Adsorption of R. meliloti was 10 times
greater on root hairs of its host, alfalfa, than on clover. These results
represent the first direct indication that preferential adsorption of
infective Rhizobium cells to root hairs of its homologous host is an early
expression of host specificity that preceeds infection.

The cell surfaces of R. trifolii and root hairs of its specific host
clover were studied. Various immunochemical techniques, including quantitative tube agglutination, immunofluorescence with reciprocal cell adsorptions,
and radioimmunoassay (I^{125}-labelled immunoglobulin) clearly indicated the
presence of a surface antigen on R. trifolii which was cross reactive with
an antigen on clover roots (Dazzo & Hubbell, 1975c).* The antigen was
immunochemically unique to infective strains of R. trifolii since it was
not detected on Rhizobium strains incapable of infecting clover
(R. leguminosarum, R. meliloti, R. japonicum, R. phaseoli, and several
"cowpea" rhizobia). Recent studies have indicated that the cross reactive
antigen is immunochemically distinct from surface antigens on non-infective
R. trifolii mutants. The cross reactive antigen was isolated from
R. trifolii strain 0403 and was characterized as a high molecular weight
(> 4.6 x 10^6 daltons), β-linked, acidic heteropolysaccharide containing
2-deoxyglucose, glucose, galactose, and glucuronic acid. The sugar
2-deoxyglucose had never before been reported in Rhizobium polysaccharides.
The discovery of this immunochemically unique polysaccharide (Dazzo &
Hubbell, 1975c) represents the first identification of a Rhizobium product
which displays a characteristic unique to the infective phenotype of the
bacterium Rhizobium trifolii. The structural integrity of this surface
polysaccharide is acid-labile with 63, 89, and 100% loss of antibody
binding characteristics after treatment at pH 5, 4 and 3, respectively
(Dazzo & Hubbell, 1975c).

Recent emphasis has been placed on the finding of legume proteins
that interact specifically with surface carbohydrate receptors of homologous

---

*DeVay has reviewed the literature on the occurrence of cross reactive
antigens shared between phytopathogenic microbes and their plant hosts
(DeVay & Adler, 1976). Cross reactive antigens between other rhizotropic
nitrogen-fixing bacteria and roots of the associated plants have been
described. These include Rhizobium sp., HR1 and joint vetch (Aeschynomene
americana, Napoli, Dazzo & Hubbell, 1975b), Spirillum lipoferum 13t and
guinea grass (Panicum maximum, Dazzo & Milam, 1976), and possibly
R. japonicum 138 and soybean (Glycine max, Takats, personal communication).

rhizobia (Hamblin & Kent, 1973; Bohlool & Schmidt, 1974; Dazzo & Hubbell, 1975c; Dazzo & Brill, 1976; Wolpert & Albersheim, 1976). However, fluorescent microscopic studies with purified FITC-Concanavalin A from jackbean indicated that it would interact with complementary cell surface receptors of some 19 strains of Rhizobium (R. trifolii, R. japonicum, R. meliloti, R. phaseoli, R. leguminosarum, and several "cowpea" rhizobia), only 2 strains of which (strains CE22A1 and 127E10 from J.C. Burton) were successful in nodulating jackbean (Dazzo & Hubbell, 1975b). Con A could be specifically eluted from the cells by competitive displacement with α-methylglucoside or mannoside indicating that the jackbean protein interacts with the Rhizobium cell surface receptors through its carbohydrate binding site. This work showed that interactions between certain legume lectins and Rhizobium cells may not always account for the specificity expressed by the nodule bacteria for their respective legume hosts. Other anomalous results also have been reported for the interaction of the galactosyl-specific soybean lectin and nodulating strains of R. japonicum (Bohlool & Schmidt, 1974; Brethauer & Paxton, 1975; Maier, Dazzo & Brill, unpublished observation). These results emphasize that investigators should not be prejudiced toward considering only the lectins classically described and should not overlook hitherto unrecognized legume proteins that may bind to and distinguish nodulating and non-nodulating strains of Rhizobium.

A protein from white clover has been found which binds specifically to the unique surface polysaccharide of R. trifolii (Dazzo & Hubbell, 1975c). The Rhizobium binding protein from clover (hereafter called RBP) agglutinates infective strains of R. trifolii (including wild type infectives, infective revertants, and "promiscuous" rhizobia which infect clover outside of the normal cross inoculation group). The RBP does not agglutinate cells of non-infective R. trifolii, or R. leguminosarum, R. meliloti, R. phaseoli, R. japonicum, and several "cowpea" rhizobia incapable of infecting clover. The purification of RBP is in progress. The purification is followed by an increase in specific activity (endpoint agglutinating units/mg protein). The quantitative agglutination test at the present can detect less than 1 μg agglutinating protein. [Agglutination of Rhizobium cells has been reported in situ in simulated rhizosphere environments (Napoli, Dazzo & Hubbell, 1975a), and has been confirmed by others (Solheim, personal communication). Agglutination may influence the frequency of root hair infections. Analysis of 231 infected root hairs indicated that 210 were in physical contact with agglutinated rhizobia in comparison with 21 infected root hairs in contact with single cells only (Napoli, Dazzo & Hubbell, 1975a).] The protein has a 2-deoxyglucose binding specificity on R. trifolii cells based on inhibition of direct cell agglutination. Binding specificity which is 2-deoxyglucose inhibited has been confirmed using passive hemagglutination of erythrocytes coated with the immunochemically unique surface polysaccharide. Agglutination of cells by RBP is different from cellulose mediated flocculation (Napoli, Dazzo & Hubbell, 1975a) since the former can occur in the presence of cellulase (Dazzo & Hubbell, 1975c).

The possible role of RBP in the adsorption of R. trifolii to clover roots is being examined. Precoating infective cells of R. trifolii with RBP from clover increases their firm adsorption to clover roots (Dazzo, Napoli & Hubbell, 1976). Similar enhancement of bacterial adsorption has been observed for Vicia lectin-coated Rhizobium on Vicia roots (Solheim, 1975).

The sugar 2-deoxyglucose was the only sugar component of the unique R. trifolii antigen which was capable of blocking its interaction with the RBP (Dazzo & Hubbell, 1975c). It was hypothesized that this sugar may interfere with the bacterial adsorption to clover roots by occupying all the binding sites. Lack of inhibition by this sugar would therefore indicate independence of R. trifolii attachment and an attachment role of RBP. The results indicated that 2-deoxyglucose inhibited the adsorption of R. trifolii to clover root hairs but not the adsorption of R. meliloti to alfalfa root hairs (Dazzo, Napoli & Hubbell, 1976). The inhibition of bacterial adsorption was specific for 2-deoxyglucose and was not observed with 2-deoxygalactose or α-D-glucose.

Clover root hairs contain surface receptors which specifically bind the unique surface polysaccharide of R. trifolii (Dazzo & Brill, 1976). The binding of the polysaccharide can be specifically inhibited by 2-deoxyglucose. Examination of clover root washings indicated the presence of a protein with identical electrophoretic and sugar inhibitory characteristics as that of RBP. The specific activity of RBP was 30 times greater in concentrated root eluants obtained with 2-deoxyglucose as compared with α-D-glucose. The eluted protein can bind to the capsules of R. trifolii strain 0403. These results suggest that the R. trifolii binding receptor site on clover root hairs contains a Rhizobium binding protein that is non-covalently bound through a 2-deoxyglucose receptor site.

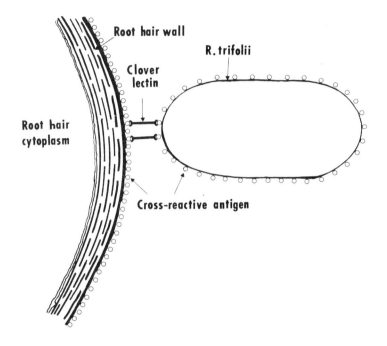

Figure 1.   Schematic diagram showing the proposed cross-bridging of the
cross-reactive antigens of Rhizobium trifolii and clover root
hairs with a clover lectin.

These data are consistent with the model (Figure 1) to explain host
specificity based on the preferential adsorption of infective R. trifolii
cells to clover roots by a cross-bridging of their common surface antigens
with the multivalent, 2-deoxyglucose specific clover lectin (Dazzo &
Hubbell, 1975c).

Acknowledgements

This research was conducted in collaboration with Dr. David H. Hubbell,
University of Florida, Gainesville, and Dr. Winston J. Brill, University of
Wisconsin, Madison. The research was supported by NSF grant GB-31307,
DEB-75-14043, and BMS-74-01134, and by a Grant-in-Aid for Research by
Sigma Xi, the Scientific Research Society of North America.

Dazzo - Cross reactive antigens

References

Bohlool, B.B. & Schmidt, E.L. (1974). Lectins: a possible basis for specificity in the Rhizobium-legume root nodule symbiosis. Science 185, 269-271.

Brethauer, T.S. & Paxton, J.D. (1975). Interaction of soybean lectin with extracellular material from Rhizobium bacteria. Proceedings of the American Phytopathological Society 2, 83.

Dazzo, F.B. & Brill, W.J. (1976). The receptor site for Rhizobium trifolii on clover roots. (Submitted for publication)

Dazzo, F.B. & Hubbell, D.H. (1975a). Antigenic differences between infective and noninfective strains of Rhizobium trifolii. Applied Microbiology 30, 172-177.

Dazzo, F.B. & Hubbell, D.H. (1975b). Concanavalin A: lack of correlation between binding to Rhizobium and specificity in the Rhizobium-legume symbiosis. Plant and Soil 43, 713-717.

Dazzo, F.B. & Hubbell, D.H. (1975c). Cross reactive antigens and lectin as determinants of symbiotic specificity in the Rhizobium-clover association. Applied Microbiology 30, 1017-1033.

Dazzo, F.B., Napoli, C.A. & Hubbell, D.H. (1976). Adsorption of bacteria to roots related to host specificity in the Rhizobium-clover symbiosis. Applied and Environmental Microbiology 31 (in press, July 1976 issue).

Dazzo, F.B. & Milam, J.R. (1976). Serological Studies of Spirillum lipoferum. Proceedings of the Florida Soil and Crop Science Society 35 (in press).

DeVay, J.E. & Adler, H.E. (1976). Antigens common to hosts and parasites. Annual Review of Microbiology 30 (in press).

Fahraeus, G. (1957). The infection of clover root hairs by nodule bacteria studied by a simple glass slide technique. Journal of General Microbiology 16, 374-381.

Hamblin, J. & Kent, S.P. (1973). Possible role of phytohemagglutinin in Phaseolus vulgaris L. Nature, New Biology (London) 245, 28-30.

Li, D. & Hubbell, D.H. (1969). Infection thread formation as a basis of nodulation specificity in Rhizobium-strawberry clover associations. Canadian Journal of Microbiology 15, 1133-1136.

Napoli, C.A. & Hubbell, D.H. (1975). Ultrastructure of Rhizobium-induced infection threads in clover root hairs. Applied Microbiology 30,

1003-1009.

Napoli, C.A., Dazzo, F.B. & Hubbell, D.H. (1975a). Production of cellulose microfibrils by Rhizobium. Applied Microbiology 30, 123-131.

Napoli, C.A., Dazzo, F.B. & Hubbell, D.H. (1975b). Ultrastructure of infection and "common antigen" relationships in Aeschynomene. Proceedings of the 5th Australian Legume Conference, March 18-21, 1975, Brisbane, Australia.

Solheim, B. (1975). Possible role of lectins in the infection of legumes by rhizobia. National Alliance Treaty Organization Conference on Specificity in Plant Diseases, Advanced Study Institute, Sardinia. May 4-16, 1975.

Wolpert, J.S. & Albersheim, P. (1976). Host-symbiont interactions. I. The lectins of legumes interact with the O-antigen containing lipo-polysaccharides of their symbiont Rhizobia. Biochemistry and Biophysics Research Communications (in press).

DISCUSSION

Chairman: J. Raa

Albersheim: We heard this morning a lot about rhizobia and I don't think we ever described the fact that rhizobia have an exopolysaccharide. We assumed everybody knew this, as well that they have, of course, the classical lipopolysaccharide of gram-negative bacteria. We did hear earlier that the exopolysaccharide has a uronic acid and glucose and some mannose, and has been studied in great detail. There has been excellent chemical work characterizing the exopolysaccharide. The best, and probably some of the most recent came out of Bengt Lindberg's laboratory in Sweden. They found no significant differences between the exopolysaccharides of several Rhizobium species. There are two major exopolysaccharide types for many Rhizobium species. These results agree well with previous findings on the exopolysaccharides. There has been a number of immunological studies of the exopolysaccharides. People thought that the exopolysaccharides must repre-sent the serological difference between the species, but they did not find it. Instead, they found on the surface of the bacteria tremendous serolo-gical differences which could account not only for species differences but

also for strain differences. I don't know of any evidence to suggest that
exopolysaccharide and the capsular polysaccharide are different, as far as
I know they are the same molecule. So Dazzo's suggestion this morning that
the capsular of exopolysaccharide is species specific is really in
opposition to what I consider a large amount of evidence in the literature
suggesting that these do not vary from species to species. We do know that
there are capsular polysaccharides in some bacteria which are species
specific. But it was not established for rhizobia, in fact shown not to be
the case. Lipopolysaccharides have always been shown to be species specific.
How does this account for Dazzo's data? I can't account for it. However, I
do not think that exopolysaccharide or the capsular polysaccharide was purified to any degree by the method used. There was no chromatography or
gradient elution from a column that would have separated it from other
components. There are probably other polysaccharides present in the exopolysaccharide preparations in minor amounts, probably O-antigen or intact
lipopolysaccharides. We know, again by the literature, that curling factors
are secreted by rhizobia which, when applied to roots, cause the root hairs
to curl. Maybe Solheim will comment more about this because he has done a
lot of work in this area. We don't really know what the curling factors are,
maybe they are the capsular polysaccharides, but that to me seems unlikely.
They wouldn't be species specific. Curling factors are species specific.
There could be quantitatively minor components or hormone level components
secreted by the bacteria which cause changes in the surface of the root
hairs and which then cause attachments of the bacteria. These minor
components could certainly contaminate the capsular polysaccharide. What I
am trying to say is that there is a lot we don't understand about all of
this and that there is a lot of conflicting data in the literature and I
wanted to make sure that you all understood it wasn't clear to all of us who
work in the field that the exopolysaccharide or capsular polysaccharide,
which Dazzo presented a lot of data for this morning, is in fact likely to
be the species specific component. At least it is not in my mind.

Dazzo: The immunochemical data that have been presented in the past
involved the use of antiserum prepared against the bacteria. The immunochemical data that I presented utilized antiserum prepared against clover
roots. I see no reason why the antibodies prepared against the bacteria
have to bind to the same immunodominant antigenic determinant. It is well

known that gram-negative bacteria have on their surfaces a mosaic of many
antigenic possibilities and I am only saying that I can find in the capsular
antigen preparation some antigenic determinant which appears to be unique to
the species Rhizobium trifolii. In regard to finding no differences in
carbohydrate composition, I would like to point out a paper that has been
recently published by Professor W.F. Dudman (Carb. Res. 46, 97-110, 1976),
who has shown that in Rhizobium japonicum exopolysaccharides, there are some
sugars, e.g. galacturonic acid, 4-O methyl galactose, that have never before
been described in Rhizobium. There are several papers that Dr. Albersheim
mentioned which utilize sophisticated carbohydrate analytical techniques.
In considering Dudman's information, a new controversy is what went wrong
with the earlier analyses which looked at R. japonicum polysaccharides and
didn't detect the galacturonic acid? I would also like to point out that
Dr. Peter Graham published a series of studies (Graham, P.H., Antonie van
Leeuwenhoek 29, 281-291, 1963; Graham, P., Antonie van Leeuwenhoek 31, 349-
354, 1965; Graham, P. & O'Brien, M., Antonie van Leeuwenhoek 34, 326-330,
1968; Graham, P., Analytical serology of microorganisms 2, 353-378, 1969)
where he examined the exopolysaccharide and lipopolysaccharide contents of
rhizobia representing several cross inoculation groups. He found that he
could not explain speciation in Rhizobium based on the sugar composition of
these exopolysaccharides and lipopolysaccharides. When I presented my work
at the Rhizobium Conference in Raleigh, North Carolina (1975), his comment
was that the lipopolysaccharide composition of Rhizobium trifolii,
R. phaseoli, and R. leguminosarum appears to be the same as the composition
of what I presented on the R. trifolii capsular polysaccharide, again
showing no unique differences in sugar composition. The techniques that I
used for the extraction and the purification of the capsular polysaccharides
did not utilize any chromatographic procedures. Capsular polysaccharides
can be isolated from pelleted cells by the method of Nowotny (Basic exercises
in immunochemistry, Springer Verlag, N.Y., 1969). I'd like to know if
your published techniques can be applied in general to the separation of
Rhizobium exopolysaccharides and lipopolysaccharides. If they can, then I
will use your techniques on the material that I have where I left off.
Then I will examine whether or not capsular polysaccharides bind to the
Rhizobium binding protein from clover.

Albersheim: Clearly I can't predict that the method I use will work for all rhizobia. I think that you have to look at the composition of these things as you are going along and see whether you have a composition of LPS or exopolysaccharide and see whether you get these things purified. What we published in Biochem. Biophys. Res. Commun. (70, 729-737, 1976), which is the only one we have out on this subject, was not nucleic acid free, it was exo-free. I think that it is important to get your thing totally purified and you should certainly do an ion exchange chromatography on your material in addition to what we had on that paper.

Ballou: Immunochemical cross reactivity is not usually good evidence for structural identity since we know that antibodies recognize usually single sugars or small parts of the whole molecule. On the other hand, the kind of chemical structural analyses that Lindberg and even I do is not a very good test for differences of immunochemical types. That is, small differences that are recognized by antibodies are not always easily found by chemical structural analyses. As far as looking for the possible role of exopolysaccharide contamination of the LPS preparations, I should think one could use strains that don't make the capsule and isolate the LPS there. Then there should be no question as to whether they are free of the exopolysaccharide.

Dazzo: I would also like to state that when I go back and use Albersheim's procedures to try to separate R. trifolii capsular polysaccharides from lipopolysaccharides, in addition to using an anthrone test to measure the carbohydrate material coming off the column I am going to incorporate specific tests which can distinguish "lipopolysaccharides" from what you call "exopolysaccharides". One test that is extremely sensitive is the limulus lysate assay, and the other test than can distinguish lipid-bound vs. lipid-free polysaccharide is the presence of fatty acids. Dr. Albersheim reported neither a reaction with the limulus lysate nor the presence of fatty acids in his LPS preparation. I will strive to purify a capsular polysaccharide until it is limulus lysate negative and when I do that I will be very confident that it does not contain lipopolysaccharides.

Jarvis: Is the bending of the clover root hairs by the application of capsular material accompanied by any visible cytoplasmic changes in the hair? Cytoplasmic changes like that might well not be visible, but if any of them were happening then how does the information to start such a change

get to the plasmalemma through the wall? If these polysaccharides are as
large as they seem to be, they would presumably have some difficulty
reaching the plasmalemma. Alternatively if what we are thinking about is
happening only in the wall, then it would be interesting to hear any
speculations as to why bending might occur, bearing in mind that there is
serological similarity demonstrated between the capsular polysaccharides
and the wall. If not the wall, then at least the surface of it.

Dazzo: It is an interesting observation that many lectins have a variety of
interactions with mammalian cell membranes and trigger intracellular events
without necessarily being transported in. Sooner or later they may get in
but there are examples where lectins bind to membranes and in a short period
of time the cell begins to differentiate with blast formation and there is
incorporation of tritium labelled thymidine to a higher level. A good
example of a compound that binds to the cell membrane and triggers intra-
cellular events without necessarily getting in is with the hormone
epinepherine. There is a receptor site for epinepherine on the outer
surface of membranes which accepts epinepherine. Beneath the epinepherine
receptor on the inner surface of the membrane lies adenyl cyclase. When
epinepherine binds to its outer membrane receptor, membrane-bound adenyl
cyclase is activated to catalytically convert ATP to cyclic AMP. Dr. Jarvis,
I cannot answer your question as to whether or not capsular polysaccharides
do get in but it is interesting to speculate that if there are chemical
similarities based on immunochemical evidence, then perhaps the capsular
polysaccharides may be incorporated into the growing root hair cell wall
and in one way or another alter the cell wall synthesis resulting in a
three dimensional change which is observed as the root hair deformation.
Dr. Solheim and Dr. Raa have done much more work with the curling factor
and perhaps they can expand on this in terms of the mechanism.

Raa: The only way we could interpret our research on the curling factor
was that a heat unstable compound was released from the bacteria. This
was stabilized by a compound released from the host. The complex was more
active in causing curling reaction than the compound released from the
bacteria itself. This fits into the model you suggested, of course, if we
say that the stabilizing factor is identical to your lectin.

Dazzo: We ran stability studies of the clover agglutinating factor at 56° C and 80°C for 10 minutes followed by rapid cooling. There is a considerable amount of denaturation of seed material which is removed by centrifugation. Then we run an endpoint agglutination titer using infective R. trifolii cells following the heat treatment and find that the titer is reduced down to a non-detectable level. Regarding enhanced heat stability of macro-molecules by lectin binding, I would like to point out a paper presented in the symposium on Con A (A. Surolia et al., Adv. Exp. Med. Biol. 55, 95-115, 1975) which was recently published showing one example of a lectin which binds to a glycoprotein enzyme called arylsulphatase which was isolated from chicken brains. The interaction of the lectin Con A with the carbohydrate portion of this arylsulphatase glycoprotein tremendously increased the enzyme's heat stability. So there is information in the literature to support the concept that perhaps materials with lectin-like properties may heat stabilize other components to which the lectins bind. The Rhizobium-binding protein that I studied from clover seeds was inactivated by heat. I have not yet examined the heat-stability of the Rhizobium-binding protein obtained from clover roots.

Solheim: I want to add that both the curling factor and the stabilizing factor are heat labile. It is the complex which is heat stable (J. Gen. Microbiol. 77, 241-247, 1973). It has been reported that the curling factor is heat stable, that it can be autoclaved and still curls the root hairs. In my experience this is not so. This might be an argument against polysaccharide being the curling factor, because polysaccharides are usually heat stable. I have not yet purified the curling factors, and do not really know the chemical nature of them.

ADSORPTION OF PROTEINS FROM TROPICAL LEGUMES TO RHIZOBIUM

P.J. Robinson, C.S.I.R.O., Division of Tropical Agronomy, Privat Bag,
Townsville, Quinsland 4810, Australia

The genus Stylosanthes, native to Central and South America, is of
considerable importance as a forage legume in northern Australia, and some
700 accessions have been introduced for agronomic evaluation. As a
considerable number of these introductions will not successfully nodulate
or fix nitrogen with native Australian Rhizobium strains, considerable
effort has been spent on strain selection. It has been shown that in the
large and variable species S. guyanensis, accessions requiring specific
strains possessed seed acid phosphatase and esterase isozyme patterns which
differed from accessions with promiscuous nodulating habits. This association was fortuitous but raised the question of whether the different
isozymes (proteins) some of which are present in roots, bound specifically
to nodulating strains, as had been shown for soybean lectin and nodulating
strains of Rhizobium japonicum.
    Seven accessions of S. guyanensis were selected for further study,
based on their nodulation performance with three strains of "cowpea"
Rhizobium; accession-strain combinations ranged from non-nodulating, through
nodulating but with ineffective nitrogen fixation, to high effective
nitrogen fixing symbioses. Crude protein extracts of seed and root were
incubated with cell suspensions of the Rhizobium strains (both motile cells
and cells which had been heat treated to destroy the flagellar antigens).
The binding of protein to cells was monitored by polyacrylamide gel
electrophoresis, nitrogen analysis of the cells, and reduction in protein
N and acid phosphatase activity after removal of the cells. Tube aggluti-
nations were also carried out and a range of sugars examined for inhibition
of the agglutination reaction.
    The optimum ratio for binding of seed protein to bacterial cells was

found to be 1 mg protein N/ml to 10^{10} cells/ml. Numbers of bacteria either more or less than this resulted in a decrease in removal of protein from the solution. The amount of N bound to motile cells of CB_{756}, for example, ranged from 0.9 to 14.1 µgN per 10^9 cells for the 7 guyanensis accessions studied, and was not correlated with nodulation data. Cells of CB_{756} also bound 18.3 µgN/10^9 cells when incubated with a seed extract of Medicago sativa, which is not nodulated by this strain. E. coli cells (ATCC-25922-1) bound 10.1 µgN/10^9 cells, when incubated with a seed extract of S. guyanensis CP1.38754.

The concentration of protein N in the Stylosanthes seed and root extracts required to agglutinate suspensions of the Rhizobium cells also did not correlate with nodulation data. Concanavalin A and root extracts of alfalfa, white clover and Urochloa mozambicensis (a grass) also agglutinated the Rhizobium cells. Agglutination of CB_{756} by extracts of CP1 38754 could be inhibited by glucosamine, acetylglucosamine, galactosamine and acetylglucosamine, but not by glucose, galactose, fucose or mannose.

Electrophoresis revealed that the bacteria bound selectively to some proteins and esterase and acid phosphatase isozymes in the seed extracts. Preliminary work reveals that the bands which are bound may be glycoproteins, while proteins without sugar residues are not removed by the Rhizobium cells. Whether the isozymes are bound because of this enzymic activity or because of their glycoproteinic nature is unknown at this stage. It is also unknown whether the various individual bands have any haemagglutinating activity. A possible explanation for the non-specific binding of Stylosanthes proteins to both nodulating and non-nodulating strains is that the "cowpea" Rhizobium species is reputably far less selective in its symbioses than other Rhizobium species.

PEA LECTIN PURIFICATION

J.W. Kijne, K. Planqué and P.P.H. Swinkels, Botanisch Laboratorium,
Rijksuniversiteit, Leiden, Nederland

Entlicher et al. (1969, 1970) isolated and characterized two lectins
from the seeds of Pisum sativum var. Pyram. We examined the number of
lectins present in Pisum sativum cv. Rondo-seeds, and their separation.
Hemagglutinating activity of the various fractions was assayed in a double
dilution test using a 2% v/v saline suspension of washed human A-
erythrocytes. The active protein fraction of a crude water extract of
fine pea seed meal was loaded on a Sephadex G100 or G150 column, whether
or not after $(NH_4)_2SO_4$-precipitation and following dialysis and lyophilization. Elution with 0.01 M glycine-HCl (pH 2.0) gave one broad absorbance
peak at 280 nm. After dialysis and lyophilization this peak material was

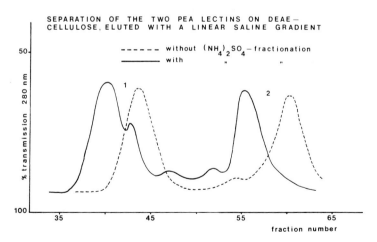

Figure 1. Separation of the two pea lectins on DEAE-cellulose, eluted
with a linear saline gradient.

seaparated on DEAE-cellulose, resulting in the revealing of two main components both active in the agglutination of red cells and Rhizobium
leguminosarum A171. Elution with a linear NaCl-gradient (25 nM to 150 nM
in tris-HCl pH 8.4) gave a far better separation than elution with 150 nM
NaCl in the same buffer. Besides, material subjected to ammonium sulphate
fractionation yielded some additional small peaks compared with non-
fractionated material (Figure 1). Rechromatography of peak 1 after dialysis
and lyophilization yielded two peaks, one of them being inactive. Polyacrylamide electrophoresis of active fractions invariably showed only one
band in acidic gels (Reisfeld et al., 1962) and two bands in alkaline gels
(Steward et al., 1965). We suggest that pea "Rondo"-seeds contain two
lectins, and that dialysis against distilled water and lyophilization could
raise artefacts during their isolation.

References

Entlicher, G., Tichá, M., Kostír, J.V. & Kocourek, J. (1969). Studies on
    phytohemagglutinins. II. Phytohemagglutinins of Pisum sativum L. and
    Lens esculenta Moench: specific interactions with carbohydrates.
    Experientia 25, 17-19.
Entlicher, G., Kostír, J.V. & Kocourek, J. (1970). Studies on phyto-
    hemagglutinins. III. Isolation and characterization of hemagglutinins
    from the pea (Pisum sativum L.). Biochimica et Biophysica Acta 221,
    272-281.
Reisfeld, R.A., Lewis, U.J. & Williams, D.E. (1962). Disk electrophoresis
    of basic proteins and peptides on polyacrylamide gels. Nature 195,
    281-283.
Steward, F.C., Lyndon, R.F. & Barber, J.T. (1965). Acrylamide gel electro-
    phoresis of soluble plant proteins: study on pea seedlings in relation
    to development. American Journal of Botany 52, 155-164.

Kijne et al. - Pea lectin purification

DISCUSSION

Chairman: D. Delmer

Dazzo: What was your recovery of total agglutinating activity in your
ammonium sulfate fractionated material as compared to material prior to
ammonium sulfate?

Kijne: You mean the titer or the amount of excess protein?

Dazzo: Did you look at the titer before and after ammonium sulfate
fractionation? In our experience in working with trying to purify clover
lectin we found that ammonium sulfate fractionation would indeed precipitate our lectin in a 20-40% range but that the recovery of total activity
was only about 3%. That is, we could only recover about 3% of the activity
that we had when we started prior ammonium sulfate precipitation. So
we've abandoned that technique. You might compare this total activity
prior to the ammonium sulfate precipitation and follow the original
extraction as your 100% activity and calculate it based on that.

Kijne: The titer of filtrate before this ammonium sulfate fractionation
was 16, or 32. After the ammonium sulfate precipitation there is no
activity left in the remaining fraction. So we indeed, supposedly got
all of the existing activity, certainly not 3%.

Dazzo: O.K., well, maybe this lectin behaves in a different way. But if
you look at the purification schemes of many of the lectins, ammonium
sulfate is a commonly used procedure initially and many of these schemes
do not examine the recovery of activity following that first step.

Kijne: Well, we found that enrichment of our activity by this step was
slightly better.

Paxton: Does your pea lectin have a specificity for glucose?

Kijne: For glucose and for mannose and for piccarose. It's rather
unspecific.

Paxton: We avoided the ammonium sulfate precipitation step all together
when we make an affinity column and simply run the crude extract through
that. It saves you a lot of trouble.

Kijne et al. - Pea lectin purification

Kijne: That's what we are doing now.

Paxton: And then you can elute it with the sugars. It might be a good
idea to try adding a sugar to your initial extract to try to keep the
lectins from binding to those residues in cell walls and other materials
in your crude preparation when you first do the extraction.

Kijne: I think that's a good suggestion. We didn't work with elution
with sugar solution so far. I think it's much milder than glycine-
hydrochloric acid and should be preferred, but I don't think the amount of
activity coming off with sugar elution is as high as when you use acidic
conditions.

Paxton: It is.

Delmer: I was wondering, if you treat bacteroids with the haptens (glucose
or mannose) and wash them well, will they react now with the lectin? This
could be a possible way to remove residual bound lectin.

Kijne: This is certainly one of the things we are going to do.

ULTRASTRUCTURE OF RHIZOBIUM-INDUCED INFECTION THREADS IN CLOVER ROOT HAIRS

C.A. Napoli and D.H. Hubbell, Departments of Microbiology and Soil Science,
University of Florida, Gainesville, Florida 32611, USA

The first stage in the establishment of the Rhizobium-legume N_2-fixing
symbiosis is the infection of the host legume by the appropriate Rhizobium
species. This is a highly specific interaction as each strain of Rhizobium
is restricted to a particular groups of legumes it can infect. These
restrictions form the basis for speciation in Rhizobium and the so-called
cross inoculation groups of host legumes and nodulating rhizobia.

Root hairs are the site of infection by Rhizobium in a large number of
legume species, particularly those of the families Trifolieae and Viciae.
In the aquatic legume Neptunia oleracia, Schaede did not find root hairs
and proposed entry of the bacteria through the epidermal cells (Schaede,
1940). Another important route of entry of rhizobia is the point of lateral
root emergence (Allen & Allen, 1940; Napoli, Dazzo & Hubbell, 1975). A
recent and complete review of the infection process has been given by Dart
(1974).

The first microscopically visible indication of the bacteria-plant
interaction is deformation and curling of the normally straight root hairs.
A characteristic deformation is a curling of the root hair tip to produce
a "shepherd's crook" (Fahraeus, 1957). The bacteria enter the root hair and
are enclosed in a tubular structure, the infection thread, which is the
first microscopically visible sign of a successful infection (Li & Hubbell,
1969). The majority of infected root hairs have the shepherd's crook at
the infection thread origin, but exceptions exist (Fahraeus, 1957; Nutman,
1956; Napoli & Hubbell, 1975). However, not all deformed root hairs contain
infection threads.

Several theories have been proposed regarding the entry of the bacteria
into the root. When the rhizobia were first isolated in pure culture, it

- 407 -

was determined that the bacteria did not hydrolyze cellulose (Beijerinck, 1888). McCoy (1932) confirmed that the rhizobia did not hydrolyze cellulose or pectin when cultured in media containing these substrates. The large scale production of hydrolytic enzymes would be detrimental to the symbiosis as the rhizobia would have the potential to be pathogenic. The infection of the legume by the rhizobia must proceed in such a manner that the physiology of the plant is affected as little as possible.

Nutman (1956) had advanced the hypothesis of root hair cell wall invagination. An invagination results from the redirection of plant cell wall growth at a localized point, resulting in the wall growing back into the root hair to form the tubular infection thread. There is no penetration through the wall at the point of entry, and the bacteria remain extra-cellular, i.e. there is no direct contact with the host cytoplasm.

Nutman's theory of invagination has been challenged on several points. First, how the cell wall invaginates against the high hydrostatic pressure of the root hair is unknown (Dixon, 1969). Secondly, invagination would form an open pore, which had not been shown in earlier electron micrographs (Higashi, 1966; Sahlman & Fahraeus, 1963). However, serial sections of root hairs were not used in these studies. Additionally, an open pore would allow simultaneous entry of different cell types which would result in several Rhizobium strains being isolated from one nodule. Early studies indicated that only one strain of Rhizobium was isolated from one nodule when the host had been inoculated with a mixture of infective rhizobia differentially marked by antibiotic resistance (MacGregor & Alexander, 1972) or serological types (Hughes & Vincent, 1972; Jones & Russell, 1971). However, recent studies (Johnston & Beringer, 1975; Lindemann, Schmidt & Ham, 1974) have shown that several strains can be isolated from one nodule.

Ljunggren & Fahraeus (1961) have proposed a "polygalacturonase" hypothesis in which the rhizobial exopolysaccharide increased plant pectic enzyme activity and a single bacterial cell softened and subsequently penetrated the plant cell wall without pronounced structural disruption. The infection thread is presumably initiated once the bacterium penetrates to the plant plasmalemma.

In support of this theory these workers demonstrated that a crude preparation of extracellular polysaccharide of infective rhizobia increased the activity or de novo synthesis of plant produced pectinolytic enzymes. This activity was strain specific in that it correlated with the plant-bacterium specificity. Munns (1969) provided evidence to support this theory and demonstrated that the induction of pectinase (pectin transeliminase) was acid sensitive. Bonish (1973) found pectinolytic enzyme activity was not correlated with infectivity of strains. In addition, other workers (Lillich & Elkan, 1968; MacMillan & Cooke, 1969; Solheim & Raa, 1971) have not been able to verify this hypothesis.

Few workers have attempted ultrastructural studies of infected root hairs. Sahlman & Fahraeus (1963) and Higashi (1966) examined infected root hairs under the electron microscope. These authors did not section infected root hairs through the origin of the infection thread but did offer their micrographs as support of Nutman's theory of invagination. Dart (1971) examined root hairs under the scanning electron microscope. He reported that root hairs and epidermal cells were coated with many bacteria, some of which appeared to be embedded in the wall. The root hair tips were often smooth but some older root hair surfaces had a fibrillar meshwork pattern.

The light microscope has been invaluable in studying the growth of the infection thread through the root hair. However, the point of entry of the bacteria into the root hair, and thus the mechanism of infection, cannot be observed with the light microscope. Infections are initiated in tightly curled root hair tips, in areas where root hairs touch, or in areas covered by bacterial flocs. The electron microscopic examination of ultrathin serial sections was used as an approach in resolving this problem. The technique of serial sectioning was used so the root could be seen in its entirety.

Three and seven day old inoculated clover seedlings were glutaraldehyde osmium fixed and flat embedded in Spurr plastic. The embedded seedlings were viewed under phase-contrast microscopy, and areas containing infection threads were selected for sectioning. Serial sections were cut and picked up on Formvar-coated, one-hole grids and stained with Reynold's lead citrate.

Five infected root hairs having the shepherd's crook at the infection thread origin were serially sectioned, and in every case the root hair wall was invaginated (Napoli & Hubbell, 1975). Serial sectioning showed that there were no breaks in the root hair cell wall at the point of entry and the root hair cell wall was continuous with the wall of the infection thread. The infection thread walls at the point of invagination were difficult to see. However, the infection thread walls away from the point

of invagination were clearly recognizable. This may have reflected a
physical and/or chemical alteration of the cell wall structure at the
invagination origin, where the specific bacteria-plant interactions resulted
in the initiation of an infection thread.

The invagination process resulted in the formation of a pore.
Infection threads on seedlings inoculated for seven days were examined and
the pores were open. At this time it is unknown if the pores remain open
for a longer period of time. The pore was not directly exposed to the
rhizosphere but was either encircled by the shepherd's crook or plugged
with a floc of bacteria. In this way, entry into the infection thread
would be limited to those bacteria trapped in the shepherd's crook or the
bacteria comprising the floc. This may explain the infrequent occurrence
of isolation of more than one strain of Rhizobium from one nodule.

Serial sections through a root hair with an infection thread which
did not originate in the tightly curled shepherd's crook showed an
invagination which resulted in the formation of a pore. A floc of bacteria
was attached to the root hair at the origin of the infection thread.

The cytoplasm surrounding the infection thread appeared to be meta-
bolically active. There was an accumulation of mitochondria, dictyosomes,
and rough endoplasmic reticulum. The infection thread was composed of two
distinct layers, an outer layer and an inner amorphous layer which
surrounded the bacteria and was unstable in the electron beam. The outer
layer of the infection thread and the plant cell wall stained positive
for polysaccharide with the periodic acid-silver hexamine stain. However,
the inner layer did not stain. If this layer contained polysaccharide, it
was not sensitive to periodic acid oxidation. There was no discontinuity
in infection threads crossing from one cell into another in the root
cortex. The wall of the infection thread and the plasmalemma of each cell
were continuous which suggested a repetition of the invagination process.

Infection threads adjacent to plant cell nuclei were sectioned and in
each instance there was an accumulation of darkly staining, diffuse material
between the nucleolus and the nuclear envelope. This may have represented
the transfer of ribonucleoprotein from the nucleus to the cytoplasm. It has
been thought by several workers (Fahraeus, 1957; Nutman 1969) that there is
a direct biochemical communication between the plant cell nucleus and the
growing infection thread as the nucleus precedes the growing tip into the
base of the root hair cells. Nuclei positioned beside and at the tip of

infection threads had numerous nuclear pores.

Ultrastructural evidence presented by this research showed that the bacteria entered the root hair by a process of invagination. The bacteria redirected the growth of the plant cell wall so that the root hair grew back into itself. In this way, the bacteria entered the root through a bacterial induced, plant synthesized infection thread which resulted from the invagination process. The bacteria remained extracellular while inside the root hair.

The majority of infected root hairs have the tightly curled root hair tip, but infection will occasionally occur in a relatively undeformed root hair. It is believed that the attached bacterial floc on the slightly curled root hair served the same function of the tightly curled root hair tip, localizing and concentrating the biochemical interactions of the plant and bacteria to a threshold level which is required for initiation of infection.

Future studies on the infection of legumes by Rhizobium should concentrate on the specific biochemical interactions which allow infection to take place and which result in the invagination of the plant cell wall.

Acknowledgements

This research was supported by the National Science Foundation Grants GB 31307 and DEB 75-14043 and a Grant-in-Aid for Research from Sigma Xi, the Scientific Research Society of North America.

References

Allen, O.N. & Allen, E.K. (1940). Response of the peanut plant to inoculation with rhizobia, with special reference to morphological development of the nodules. Botanical Gazette 102, 121-142.

Beijerinck, M.W. (1888). Die Bacterien der Papilionaceenknollchen. Botanische Zeitung 46, 725. (Translated in Brock, T.D., Milestones in Microbiology, 1961. Prentice-Hall Int. Inc., London.)

Bonish, P.M. (1973). Pectolytic enzymes in inoculated and uninoculated red clover seedlings. Plant and Soil 39, 319-328.

Dart, P.J. (1971). Scanning electron microscopy of plant roots. Journal of Experimental Biology 22, 163-168.

Dart, P.J. (1974). The infection process. pp. 381-429 in The biology of nitrogen fixation, ed. by A. Quispel. North Holland Publishing Co., Amsterdam.

Dixon, R.O.D. (1969). Rhizobia (with particular reference to relationships with host plants). Annual Reviews of Microbiology 23, 137-157.

Fahraeus, G. (1957). The infection of clover root hairs by nodule bacteria studied by a simple glass slide technique. Journal of General Microbiology 14, 374-381.

Hughes, D.Q. & Vincent, J.M. (1972). Serological studies of the root-nodule bacteria. III. Test of neighboring strains of the same species.. Proceedings of the Linnean Society of New South Wales 67, 142.

Higashi, S. (1966). Electron microscopic studies on the infection thread developing in the root hair of Trifolium repens L. infected with Rhizobium trifolii. Journal of General Microbiology 13, 391-403.

Johnston, A.W.B. & Beringer, J.E. (1975). Identification of the Rhizobium strains in pea root nodules using genetic markers. Journal of General Microbiology 87, 343-350.

Jones, D.G. & Russell, P.E. (1971). The application of immonofluorescence techniques to host plant/nodule bacteria selectively using Trifolium repens. Soil Biology and Biochemistry 4, 277-282.

Li, D. & Hubbell, D.H. (1969). Infection thread formation as a basis of specificity in Rhizobium-strawberry clover associations. Canadian Journal of Microbiology 15, 1133-1136.

Lillich, T.T. & Elkan, G.H. (1968). Evidence countering the role of polygalacturonase in invasion of root hairs of leguminous plants by Rhizobium spp. Canadian Journal of Microbiology 14, 617-625.

Lindemann, W.C., Schmidt, E.L. & Ham, G.E. (1974). Evidence for double infection within soybean nodules. Soil Science 118, 274-279.

Ljunggren, H. & Fahraeus, G. (1961). The role of polygalacturonase in root hair invasion by nodule bacteria. Journal of General Microbiology 26, 521-528.

MacGregor, A.N. & Alexander, M. (1972). Comparison of nodulating and non-nodulating strains of Rhizobium trifolii. Plant and Soil 36, 129-139.

MacMillan, J.D. & Cooke, R.O. (1969). Evidence against involvement of pectic enzymes in the invasion of root hairs by Rhizobium trifolii. Canadian Journal of Microbiology 15, 543-645.

McCoy, E. (1932). Infection by Bact. radicicola in relation to the microchemistry of the host's cell walls. Proceedings of the Royal Society of London, Series B 110, 514-533.

Munns, D.N. (1969). Enzymatic breakdown of pectin and acid-inhibition of the infection of Medicago by Rhizobium. Plant and Soil 30, 117-120.

Napoli, C.A., Dazzo, F.B. & Hubbell, D.H. (1975). Ultrastructure of infection and "common antigen" relationships in Aeschynomene. Proceedings of the 5th Australian Legume Conference, March 18-21, Brisbane, Australia.

Napoli, C.A. & Hubbell, D.H. (1975). Ultrastructure of Rhizobium-induced infection threads in clover root hairs. Applied Microbiology 30, 1003-1009.

Nutman, P.S. (1956). The influence of the legume in root nodule symbiosis. A comparative study of host determinants and functions. Biological Reviews of the Cambridge Philosophical Society 31, 109-151.

Nutman, P.S. (1959). Some observations on root hair infection by nodule bacteria. Journal of Experimental Botany 10, 250-263.

Sahlman, K. & Fahraeus, G. (1963). An electron microscope study of root hair infection by Rhizobium. Journal of General Microbiology 33, 425-427.

Schaede, R. (1940). Die Knollchen der adventiven Wasserwurzein von Neptunia oleracea und ihre Bakterien Symbiose. Planta 31, 1-21.

Solheim, B. & Raa, J. (1971). Evidence countering the theory of specific induction of pectin degrading enzymes as basis for specificity in Rhizobium-Leguminosae associations. Plant and Soil 35, 275-280.

CELL PROLIFERATION IN PEA ROOT EXPLANTS SUPPLIED WITH RHIZOBIAL LPS

J.W. Kijne, S.D. Adhin and K. Planqué, Botanisch Laboratorium,
Rijksuniversiteit, Leiden, Nederland

The lipopolysaccharides (LPS) in the outer membrane of various gram-negative bacteria are biologically active, e.g. in the vertebrate immune system. The action of LPS as a mitogen, and a trigger of differentiation (Kearny & Lawton, 1975) is most intriguing. Although most plant pathogenic bacteria are gram-negative, no reports are known to us in which any role of purified LPS in plant pathogenesis is established. We investigated the influence of the LPS of the root nodule inducing Rhizobium leguminosarum A171 on cell division in root explants of Pisum sativum cv. Rondo. Main-root explants of 7 days-old plants (with or without the stele) were cultured on nutrient agar with the basal medium of Shigemura (BS) (1958) in case supplied with 10^{-6}M 2,4-D and $3,2 \times 10^{-6}$M kinetin, following methods of Libbenga (Libbenga et al., 1973). Purified Rhizobial LPS was subjected to DNase and RNase treatment, dialyzed, sterilized by filtration or autoclavation, and a drop of a 0.25% aqueous solution was placed on explants cultured on BS + 2.4-D. After 7 days the cell proliferation pattern on the upper surface of complete explants on BS + 2.4-D + kinetin and BS + 2.4-D + LPS was the same, and differed notably from the pattern in BS- and BS + 2.4-D- cultured explants (Figure 1). The cytokinin-like LPS-effect was much less pronounced or absent in root cortical explants. The results indicate that in the presence of the stele the addition of LPS has the same effect as the addition of kinetin. In another system, the Amaranthus-test on cytokinin activity according to Biddington & Thomas (1973), the 0.25% LPS-solution displayed an activity comparable with $1-3 \times 10^{-8}$M kinetin.

SCHEMATIC SURFACE VIEW ON 7 DAYS—
OLD COMPLETE PEA ROOT EXPLANTS,
SHOWING DOTTED AREAS WITH CELL
PROLIFERATION. ___ = endodermis; ⅄ = xylem

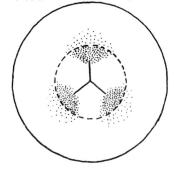 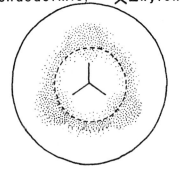

BS + 2,4 D (and BS)   BS + 2,4 D + KIN or LPS

Figure 1.

References

Biddington, N.R. & Thomas, T.H. (1973). A modified Amaranthus bioassay
    for the rapid determination of cytokinins in plant extracts. Planta
    Berlin 111, 183-186.

Kearney, J.F. & Lawton, A.R. (1975). B lymphocyte differentiation induced
    by lipopolysaccharide. I. Generation of cells synthesizing four major
    immunoglobin classes. Journal of Immunology 115, 671.

Libbenga, K.R., Van Iren, F., Bogers, R.J. & Schraag-Lamers, M.F. (1973).
    The role of hormones and gradients in the initiation of cortex
    proliferation and nodule formation in Pisum sativum L. Planta Berlin
    114, 29-39.

Shigemura, Y. (1958). The nutritional and auxin requirements for the growth
    of pea root callus tissue. Ph.D. Thesis, University of California.

INTERACTION BETWEEN PLANT LECTINS AND CELL WALL COMPONENTS OF PSEUDOMONAS
SOLANACEARUM: ROLE IN PATHOGENICITY AND INDUCED DISEASE RESISTANCE

T.L. Graham and L. Sequeira, Department of Plant Pathology, University
of Wisconsin, Madison, Wisconsin 53706, USA

Ultrastructural Evidence

When tobacco leaves are infiltrated with suspensions of avirulent (B1)
variants of Pseudomonas solanacearum, the bacterial cells attach readily to
mesophyll cell walls (Sequeira et al., 1976). By 4 hr after infiltration,
fibrillar and granular material envelops the attached bacteria. In
contrast, virulent strains of the pathogen (e.g. isolate K60) are not
attached and multiply freely in the intercellular fluid, eventually
becoming systemic. Attachment and envelopment seem to be of fundamental
significance in the early interactions between host and pathogen; they may
determine the ultimate fate of the bacteria that reach potential sites of
infection.

Agglutination of Bacterial Cells by Lectins

The initial attachment of avirulent cells to host cell walls apparently
involves recognition between molecular components of the bacterial and host
cell walls. Our working hypothesis is that recognition involves bacterial
lipopolysaccharide (LPS) and lectins on the host cell wall. A lectin has
been extracted from the roots and leaves of tobacco (Nicotiana tabacum L.
"Bottom Special") that agglutinates avirulent (B1), but not virulent (K60)
cells. This lectin appears to be present on mesophyll cell walls; thus, it
may be important in the attachment and immobilization of incompatible
bacteria described above.

Because the procedures for purification of potato lectin are established,
most of our work has centered on the interaction of potato lectin and

P. solanacearum. Potato leaves react to strains of P. solanacearum in a manner similar to tobacco. Fifty-five virulent isolates and 33 avirulent variants of P. solanacearum from different geographic locations and representing all major races and biotypes were tested for their ability to bind to potato lectin. The lectin was purified to homogeneity from "Katahdin" potato tubers by modifications of the method of Marinkovitch (1964). All avirulent isolates agglutinated in the presence of the lectin whereas virulent isolates (except for two isolates from potato in Australia, which gave variable results) failed to agglutinate. Potato lectin conjugated to fluorescein isothiocyanate bound to rabbit erythrocytes and to avirulent, but not to virulent cells of P. solanacearum. Failure to bind to the lectin by bacterial cells was correlated with the presence of extracellular poly-saccharide (EPS) which is formed by virulent, but not by avirulent cells. Virulent cells (K60) were agglutinated by potato lectin after EPS was removed by repeated washing of the cells. Moreover, binding to avirulent cells (B1) was prevented when EPS was added to the cells prior to addition of the lectin.

Purified lipopolysaccharide (LPS), extracted from B1 cells with a mixture of phenol, chloroform, and petroleum ether (Galanos et al., 1969) precipitated when mixed with potato lectin, and may, therefore, represent the binding site of the lectin on B1 or washed K60 cells. Binding sites for the lectin are present in the Lipid A section of the LPS molecule, but may be present in other sections as well. The LPS molecule consists of three major sections: the O-antigen polysaccharide, the R-core, and the Lipid A backbone (Luderitz et al., 1971).

The pattern for recognition of compatible or incompatible bacteria that emerges from this work, then, involves the interaction of three molecular components: a) bacterial LPS, which may represent the binding site for the lectin; b) bacterial EPS, which interferes with binding; and c) host cell wall lectin. Because slight changes in the presence, chemical structure or steric accessibility of each of these components can affect binding to one another, the potential for variability is very high. Whether the EPS of virulent cells prevents agglutination by presenting a simple physical barrier or by competing in its free, soluble state for lectin binding sites, is the subject of current study.

Role of LPS in Induced Resistance

Attachment of bacteria to host cells and induction of disease resistance may be related phenomena. Infection of tobacco leaves by P. solanacearum can be totally prevented if leaves are previously infiltrated with heat-killed cells (Lozano & Sequeira, 1970), which attach to the cell wall in the same way as live cells (Sequeira et al., 1976). The induction of disease resistance or protection may involve, as a first step, the interaction between host cell wall lectin and the same components of the bacterial cell wall involved in recognition and attachment.

The protection inducer in crude preparations is highly stable at neutral or alkaline pH at temperatures up to $150^{o}C$ for 30 min. However, at avid pH, the inducer loses all of its activity within 15 min at $50^{o}C$. Although unaffected by proteases and nucleases, activity is totally destroyed by a mixture of β-glycosidases. These initial observations suggested that the inducer was probably carbohydrate in nature. Purified peptidoglycan from P. solanacearum does not induce protection (Wacek & Sequeira, 1973). Of the many gram-negative and gram-positive bacteria examined, only gram-negative species induced protection, suggesting that the inducer was possibly LPS.

LPS was isolated by a variety of methods from both compatible and incompatible strains of P. solanacearum as well as from smooth and rough forms of E. coli. LPS is frequently associated in these preparations with cell wall protein and a non-covalently linked lipid. Purified LPS preparations, free of protein and non-covalently bound lipid, are highly active as protection inducers at concentrations of about 50 µg/ml. Preparations from rough mutants, which lack the O-antigen polysaccharide, are also active. The remaining R-core: Lipid A region of the LPS thus seems responsible for activity. Mild, specific chemical hydrolyses indicate that the linkage between the R-core and Lipid A is necessary for activity, and that fatty acid esterification of Lipid A is required.

Since, as discussed above, LPS may also be involved in the attachment of incompatible bacteria to the tobacco mesophyll cell wall, it is possible that the same interaction that results in attachment of the bacteria is a first step in disease resistance.

References

Galanos, C., Luderitz, O. & Westphal, O. (1969). A new method for
     extraction of R lipopolysaccharides. European Journal of Biochemistry
     9, 245-249.
Lozano, J.C. & Sequeira, L. (1970). Prevention of the hypersensitive
     reaction in tobacco leaves by heat-killed bacterial cells. Phyto-
     pathotlogy 60, 875-879.
Luderitz, O., Westphal, O., Staub, A.M. & Nikaido, H. (1971). Isolation
     and chemical and immunological characterizations of bacterial lipo-
     polysaccharides. In Microbial Toxins, ed. by G. Weinbaum, S. Kadis
     & S.J. Ajl. Vol. 4, Academic Press, New York.
Marinkovitch, V.A. (1964). Purification and characterization of the
     hemagglutinin present in potatoes. Journal of Immunology 93,
     732-741.
Sequeira, L., Gaard, G. & de Zoten, G.A. (1976). Attachment of bacteria
     to host cell walls; its relation to mechanisms of induced resistance.
     (In press).
Wacek, T.J. & Sequeira, L. (1973). The peptidoglycan of Pseudomonas
     solanacearum: chemical composition and biological activity in relation
     to the hypersensitive reaction in tobacco. Physiological Plant
     Pathology 3, 363-369.

DISCUSSION

Chairman: J. Raa

Albersheim: I would like to ask you whether your protection using the
lipid A involved an induction of phytoalexin or do you have any information
on this?

Graham: We don't know at this time. The bacteria in the hypersensitive
reaction actually seemed to be killed but the problem is that you have both
the hypersensitive reaction and the protective response going on at the
same time. Peroxidase may be involved as a protective factor, but we really
don't know.

Albersheim: Are there phytoalexins known in the tobacco system?

Graham: I don't believe so.

Albersheim: Has anyone looked?

Graham: I don't believe so.

Brown: You seem to have a paradox here, if I understand correctly. The specificity appears to lie in the R region which is supposed to be buried in the membrane and you are still getting agglutination. You seem to get good correlation between isolates, virulent and avirulent, from a whole range of plants and you are testing them with only the potato lectin.

Graham: I don't believe that the R core is buried in the cell wall. There is a common antigen which is common to E. coli for instance. Different strains have common antigens which are based on the R core, so I believe that they are exposed at least enough to elicit an antibody.

Brown: What about potato lectins versus all these isolates?

Graham: Well, we haven't checked any of the other lectins. That would be a monumental task. But these results do fit the virulence or avirulence of these particular isolates on tobacco and that's the important thing.

Kijne: I would like to add to this last question of Dr. Brown that it seems pretty well established now that for E. coli as well as for certain pseudomonads the smooth LPS is made up of a mixture of smooth and rough LPS, so there may well be an exposure of rough LPS to the surface.

Lippincott: In our pictures of Agrobacterium infection we do see similar fibrillar materials produced and surrounding the bacteria and also we see very large haloes which have unusual shapes suggesting that the bacteria is able to digest part of this away. I thought previously, the hypersensitive reaction was believed to require bacterial attachment to induce this and now you are saying that it doesn't. Is that correct?

Graham: No! In fact we believe it does require attachment and it also seems to require it in the live cells, at least no one has been able to isolate the hypersensitive reaction inducer.

Kauss: Wheat germ agglutinin have been reported to agglutinate certain strains of bacteria without really interfering with growth. Do you know if that is true for yours too?

Graham: We have had the same problem as Bauer (this symposium) in trying to study the effect of the lectins on growth. Our lectin binds very tenaciously to glass surfaces and so we haven't been able to carry out the growth inhibition experiments yet.

Paxton: We have tried the soybean lectin on Phytophthora. Although we do see it bind indiscriminately to the hyphae we don't see any interference with growth. Of Course, Mirelman et al. (Nature 256, 414-416, 1975) did find inhibition of growth and that is why we checked our system. But it doesn't seem to work at least with the Phytophthora.

ULTRASTRUCTURAL EVIDENCE OF AN "ACTIVE" IMMOBILIZATION PROCESS OF
INCOMPATIBLE BACTERIA IN TOBACCO LEAF TISSUE: A RESISTANCE REACTION

R.N. Goodman, D.J. Politis and J.A. White, Department of Plant Pathology,
University of Missouri, Columbia, Missouri 65201, USA

Introduction

Previous studies have noted that the hypersensitive defense reaction
(HR) manifests itself in three ways: a cessation of bacterial proliferation
(Goodman, 1972), intense ultrastructural disorganization (Goodman & Plurad,
1971), and precise localization of the incompatible bacterial pathogen in
the intercellular space into which they are injected (Klement & Goodman,
1967).

For a period of time we considered that immobilization was due to the
entrapment of the injected bacteria between the mesophyll cells that
collapse as a consequence of HR (Klement & Goodman, 1967) (when the
inoculum concentration is 5×10^6 cells/ml or greater). However, a study
by Turner & Novacky (1974) revealed that injection of tobacco leaves with
the incompatible pathogen Pseudomonas pisi (a pathogen of pea) with inocula
of less than 5×10^6 cells/ml cause isolated mesophyll cell death without
the collapse of tissue. In fact they detected a 1:1 ratio between the
number of P. pisi cells introduced and the number of dead plant cells
present 6 h after inoculation. Since these inoculated bacteria were also
"localized", the collapsed tissue-hypothesis no longer appears tenable.

The immobilizing mechanism has remained obscure in ultrastructural
studies because inoculum levels used were too low. For example, an inoculum
dose of 10^8 cells/ml (our conventional experimental inoculum dose) injected
into tobacco leaf tissue places approximately 10^6 bacterial cells into a
leaf disk 1 cm in diameter. Since $\sim 4 \times 10^{11}$ bacteria can be accommodated
in the intercellular space of such a leaf disk, it is clear why a population
level of 10^6 cells/disk would rarely reveal the presence of bacteria

(Goodman & Plurad, 1971).

Methods

In the studies reported herein an inoculum dose of 10^9 cells/ml was used (this provides a population/disk 1 cm in diameter of 10^7 bacterial cells) in order to more clearly visualize the immobilization phenomenon. Since this inoculum level caused intense and rapid tissue collapse, a study was subsequently conducted using an inoculum of 10^7 cells/ml. Tobacco leaf tissue was examined electronmicroscopically 20 min., 2, 4 and 6 h after infiltration with 10^9 or 10^7 cells/ml of P. pisi. Inoculum-containing fluid usually equilibrated with the atmosphere in 30 minutes, and the bacteria consequently migrated to the cell wall surfaces bordering inter-cellular space (Figure 1). The methods used in these studies are described in detail elsewhere (Goodman, Huang & White, 1976).

Results

At 20 min. after infiltration we regularly observed a filamentous structure that appeared to be separating from the surface of parenchyma cells bordering intercellular space. This plant cell wall surface layer seemed to separate from the wall per se where bacteria were in close proximity (Figure 2). This wall surface layer which we have designated as wall cuticle has embedded in it fibrils, vesicles with an average dia-meter of 25 nm and an amorphous ground substance, probably cutin (Artz, 1933; Scott, 1950).

Two h after inoculation the wall cuticle has clearly separated from the mesophyll cell wall where bacteria are in close proximity. This may be a response to the bacterial cell surface materials per se (Figure 3). Separation of the wall cuticle causes the underlying fibrils to become raised giving the newly exposed wall surface a "fuzzy" appearance and dense-staining vesicles seem to be pulled out of the wall (Figure 4). The "lifting off" of the wall cuticle also seemed to permit additional dense-staining fibrils and vesicles to collect at the surface and as a consequence, the deeper layers of the cell wall stained more intensely. The apparent increased permeability of the wall to heavy metal stains reflects either a structural modification and/or a biochemical alteration of the wall. This is consistent with our findings concerning the nature of HR-provoked rapid plasmamembrane disintegration (Huang & Goodman, 1972). Close proximity of

bacteria to the plant cell surface also causes convolution of the plasmalemma. This is observed as early as 2 h after inoculation (Figure 5). The
saprophyte P. fluorescens also induces this effect; however, it occurs 4 h
later and other organelles are apparently unaffected (Figure 11).

Four hours after inoculation the detached wall cuticle has become
thicker with vesicles, fibrils and membrane fragments contributing to its
complexity (Figures 7 and 8). In addition, at positions opposite "immobilized" bacteria, the plasmalemma, which was earlier merely convoluted, has
now vesiculated and the cell wall is markedly broadened by a loose array of
fibrils and vesicles (Figure 6).

At 6 h after inoculation HR-evoked intense tissue collapse is apparent.
Some bacteria are trapped between collapsed mesophyll cells whereas others
are immobilized by an extremely complex wall cuticle (Figures 9 and 10).

When 10^9 cells/ml of a saprophytic species, P. fluorescens, was
injected into tobacco, there was very little evidence of wall cuticle
development until 6 h after inoculation. However, at 6 h a very thin wall
cuticle, with a minimum number of vesicles does develop and clearly localizes
the P. fluorescens cells (Figure 11). In addition, this localization occurs
without any signs of HR-evoked plant cell collapse. Apparently even saprophytic bacteria induce a mild response from plant cells which is sufficient
to cause their immobilization.

Four to six hours after inoculation of tobacco with either 10^7 or 10^9
cells, the compatible pathogen, P. tabaci, a weak wall localizing process
develops (Figure 12). However, the enclosure appears to be largely a
loose array of fibrils, without vesicles, that fails to envelop the bacteria
completely. It is also known that under these conditions the bacteria
continue to multiply rapidly and spread to uninoculated tissue.

When an inoculum dose of 10^7 cells/ml was injected into tobacco leaves,
the wall modification and tissue collapse (HR) reactions were significantly
slowed. As a consequence, wall thickening developed more prominently than
it had with an inoculum of 10^9 cells/ml, presumably because the leaf cells
remained functional for a longer period of time. The fibrillar material
appeared as a prominent feature as early as 4 h after infiltration, forming
an elliptically shaped structure that we have designated as "wall apposition". These areas of wall apposition also characteristically invaginate
the plasmalemma at points opposite the localized bacteria (Figures 6, 8 and
13). Six hours after infiltration the wall apposition may assume massive

proportions, but is still exclusively opposite the bacteria that have been enveloped by the complex wall cuticle.

Discussion

At both 4 and 6 hours after infiltration parallel profiles of dilated ER and numerous vesicles are associated with the wall apposition. Although ER may be contributing to the production of some vesicles, most are probably derived from the plasmalemma. It is also possible that there are in fact two types of vesicles, each carrying diverse substances. The former may be carrying hydrolases (Figures 7 and 8) (Matile, 1975), whereas the latter may be bearing cell wall material as shown in Figure 14. The vesicles shown in Figure 14 are very evidently double-track membrane-bound and frequently they can be found embedded within the wall apposition. It seems logical to suggest that they discharge their contents contributing thereby to the growing region of wall apposition. Willison & Cocking (1975) have recently presented convincing evidence that wall microfibril assembly occurs within the outer half of the unit membrane or within a thin coating of its outer surface.

The flattened membrane embedded deeply in the regions of wall apposition appears to further fragment between 4 and 6 hours (Figures 8 and 10). Our electron micrographs suggest that these membrane fragments migrate through the cell wall per se. They can be seen issuing from the wall as small (25-30 nm) vesicles where the cuticle (Figures 6 and 10) has pulled away in response to the presence of bacteria.

It is also tempting to speculate on the role these minute vesicles play in localizing the bacteria. Their size is almost identical to vesicles observed in apple xylem vessels that agglutinate the avirulent strain of Erwinia amylovora in vivo (Huang et al., 1975). They may also be related to the agglutinating factor observed by Goodman, Huang & Thaipanich (1974) in P. pisi infiltrated tobacco leaves. This factor would agglutinate P. pisi in vitro but not the compatible pathogen, P. tabaci. It would not be surprising if these small vesicles were glycoprotein-containing and had lectin-like qualities.

Summary

Infiltration of tobacco leaf tissue with either 10^7 or 10^9 cells/ml of the incompatible bacterial pathogen, P. pisi, very rapidly causes extensive

Goodman et al. - Immobilization of incompatible bacteria

parenchyma cell wall modification prior to the development of HR-induced microorganellar disorganization and cell collapse. Specifically, the cell wall surface cuticle peels off within 20 minutes in regions that are in close proximity to bacteria. Within 2 h the plasmalemma becomes convoluted and then vesiculates between 4 and 6 h after infiltration. These vesicles of various sizes appear to contribute to the formation of an elliptical mass of fibrillar material which we have designated as an area of wall apposition. All of this activity is confined to the region wall that is near or in contact with one or more bacterial cells. These initially very localized responses will at 6 h after infiltration result in general ultrastructural destruction and cell collapse.

The numerous vesicles of plasmalemma origin appear to fragment still further as they pass through the cell wall. Once in intercellular space they become integrated into the wall cuticle and form a complex immobilizing envelope around the bacteria. This process is regarded as an active resistance reaction evoked by the presence of bacteria in close proximity to or in contact with the plant cell wall surfaces.

Acknowledgements

University of Missouri Agricultural Experiment Station Journal Series Paper Number 7669. Supported in part by National Science Foundation Grant BMS-19432.

Legend to figures

Abbreviations for electronmicrograps:
B = bacteria; BL = blebs; BC = bacterial capsule; C = chloroplast; CPL = convoluted plasmalemma; CW = cell wall; DER = dilated endoplasmic reticulum; DSF = dense-staining fibrils; F = fibrils; HRT = hypersensitive reaction tissue; IS = intercellular space; M = mitochondrion; MF = membrane fragments; N = nucleus; Nu = nucleolus; PL = plasmalemma; T = tonoplast; V = vesicle; Va = vacuole; VPL = vesiculated plasmalemma; WA = wall apposition; WC = wall cuticle.

Figure 1.   Control, tobacco leaf tissue 6 h after infiltration with buffer.
            All organelles are normal, arrows point to smooth wall surfaces
            (X 13,500).

Figure 2.   Twenty minutes after inoculation with 10^9 Pseudomonas pisi
            cells/ml showing wall cuticle containing fibrils, vesicles, and
            ground substance (probably cutin). Note fibrillar surface where
            cuticle has separated from wall (arrows) (X 36,400).

Figure 3.   A bacterial cell in close proximity to a plant cell wall 2 h
            after inoculation (note bacterial capsular fibrils in contact
            with the plant wall) (X 93,000).

Figure 4.   Wall cuticle folding back from the wall with adhering vesicles
            and fibrils and underlying dense-staining fibrils (X 53,500).

Figure 5.   Clear evidence of convoluted plasmalemma at a point opposite
            bacterial cells, 2 h after infiltration with 10^9 P. pisi cells/ml
            (X 67,500).

Figure 6.   After inoculation with 10^9 P. pisi cells/ml, at 4 h the plasma-
            lemma has vesiculated. Note area of wall apposition and vesicles
            emerging from wall surface (X 67,500).

Figure 7.   Cells of P. pisi completely ensheathed by wall cuticle 4 h after
            inoculation with 10^9 P. pisi cells/ml (X 9,000).

Figure 8.   Enlargement of a portion of Figure 7 showing wall apposition,
            vesiculated plasmalemma and movement of membrane fragments
            (probably from plasmalemma) through cell wall into wall
            cuticle (X 45,000).

Figure 9.   Groups of bacteria trapped between spongy parenchyma cells as
            a consequence of HR and other bacteria completely enveloped by
            wall cuticle 6 h after inoculation with 10^9 P. pisi cells/ml
            (X 10,900).

Figure 10. Enlargement of a portion of Figure 9 showing movement of vesicles
            through the plant cell wall and other vesicles, fibrils and
            membrane fragments integrated into the wall cuticle.

Figure 11. Localization of a saprophytic bacterium by a minimal wall
            cuticle 6 h after infiltration with 10^9 P. fluorescens cells/ml.
            Note the very few vesicles in the wall cuticle, the convoluted
            plasmalemma and the dilated endoplasmic reticulum (X 42,800).

Figure 12. A juncture between tobacco leaf cells 6 h after inoculation
            with 10^9 cells/ml of the compatible pathogen P. tabaci. The

wall cuticle is incomplete, consisting of fibrils only. Note the absence of vesicles (X 42,800).

Figure 13. Two bacterial cells trapped between two plant cell walls 6 h after inoculation with 10^7 P. pisi cells/ml. Note wall apposition and vesiculated plasmalemma on either side of the bacteria (X 7,000).

Figure 14. An enlargement of a portion of Figure 13. Note wall apposition and vesiculated plasmalemma. Some vesicles contain fibrils similar to those in the region of wall apposition (X 40,000).

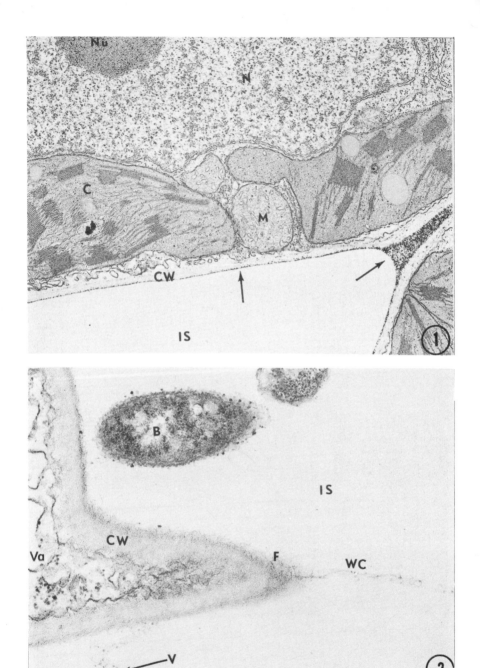

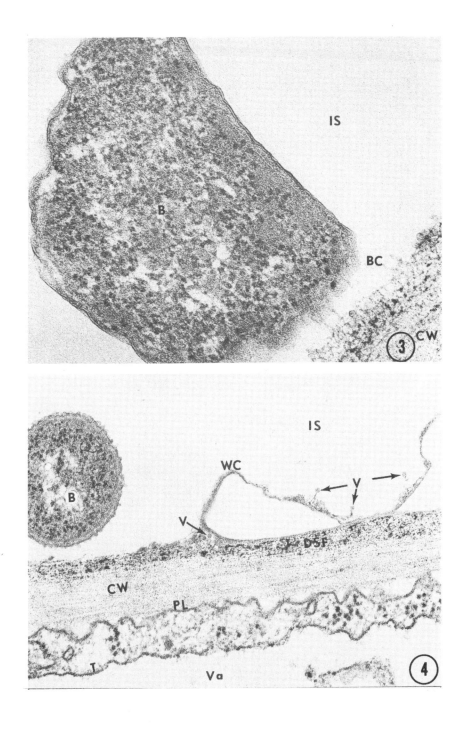

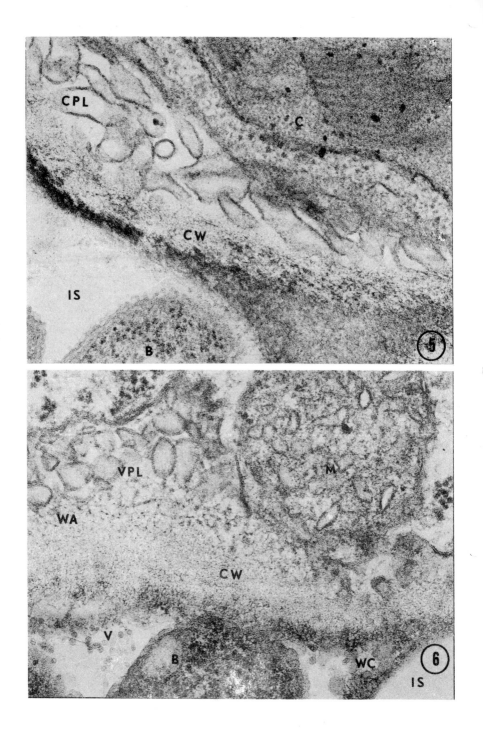

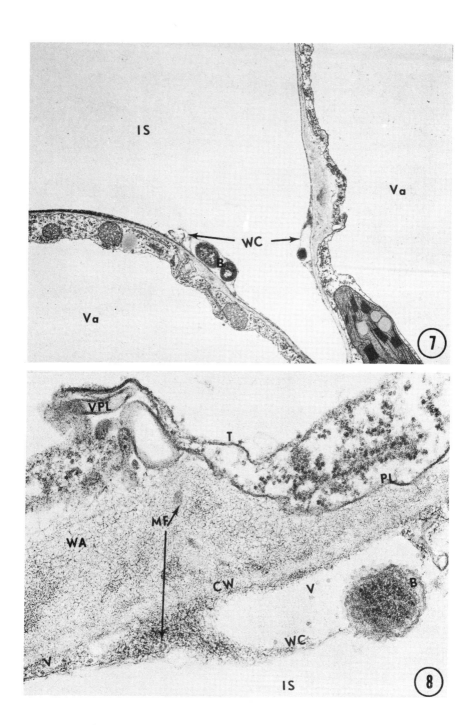

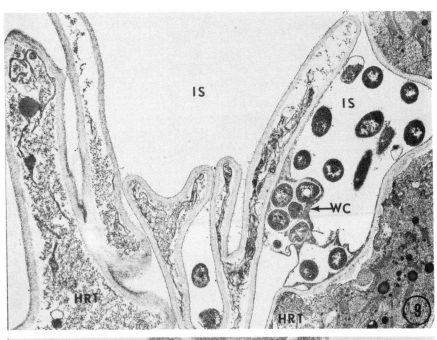

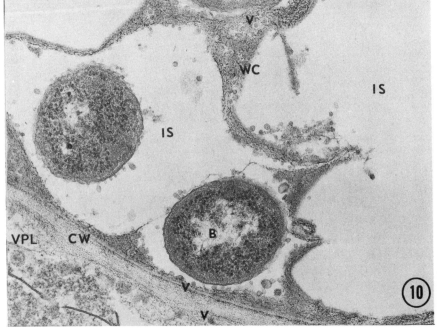

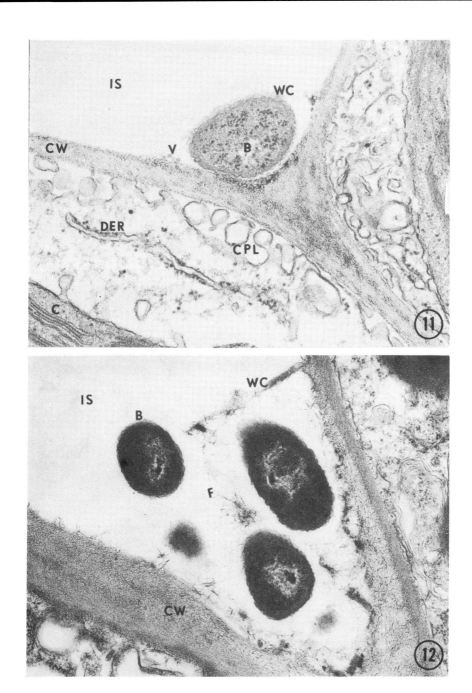

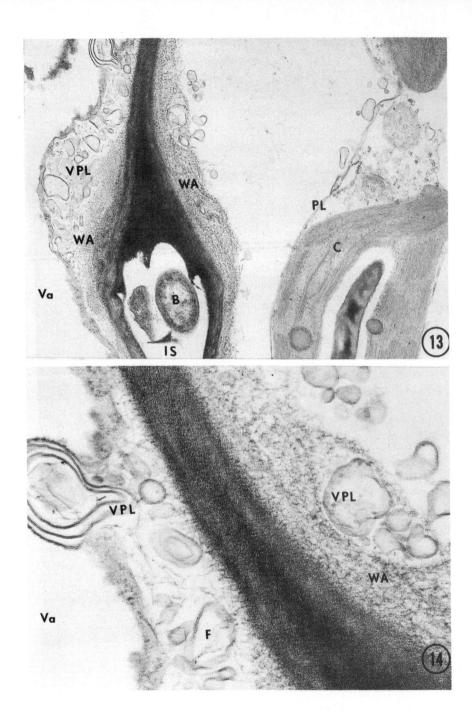

Goodman et al. - Immobilization of incompatible bacteria

References

Artz, T. (1933). Untersuchungen über die Vorkommen einer Kutikula in den Blattern dikotyler Pflanzen. Berichte der Deutschen Botanischen Gesellschaft 51, 470-500.

Goodman, R.N. (1972). Electrolyte leakage and membrane damage in relation to bacterial population, pH and ammonia production in tobacco leaf tissue inoculated with Pseudomonas pisi. Phytopathology 62, 1327-1331.

Goodman, R.N., Huang, P.Y. & Thaipanich, V. (1976). Induced resistance to bacterial infection. pp. 35-42 in Biochemistry and Cytology of Plant Parasite Interaction, ed. by K. Tomiyama et al. Kodansha LTD, Tokyo.

Goodman, R.N., Huang, P.Y. & White, J.A. (1976). Ultrastructural evidence for immobilization of an incompatible bacterium, Pseudomonas pisi, in tobacco leaf tissue. Phytopathology (in press).

Goodman, R.N. & Plurad, S.B. (1971). Ultrastructural changes in tobacco undergoing the hypersensitive reaction caused by plant pathogenic bacteria. Physiological Plant Pathology 1, 11-15.

Huang, J.S. & Goodman, R.N. (1972). Alterations in structural proteins from chloroplast membranes of bacterially induced hypersensitive tobacco leaves. Phytopathology 62, 1428-1434.

Huang, P.Y., Huang, J.S. & Goodman, R.N. (1975). Resistance mechanisms of apple shoots to an avirulent strain of Erwinia amylovora. Physiological Plant Pathology 6, 283-287.

Klement, Z. & Goodman, R.N. (1967). The hypersensitive reaction to infection by bacterial plant pathogens. Annual Review of Phytopathology 5, 17-44.

Matile, P. (1975). The lytic compartment of plant cells (see page 57). Springer-Verlag.

Scott, F.M. (1950). Internal suberization of tissues. Botanical Gazette 111, 459-475.

Turner, J.G. & Novacky, A. (1974). The quantitative relation between plant and bacterial cells involved in the hypersensitive reaction. Phytopathology 64, 885-890.

Willison, J.H.M. & Cocking, E.C. (1975). Microfibril synthesis on the surfaces of isolated tobacco mesophyll protoplasts, a freeze-etch study. Protoplasma 84, 147-159.

Goodman et al. - Immobilization of incompatible bacteria

DISCUSSION

Chairman: J. Raa

Kijne: I saw a lot of lomosomes present in pea root nodules. I never saw
any contents in the vesicles present in these lomosomes, nor did I have any
idea where they came from. I think the vesicles you showed and suggested
had some granular materials in them were merely vesicles which were
superficially cut so that you were looking for the membrane and not at
the content. Have you any idea if the presence of lomosomes is really
established now as having something to do with cell root formation?

Goodman: I think that if you read the June article in Phytopathology you
will see that I have indicated that this may well be the contorted plasmalemma or the situation of plasmalemma per se. However, I don't think that
the distance between vesicles uphold that sort of a contention any longer.
I believe that the pictures that we have here suggest very clearly, to me
anyhow, that these are vesicles that have originated in the cell and I now
think are carrying something to the wall. It remains for a histochemical
technique to identify those substances in the vesicles. At the present
time I can't tell what's in them. But that's what it looks like to me.

NATURE AND SPECIFICITY OF THE BACTERIUM-HOST ATTACHMENT IN AGROBACTERIUM
INFECTION

James A. Lippincott and Barbara B. Lippincott, Department of Biological
Sciences, Northwestern University, Evanston, Illinois 60201, USA

Introduction

Crown-gall is a tumorous disease of dicots and gymnosperms caused by
Agrobacterium tumefaciens. Certain variants of this species are the causal
agents of the cane gall and hairy root diseases. The agrobacteria are
classified in the family Rhizobiaceae and DNA hybridization studies
indicate that the rhizobia and agrobacteria are closely related (Kersters,
De Ley, Sneath & Sackin, 1973).

The ability of certain avirulent strains of Agrobacterium to inhibit
tumor initiation when inoculated with tumorigenic bacteria led to the
demonstration of a site attachment step in the crown-gall tumor induction
process (Lippincott & Lippincott, 1969; Kerr, 1969; Manigault, 1970).
Thus, some part of the host wound that is necessary to establish these
infections provides a site to which the bacterium must adhere before a
tumor can be induced. Data indicate that this site exhibits specificity
for agrobacteria and that in the small carborundum-induced bean leaf wounds
used in these experiments these sites have to be few in number and of the
size of the bacterium or smaller (Lippincott & Lippincott, 1969, 1975).

To elucidate the mechanism of crown-gall induction, therefore, it was
essential to determine the nature of both the bacterial components and
host plant components of this attachment process. The following results
provide the first evidence as to the nature of the bacterium-host attachment in Agrobacterium infections.

Methods

A quantitative infectivity assay (Lippincott & Heberlein, 1965)

employing the primary leaves of 7 day old pinto bean seedlings has been the
primary experimental tool in these investigations. The leaves are sprinkled
with ≠400 grit (0.037 in diameter) carborundum powder and 0.1 ml of bacteria
(ca. 5 x 10^7 organisms) added and softly rubbed over the surface of the
leaf. A week later the tumors are counted with the aid of a dissecting
microscope. Tumor number in this assay is proportional to bacterial concentration over a 100-fold range. Using 16 leaves per test sample, standard
errors are typically less than $\pm$ 20% of the mean of the number of tumors
obtained. The ability of isolated bacterial components to compete with
tumorigenic bacteria for infection sites, thereby reducing the number of
tumors induced, has been the principal means to follow the isolation of
the bacterial site attachment component. In the same manner, a reduction in
tumor number obtained by inoculating bacteria with isolated host components
provides an estimate of the ability of these materials to compete with the
natural wound site for the bacteria.

Bacterial component of site attachment

A priori it was anticipated that some portion of the Agrobacterium cell
envelope might contain the bacterial components of attachment since
attachment of virulent cells is complete within 15 min after inoculation
on bean leaves (Lippincott & Lippincott, 1969). Bacterial cell envelope
preparations were found to effectively inhibit tumor initiation when added
before or with virulent bacteria but had no effect when added 15 min after
the bacteria (Whatley, Bodwin, Lippincott & Lippincott, 1976). Similarly,
Boivin antigen, a protein-lipopolysaccharide complex from the outer
membrane of the Agrobacterium cell wall and the isolated lipopolysaccharide
(LPS) both gave results in infectivity tests comparable to those obtained
with whole cells or cell envelope preparations. The localization of LPS in
the outer membrane of the cell envelope of gram-negative bacteria and its
role in determining antigenic specificity of gram-negative bacteria in
general (Costerton, Ingram & Cheng, 1974) is consistent with the proposed
role for Agrobacterium LPS. The LPS is quite inhibitory, 1 ng/ml being
sufficient to reduce the number of tumors induced by one-third. After acid
hydrolysis of LPS and separation of the polysaccharide and lipid A components only the polysaccharide moiety shows inhibitory activity. LPS
isolated from non-virulent strains of Agrobacterium which do not show site
competition has no effect on tumor initiation, indicating a specificity

for LPS comparable to that observed with whole cells.

Host component of attachment

The most probable host component of the attachment process appeared to be either a portion of a plant cell wall exposed by wounding or an exposed portion of the cell membrane. Both membrane fractions and cell walls have been isolated from primary bean leaves of optimum sensitivity to Agrobacterium. Cell membrane fractions inoculated with virulent bacteria, the former at concentrations of 14 to 27 mg dry weight per ml, have no significant effect on tumor induction. Cell walls isolated by the procedure of Nevins, English & Albersheim (1967), however, reduced the number of tumors in these tests, 10 mg dry weight per ml typically reducing tumor number by 60 to 90%. A cell wall concentration of 0.1 mg/ml is sufficient in most tests to reduce tumor number by 30% or more. As shown in Table 1, these cell

Table 1. Effect of cell walls isolated from bean leaves and their time of application on tumor initiation by A. tumefaciens strains B6.

Materials inoculated[a]		No. of leaves	Mean no. of tumors per leaf	% control
With wounding	15 min later			
B6	None	16	14.3	100
B6 + CW	None	16	3.4	24
B6	CW	16	15.8	111
CW	B6	16	1.8	12

[a]Concentrations applied: B6, ca. 9×10^8 viable cells/ml; cell walls, 10 mg/ml.

walls are inhibitory when added before or with the tumorigenic bacteria but when added 15 min after the bacteria they are non-inhibitory, consistent with their proposed ability to inhibit by competing with wound sites for available bacteria. Since this apparent attachment to isolated cell walls leads to a decrease in tumor induction, it is clear that attachment per se is not sufficient to result in tumor formation, but rather, effective attachment must require a site localized on a susceptible host cell.

To demonstrate that the inhibitory effect of the plant cell walls was not due to a simple physical blockage of wound sites, the cell walls were treated with whole cells or LPS from avirulent site-binding strains (e.g. IIBNV6) for 15 min. The cell walls were then removed by low speed centrifugation, washed once, and tested for their ability to inhibit tumor initiation. Table 2 summarizes results from tests of this nature using both virulent and avirulent bacteria. Both cells and LPS of strains of Agrobacterium which compete with strains B6, as shown by a reduction in tumor number

Table 2. Ability of whole cells and LPS of various strains of Agrobacterium to neutralize the inhibitory activity of Pinto bean leaf cell walls when inoculated with strain B6.

Bacterial strain	Virulence	Compete with B6 for tumor sites in vivo	Neutralization effectiveness	
			Whole cells	LPS
A. radiobacter S1005	-	-	-	-
A. radiobacter 6467	-	-	-	-
A. radiobacter TR1	-	-	-	-
A. tumefaciens Ag19	-	+		+
A. tumefaciens NT1	-	+	+	+
A. tumefaciens IIBNV6	-	+	+	+
A. tumefaciens 15955	+	+	+ (CEP)	
A. tumefaciens C-58-3	+	+		+
A. tumefaciens B6	+	+	+	+
R. leguminosarum C56	?	+		+

CEP = cell envelope preparation

when the two strains are inoculated together, were active in neutralizing the effectiveness of the bean leaf cell walls. The cells and LPS from three strains of Agrobacterium which were not effective in the in vivo competition, however, had no effect on the activity of the plant cell walls. Rhizobium leguminosarum strains C56 LPS behaves like the LPS from virulent agrobacteria, suggesting this strain has the ability to adhere to similar attachment sites.

Nature of the host cell wall inhibitory activity

Treatment of the bean leaf cell walls with 0.05 M H_2SO_4 for 1.5 hr at $100^\circ C$ to remove pectin and hemicellulose (Montague & Ikuma, 1975) largely neutralizes their inhibitory effect. Extraction of the cell walls with EDTA + Triton X-100 (Kauss & Glaser, 1974) or sodium deoxycholate (Selvendran, 1975), however, has relatively little effect on their activity. Treatment of the cell walls with pectinase, meicelase, macerase or cellulase did not significantly alter their inhibitory activity.

Several natural products of cell wall origin (Table 3) have been tested for their effect on tumor initiation to obtain some evidence concerning the possible nature of the specific component of these cell wall preparations involved in the attachment process. Plant lectins were notably without

Table 3. Effect of various commercial products on tumor induction by A. tumefaciens strain B6 (substances tested at 10 mg/ml).

No effect on tumor induction		Inhibit tumor induction
Homogenized filter paper	Lecithin	Pectin
Cellex	Polyethylene glycol	Polygalacturonic acid
Galactan	Concanavalin A	Sodium polygalactoronate
Galacturonic acid	Wheat germ lectin	Arabinogalactan
Galactose	Kidney bean lectin	
Gum arabic		
Locust bean gum		

effect, as were cellulose and galactan. Four products were found to be inhibitory, polygalacturonic acid, sodium polygalacturonate, pectin and arabinogalactan. At a concentration of 10 mg/ml, pectin typically inhibited tumor initiation about 45%, whereas the three other compounds usually reduced tumor number by 90% or more at this concentration. Since pectin and polygalacturonic acid are common components of the middle lamella, the activity of these latter compounds provides a possible clue to the activity of the isolated cell walls. Figure 1 shows a titration of sodium poly-galacturonate and of bean leaf cell walls on tumor initiation. At high concentrations of each, the polygalacturonate is minimally about 10-fold more effective than the cell walls on a dry weight basis. The cell walls, however, lose effectiveness below 10 µg/ml but the polygalacturonate shows some inhibitory effect at 1 ng/ml and below.

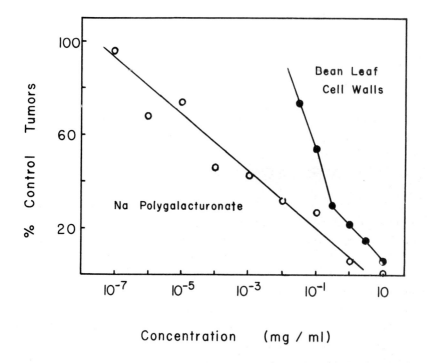

Figure 1. Titration of the inhibitory action of sodium polygalacturonate
and bean leaf cell walls on tumor initiation by A. tumefaciens
strain B6.

Inhibitory activity of cell walls from different sources

Cell walls from the different parts of 7 day old pinto bean seedlings were isolated and all proved inhibitory in tests with A. tumefaciens strain B6. Treatment of the cell walls from each of these sources with IIBNV6 significantly decreases their ability to inhibit B6 tumor initiation, as anticipated. Cell walls isolated from 3 day old pinto bean embryos (1-3 cm length), however, do not inhibit tumor initiation by strain B6 and treatment of these cell walls with strain IIBNV6 makes them slightly inhibitory, probably due to incomplete removal of the IIBNV6, which itself is a competitor for tumor initiation sites. Embryonic bean cell walls thus appear to lack components to which Agrobacterium can attach. These data suggest that. between day 3 and day 7 in the development of bean seedlings, the attachment component is either synthesized or modified to exhibit the characteristic of site attachment found in the older plant. This may explain the fact that meristematic tissues in general do not appear to respond to A. tumefaciens.

We have also compared cell walls from several species of monocots and dicots for their ability to inhibit tumor initiation and for their activity after pretreatment with strain IIBNV6. Figure 2 shows that monocot cell walls, with the exception of three varieties of barley, give little or no inhibition of tumor number when inoculated with strain B6. Treatment of these cell walls with IIBNV6 in every case increased their inhibitory activity. The cell walls from seven species of dicots, however, were all inhibitory and pretreatment with IIBNV6 greatly reduced their inhibitory activity. The monocot cell walls thus appear to lack Agrobacterium specific attachment sites and to differ significantly from dicots in general in this respect. These results may in part account for the curious ability of Agrobacterium to infect dicots and gymnosperms but not monocots.

Cell walls isolated from comparable normal and crown-gall tumor tissues were also examined for the presence of Agrobacterium binding activity. Among cultured tissues, only those from normal sources (Figure 3) yielded cell walls which were inhibitory and this activity was largely neutralized by pretreatment with IIBNV6. Tumor cell walls from three species of plant induced by five different strains of Agrobacterium were all non-inhibitory and IIBNV6 pretreatment made them somewhat inhibitory. Comparable results were obtained with cell walls from in vivo normal and tumor tissue on four species of plant. Thus, crown-gall tumors in vivo and

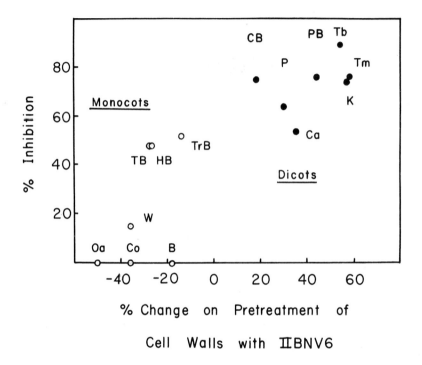

Figure 2.   Effect of cell walls isolated by the method of Nevins, English
            & Albersheim (1967) from monocots and dicots on tumor induction
            by A. tumefaciens strain B6. Ordinate: % inhibition by the cell
            walls; abscissa: difference between the % of control tumor
            initiation obtained with IIBNV6 pretreated cell walls and the %
            control tumors obtained with untreated cell walls. Symbols:
            open circles = monocots, closed circles = dicots. Abbreviations:
            CB = castor bean, P = peas, Ca = carrots, PB = pinto beans,
            K = Kalanchoe leaves, Tb = tobacco, Tm = tomato, Oa = oats,
            Co = corn, W = wheat, B = barley, Tb = trail barley, Hb =
            Himalaya barley, TrB = trophy barley.

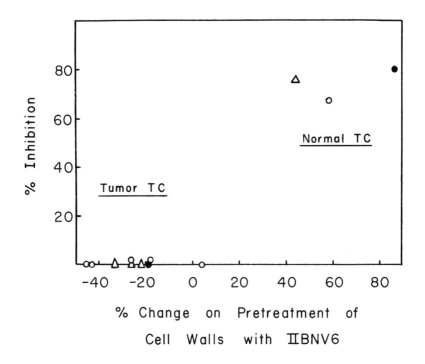

Figure 3. Comparison of cell walls isolated from normal and crown-gall
tumor tissue cultures for effects on tumor initiation by
A. tumefaciens strain B6. Symbols: open circles = Vincea
rosea, closed circles = tobacco, triangles = pinto bean.
Each separate symbol for a host species represents a separate
tumor tissue culture line induced by a different strain of
A. tumefaciens. Ordinate and abscissa as in Figure 2.

in culture have cell walls which lack the Agrobacterium attachment site, as do the cell walls of embryonic bean tissues. Regardless of the nature of this difference, it is clear that this is a persistent change maintained in tissue culture over years of subculturing. Determination of the exact nature of this change should provice a new biochemical test for crown-gall.

Pectinesterase treatment of cell walls

Based on data obtained with pectin and polygalacturonic acid, the possibility was raised that galacturonic acid residues in polysaccharides of the middle lamella might on exposure by wounding constitute the Agro-bacterium attachment site. The apparent correlation between turnover of pectin methyl groups and auxin-induced growth observed many years earlier (Ordin, Cleland & Bonner, 1955; Albersheim & Bonner, 1959) and the differences between embryonic and tumor cell walls from more mature normal cell walls also suggested that methylation of pectic compounds might account for these differences. If this were the case, treatment of these cell walls with pectinesterase to remove methyl esters should make them inhibitory. Table 4 shows results from three experiments which demonstrate that pectinesterase treatment of cell walls of corn, of cultured bean leaf tumor tissue and of embryonic bean tissue converts these cell walls to an inhibitory state. The simplest direct conclusion from these data is that the cell walls of monocots, tumors and embryonic bean tissues are sheathed with pectic substances which are sufficiently methylated that Agrobacterium does not adhere.

Summary and conclusions

The site attachment step essential in the induction of crown-gall tumors by Agrobacterium tumefaciens appears to depend on an interaction between an external cell envelope lipopolysaccharide of the bacterium and a part of a host cell wall exposed by a wound. An external portion of the host plant cell wall, most probably in middle lamella origin and consisting in part if not entirely of polygalacturonic acid appears to constitute the host site. A host site of this type which is probably common to dicots and gymnosperms is consistent with the very broad host range of the agrobacteria. Because the polysaccharide portion of the bacterial LPS effectively blocks tumor initiation, we conclude the adherence

Lippincott & Lippincott - Specificity of Agrobacterium infections

Table 4. Effect of pectinesterase[a] on the ability of certain cell walls to inhibit tumor induction by strain B6.

Materials inoculated with strain B6	No. of leaves	Mean no. of tumors per leaf	% control tumors
Expt. 1: None	16	162	100
Corn CW	16	197	122
", 0 hr enzyme treated	16	208	129
", 4 hr enzyme treated	16	63	39
Enzyme at 10^{-1} dilution	16	151	94
Expt. 2: None	12	76.2	100
Corn CW	14	47.9	63
", 4 hr enzyme treated	16	10.1	13
Pinto leaf tumor CW	14	70.1	92
", 4 hr enzyme treated	16	9.1	12
Expt. 3: None	14	15.4	100
Embryonic epicotyl CW	16	17.5	114
", 0 hr enzyme treated	14	17.7	115
", 4 hr enzyme treated	16	77.2	47

[a]Orange peel pectinesterase (Sigma Chemical Co.) at 2 mg protein and 100 units/ml.

of the bacterium to the host initiates from interactions between bacterial and plant polysaccharides. The apparent inability of agrobacteria to adhere to monocot cell walls or to "meristematic" cell walls correlates with the lack of susceptibility of these tissues to Agrobacterium in vivo. More work remains before these conclusions and their significance are unequivocally established.

Acknowledgements

These investigations were supported by Grant No. AI12149 awarded by the National Institute of Allergy and Infectious Diseases, DHEW, and by Grant No. CA05387 awarded by the National Cancer Institute, DHEW.

References

Albersheim, P. & Bonner, J. (1959). Metabolism and hormonal control of
    pectic substances. Journal of Biological Chemistry 234, 3105-3108.

Costerton, J.W., Ingram, J.M. & Cheng, K.-J. (1974). Structure and
    function of the cell envelope of gram-negative bacteria. Bacterio-
    logical Reviews 38, 87-110.

Kauss, H. & Glaser, C. (1974). Carbohydrate-binding proteins from plant
    cell walls and their possible involvement in extension growth.
    FEBS-Letters 45, 304-307.

Kerr, A. (1969). Crown gall of stone fruit. I. Isolation of Agrobacterium
    tumefaciens and related species. Australian Journal of Biological
    Sciences 22, 111-116.

Kersters, K., De Ley, J., Sneath, P.H.A. & Sackin, M. (1973). Numerical
    taxonomic analysis of Agrobacterium. Journal of General Microbiology
    78, 227-239.

Lippincott, B.B. & Lippincott, J.A. (1969). Bacterial attachment to a
    specific wound site as an essential stage in tumor initiation by
    Agrobacterium tumefaciens. Journal of Bacteriology 97, 620-628.

Lippincott, J.A. & Heberlein, G.T. (1965). The quantitative determination
    of the infectivity of Agrobacterium tumefaciens. American Journal of
    Botany 52, 856-863.

Lippincott, J.A. & Lippincott, B.B. (1975). The genus Agrobacterium and
    plant tumorigenesis. Annual Review of Microbiology 29, 377-405.

Manigault, P. (1970). Intervention dans la plaie d'inoculation de bactéries
    appartenant a différentes souches d'Agrobacterium tumefaciens (Smith
    et Town.) Conn. Annales de l'Institut Pasteur 119, 347-359.

Montague, M.J. & Ikuma, H. (1975). Regulation of cell wall synthesis in
    Avena stem segments by gibberellic acid. Plant Physiology 55,
    1043-1047.

Nevins, D.J., English, P.D. & Albersheim, P. (1967). The specific nature
    of plant cell wall polysaccharides. Plant Physiology 42, 900-906.

Ordin, L., Cleland, R. & Bonner, J. (1955). Influence of auxin on cell
    wall metabolism. Proceedings of the national Academy of Science
    (U.S.A.) 41, 1023-1029.

Selvendran, R.R. (1975). Analysis of cell wall material from plant
    tissues: extraction and purification. Phytochemistry 14, 1011-1017

Whatley, M.H., Bodwin, J.S., Lippincott, B.B. & Lippincott, J.A. (1976). Role for Agrobacterium cell envelope lipopolysaccharide in infection site attachment. Infection and Immunity 13, 1080-1083.

DEVELOPMENTAL CHANGES IN MOSS INDUCED BY ATTACHMENT OF AGROBACTERIA

Luretta D. Spiess, Lake Forest College, Lake Forest, Illinois 60045, USA

The moss plant may seem unusual for studies on host-pathogen interactions and related cell wall biochemistry, but our studies on the effects
of Agrobacterium on moss development show that bacterial-moss cell wall
interactions occur that are correlated with pathogenicity. The moss system
is excellent for quantitative studies because of the small size of the
plant (30 to 40 may be grown in one dish), it goes through distinct stages
in its development from spore to protonema to gametophore and sporophyte,
and hormonal effects on this development are well documented. The moss cell
wall is comparatively thick, yet wounding is not necessary for Agrobacterium
or Rhizobium to elicit developmental responses although adherence of the
bacteria to the moss is essential. Different strains of bacteria induce
characteristic developmental responses on the moss which vary from abnormal
to normal. As all species of moss do not respond to these bacteria,
comparative studies are possible.

    Working with the protonemal stage of the moss Pylaisiella selwynii, we
have investigated the effect of various strains of Agrobacterium and
Rhizobium on development (Spiess, 1975, 1976; Spiess et al., 1971, 1972,
1973, 1976). Only those A. tumefaciens strains which induce crown-gall
tumors on higher plants are effective in inducing a major increase in bud
formation and subsequently development into normal or abnormal gametophores
or callus. Non-pathogenic strains of A. tumefaciens and A. radiobacter have
little effect on moss. A. rubi, which induces crown-gall on higher plants,
induces copious callus and very abnormal gametophores on moss.
A. rhizogenes, the agent of the hairy root disease, induces protonemal
filaments and buds that grow out as rhizoid-like filaments. Five species
of Rhizobium all induce normal gametophore development.
    The induction of bud formation and their subsequent development

Spiess - Changes in moss induced by Agrobacteria

requires physical attachment of the bacteria with the moss (Spiess et al., 1976). Using parabiotic chambers in which moss was separated from A. tumefaciens strains B6 or Rhizobium leguminosarum strain C56 by a 0.22 μm Millipore filter, it was shown that only when bacteria were in the same chamber with moss were gametophores induced. Heat killed strain B6, which has been shown to compete with viable B6 for sites in higher plant wounds (Lippincott & Lippincott, 1969), when combined with viable B6 on moss reduced the effectiveness of viable B6 in inducing gametophores. Adding the heat-killed B6 at six or more hours after the viable bacteria, had no effect (Spiess et al., 1976). Non-site binding strains of A. radiobacter added with, before, or after A. tumefaciens had no effect on the action of the A. tumefaciens on the moss. Lipopolysaccharide (LPS) obtained from virulent A. tumefaciens was as effective as heat-killed bacteria in reducing the number of gametophores induced by B6 but LPS from non-site binding strains of A. radiobacter had no effect on this induction. These results parallel and corroborate those on the effects of LPS on site attachment in higher plants (Whatley et al., 1976).

In an attempt to visualize the bacterium-moss attachment, scanning electron microscopy (SEM) of the moss-bacterium complex was undertaken in collaboration with Dr. Jocelyn Turner and Dr. Paul Mahlberg of Indiana University. After washing and fixing, numerous bacteria still adhered to the moss protonema. A. tumefaciens appears to lie prone on the moss filament, A. rhizogenes on end, and A. radiobacter loosely connected by flagella. A. rubi seems to be partially embedded in the wall of protonemal cells, suggesting that this strain may partially degrade the moss cell wall (Figures 1-4). A. tumefaciens and A. rhizogenes are found in similar orientation on Funaria hygrometrica although this moss does not show a developmental response to these bacteria.

The mean number of bacteria on a 1 mm length of moss protonema shows no correlation with the number of gametophores induced with different strains of Agrobacterium. A correlation does exist between the number of bacteria found on the spore and germ tube and the ability of the bacteria to produce responses on the moss (Table 1). Species sites may exist on spore and germ tube where attached bacteria exert effects on development. Some messenger molecule may then transfer information to the cells where the actual developmental changes occur.

Structural differences in the cell wall of different parts of the moss

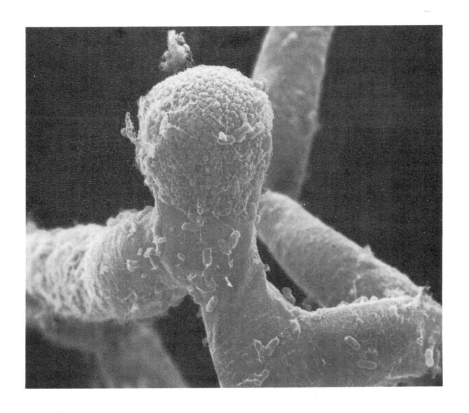

Figure 1. P. selwynii spore, germ tube and filament with A. tumefaciens
prone on the moss cells. SEM x 2000. Figure by Dr. Jocelyn
Turner and Dr. Paul Mahlberg.

Figure 2.   P. selwynii filament cell with A. rhizogenes positioned on
end to moss cell. SEM x 3600. Figure by Dr. Jocelyn Turner
and Dr. Paul Mahlberg.

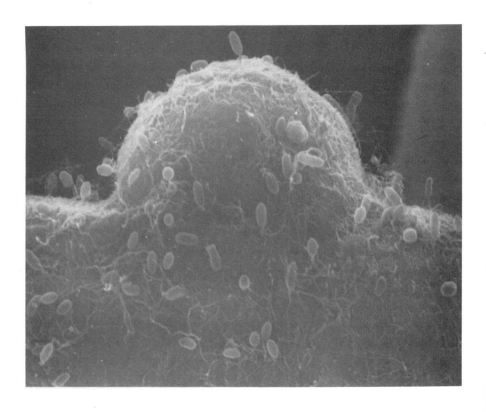

Figure 3. P. selwynii cells with A. radiobacter loosely attached by flagella. SEM x 5250. Figure by Dr. Jocelyn Turner and Dr. Paul Mahlberg.

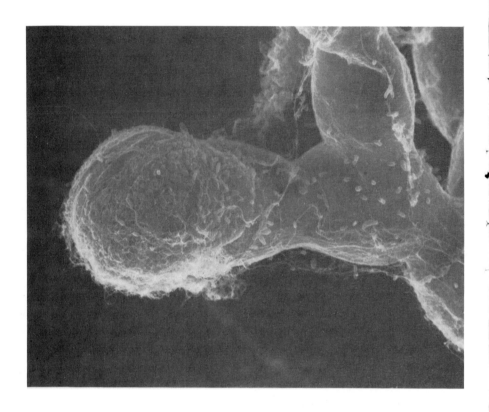

Figure 4. P. selwynii spore, germ tube and filament with A. rubi. Some of the bacteria seem to be partially embedded in the moss cell wall. SEM x 2040. Figure by Dr. Jocelyn Turner and Dr. Paul Mahlberg.

Table 1. Distribution of bacteria on moss.

	Bacteria/mm filament (mean-S.E.)	Bacteria on spore	Bacteria on germ tube	Concentration bacteria ($\times 10^{-7}$/ml)
P. selwynii				
A. tumefaciens	63.6 ± 6.5	12.0 ± 1.2	15.2 ± 3.49	9
A. rhizogenes	70.0 ± 6.2	5.1 ± 1.1	4.5 ± 0.72	6
A. radiobacter	134.4 ± 10.0	5.6 ± 0.75	4.9 ± 1.23	3.9
A. rubi	161.6 ± 12.1	32.1 ± 5.3	21.3 ± 7.61	1.3
F. hygrometrica				
A. tumefaciens	58.0 ± 5.5	4.9 ± 0.77	3.4 ± 0.95	9
A. rhizogenes	50.4 ± 7.1	4.7 ± 0.99	1.4 ± 0.20	6

can be seen in the SEM pictures. Although the protonemal cell wall surface of P. selwynii appears rough and that of F. hygrometrica smooth, there appears to be a similar number of bacteria associated with each. The cell walls of P. selwynii and F. hygrometrica gametophores both appear smooth and no bacteria are found attached to these walls. The germ tube cell walls also appear smooth but only in P. selwynii are many bacteria found attaching to this part of the moss. These results suggest that the surface composition of the moss cell walls varies with development as it does in higher plants and some of these changes are undoubtedly of significance to the bacterial interaction with the moss.

Experiments with A. tumefaciens and P. selwynii in which the material was fixed for SEM at 2, 6 or 26 hr after addition of the bacteria show a shift in the number of adhering bacteria from 156 ± 5.3 per mm of protonema at 2 hr to 92.8 ± 7.7 at 6 hr and 99.2 ± 9.8 at 26 hr. Site competition experiments have shown that site attachment effective in promoting development is complete after 6 hr (Spiess et al., 1976). This apparent reduction in number of attached bacteria seen in the SEM after 2 hr may indicate that some of the bacteria which initially attach lyse in the interval between 2 and 6 hr or the cell wall structure is altered so bacteria do not adhere as readily. This phenomenon may be of importance in the activity of the agrobacteria.

Glucose, when added with A. tumefaciens, neutralizes the effect of the bacteria on the moss. Galactose at similar concentrations is ineffective. Such results may indicate that glucose competes for either the moss or bacterial components of site attachment.

From these studies we propose that the developmental changes in the moss induced by agrobacteria or rhizobia are initiated following adherence via the lipopolysaccharide of the bacterial cell envelope to specific sites on the moss cell wall. Production of specific cytokinins such as ribosyl-zeatin by the bacteria or the initiation of their synthesis in the plant cell may control some of the developmental responses. Moss such as Funaria which do not respond to agrobacteria also fail to respond to this cytokinin (Spiess, 1975, 1976). Because ribosyl-zeatin induces only callus in responsive moss, other substances such as lysopine and octopine, which are induced by agrobacteria in higher plants, and also affect moss development may be involved in these bacterial effects (Spiess et al., 1972). Also structures similar to those induced by bacteria have been produced by adding ribosyl-zeatin, cyclic AMP and indoleacetic acid to the moss (Spiess, unpublished data). The full role of the bacterial attachment in directly contributing to some of these metabolic changes remains to be determined.

References

Lippincott, J.A. & Lippincott, B.B. (1969). Bacterial attachment to a specific wound site as an essential stage in tumor initiation by Agrobacterium tumefaciens. Journal of Bateriology 97, 620-628.

Spiess, L.D. (1975). Comparative activity of isomers of zeatin and ribosyl-zeatin on Funaria hygrometrica. Plant Physiology 55, 583-585.

Spiess, L.D. (1976). Developmental effects of zeatin, ribosyl-zeatin and Agrobacterium tumefaciens B6 on certain mosses. Plant Physiology 58 (in press).

Spiess, L.D., Lippincott, B.B. & Lippincott, J.A. (1971). Development and gametophore induction in the moss Pylaisiella selwynii as influenced by Agrobacterium tumefaciens. American Journal of Botany 58, 726-731.

Spiess, L.D., Lippincott, B.B. & Lippincott, J.A. (1972). Influence of certain plant growth regulators and crown-gall related substances on bud formation and gametophore development of the moss Pylaisiella selwynii. American Journal of Botany 59, 233-241.

Spiess, L.D., Lippincott, B.B. & Lippincott, J.A. (1973). Effect of hormones and vitamin B$_{12}$ on gametophore development in the moss Pylaisiella selwynii. American Journal of Botany 60, 708-716.

Spiess, L.D., Lippincott, B.B. & Lippincott, J.A. (1976). The requirement of physical contact for moss gametophore induction by Agrobacterium tumefaciens. American Journal of Botany 63, 324-328.

Whatley, M.H., Bodwin, J.S., Lippincott, B.B. & Lippincott, J.A. (1976). Role for Agrobacterium cell envelope lipopolysaccharide in infection site attachment. Infection and Immunity 13, 1080-1083.

DISCUSSION

Chairman: J. Raa

Dazzo: If you examine the attachment of Rhizobium to root hairs in either suspension tissue culture or using a like technique you find that the bacteria are polarly attached. If you prepare scanning electron micrographs of the bacteria attached to the surface of the root hair you find that they are, by and large, lying in a prone position. I'd like to ask, in your situation, do you get your same positioning of attachment using scanning electronmicroscopy as you would had you used a suspension or in some way examined it without a fixation treatment or any treatment for electron microscopy?

Spiess: I can't answer that because we can't see the actual bacteria attachment. We have never been able to see the actual position. We have not looked at Rhizobium, this is one thing we want to look at, to see how it is attached. These, that I showed here were all fixed at the same time through the same procedure. The implication is that the difference is because of the difference in bacteria.

Dazzo: Can you propose a mechanism whereby the mini bacteria that you have observed that attach at two hours are released?

Spiess: We know that when the bacteria attach, whether it is at the germ

tube or wherever the active sites for initiation may be, some message must
be communicated from that region through the rest of the plant body, because
the gametophores or callus do not develop at that point. The whole plant
has been programmed to respond to the condition implied by the bacteria.
So we would then propose at this time that some process has begun which is
altering the bacteria in association with it. Whether it is lysis or
changes in the cell wall I can't say at this point.

Raa: Moss is a soft and tender plant that normally lives under humid
conditions. I wonder if mosses are more susceptible to disease or attack
by microorganisms than other plants. Do they for instance react hypersensitively to the presence of microorganisms?

Spiess: Yes, the plant is killed and there is plasmolysis of the moss
cell. I don't know if this is due to the bacterial action or whether in
growing them in an enclosed environment there are toxic products from the
bacteria. As far as moss being a tender plant; if you have ever handled
one of these you will find that they are very tough and you can push them
around. In some respects I think they may be even more resilent than the
higher plant tissue.

LIST OF PARTICIPANTS AND ACCOMPANYING PERSONS

ALBERSHEIM, Peter	University of Colorado, Department of Chemistry, Boulder, Colorado 80302, USA
AYERS, Arthur R.	Swedish Forest Products Research Laboratory, Department of Chemistry, Box 5604, S-114 86 Stockholm, Sweden
BAKKERUD, Kari Gudrun	Botanical Laboratory, University of Bergen, Allégt. 70, N-5014 Bergen, Norway
BAKKERUD, Jon	As above
BALLOU, Clinton	University of California, Department of Biochemistry, Berkeley, California 94720, USA
BALLOU, Lun	As above
BATEMAN, Durward F.	Cornell University, Department of Plant Pathology, Ithaca, New York 14850, USA
BAUER, Wolfgang D.	C.F. Kettering Research Laboratory, 150 East South college St., Yellow Springs, Ohio 45387, USA
BROWN, Robert	Dalhousie University, Department of Biology, Halifax, Nova Scotia B3H 4J1, Canada
BYRDE, Robert J.W.	University of Bristol, Long Ashton Research Station, Bristol BS18 9AF, England
BYRDE, Joyce H.	As above
COOPER, Richard M.	University of Bath, School of

List of participants

	Biological Sciences, Claverton Down, Bath BA2 7AY, Avon, England
DAZZO, Frank B.	University of Wisconsin, Department of Bacteriology, Madison, Wisconsin 53706, USA
DAZZO, Olagalina G.	As above
DEFAGO, Geneviêve	Institut für spezielle Botanik der eigd. Technische Hochschule, Universi- tätstr. 2, 8006 ZH, Switzerland
DELMER, Deborah P.	Michigan State University, MSU/ERDA Plant Research Laboratory, East Lansing, Michigan 48824, USA
ERIKSSON, Karl-Erik	Swedish Forest Products Research Laboratory, Department of Chemistry, Box 5604, S-114 86 Stockholm, Sweden
ESQUERRÉ-TUGAYÉ, Marie-Thêresè	Université Paul Sabatier, Centre de Physiologie Végétale, 118, route de Narbonne, 31077 Toulouse Cédex, France
GOODMAN, Robert N.	University of Missouri, Department of Plant Pathology, Columbia, Missouri 65201, USA
GOODMAN, Phoebe	As above
GRAHAM, Terrence	University of Wisconsin, Department of Plant Pathology, Madison, Wisconsin 53706, USA
HIRUKI, Chuji	University of Alberta, Department of Plant Science, Edmonton, Alberta T6G 2E3, Canada
HJELMELAND, Knut	University of Tromsø, Institute of Biology and Geology, P.O.Box 790, N-9001 Tromsø, Norway
HUBBELL, David H.	University of Florida, Institute of Food and Agricultural Sciences, Soil Science

List of participants

	Department, 2169 McCarty Hall, Gainesville, Florida 32611, USA
JARVIS, Michael	University of Hull, Department of Plant Biology, Hull HU6 7RX, England
JOHNSEN, Arne G.	University of Tromsø, Institute of Biology and Geology, P.O.Box 790, N-9001 Tromsø, Norway
KAUSS, Heinrich	Universität Kaiserlautern Biologi, 6750 Kaiserlautern Pfaffenbergstrasse, Germany
KIER, Ilse	University of Copenhagen, Institute of Plant Physiology, Øster Farimagsgade 2A, DK 1353 København K, Danmark
KIJNE, Jan	Botanisch Laboratorium der Rijksuniversiteit, Nonnensteeg 3, Leyden, Nederland
KIMMINS, Warwick C.	Dalhousie University, Department of Biology, Halifax, N.S. B3H 4J1, Canada
KNEE, Michael	East Malling Research Station, East Malling, Maidstone, Kent, ME 19 6BJ, England
LIPPINCOTT, Barbara B.	Northwestern University, Department of Biological Sciences, Evaston, Illinois 60201, USA
LIPPINCOTT, James A.	As above
MATHISEN, Inger	University of Tromsø, Institute of Biology and Geology, P.O.Box 790, N-9001 Tromsø, Norway
MUSSELL, Harry	Boyce Thompson Institute for Plant Research, Inc., 1086 North Broadway, Yonkers, New York 10701, USA
MUSSELL, Gale	As above
NISSEN, Per	University of Bergen, Botanical Laboratory, Allégt. 70, N-5014 Bergen, Norway

List of participants

PAXTON, Jack
University of Illinois, Department of Plant Pathology, 248 Davenport Hall, Urbana, Illinois 61801, USA

PAXTON, Sarah
As above

PEGG, George F.
Wye College, Wye, near Ashford, Kent TN25 5AH, England

RAA, Jan
University of Tromsø, Institute of Biology and Geology, P.O.Box 790, N-9001 Tromsø, Norway

ROBERTSEN, Børre
University of Tromsø, Institute of Biology and Geology, P.O.Box 790, N-9001 Tromsø, Norway

ROBINSON, Peter J.
CSIRO, Davies Laboratory, Division of Tropical Agronomy, Privat Bag, Townsville, Quinsland 4810, University Road, Australia

ROBINSON, Helen
As above

RØDSÆTHER, Marianne C.
University of Tromsø, Institute of Biology and Geology, P.O.Box 790, N-9001 Tromsø, Norway

SELVENDRAN, Robert R.
Food Research Institute, Colney Lane, Norwich NR4 7UA, England

SOLHEIM, Bjørn
University of Tromsø, Institute of Biology and Geology, P.O.Box 790, N-9001 Tromsø, Norway

SPENCER, Susan L.
University of California, College of Natural Resources, Agricultural Experiment Station, Department of Plant Pathology, Berkeley, California 94720, USA

SPIESS, Luretta D.
Lake Forest College, Lake Forest, Illinois 60045, USA

SPIESS, Eliot B.
As above

List of participants

STRØM, Arne	University of Tromsø, Institute of Biology and Geology, P.O.Box 790, N-9001 Tromsø, Norway
STUSHNOFF, Cecil	University of Minnesota, Department of Horticultural Science, St. Paul, Minnesota 55101, USA
SUNDHEIM, Leif	Agricultural University of Norway, P.O.Box 70, N-1432 Ås-NLH, Norway
SØMME, Randi	Agricultural University of Norway, Institute of Chemistry, N-1432 Ås-NLH, Norway
TOUZÉ, André	Université Paul Sabatier, Centre de Physiologie Végétale, 118 route de Narbonne, 31077 Toulouse Cédex, France
TOUZÉ, Jane-Marie	As above
TRIPATHI, Ram K.	Department of Plant Pathology, G.B. Pant University of Agriculture & Technology, Pantnagar, India. P.t. Justus Liebig University, Institute of Phytopathology, Giessen, Germany
TRONSMO, Arne	University of Tromsø, Institute of Biology and Geology, P.O.Box 790, N-9001 Tromsø, Norway
TRONSMO, Anne M.	As above
UNESTAM, Torgny	Institute of Physiological Botany, University of Uppsala, Box 540, S-751 21 Uppsala, Sweden
VALENT, Barbara	University of Colorado, Department of Chemistry, Boulder, Colorado 80302, USA

AUTHOR INDEX

Numbers underlined indicate the pages on which references are listed in full.

Furcht, L.T. 349,355
Fushtey, S.G. 32,35,58

Gaard, G. 417,419,420
Galanos, C. 418,420
Galbraith, W. 347,356
Galum, E. 354,357
Galun, E. 12,23,26
Gamborg, O.L. 362,365
Garbo 256
Garibaldi, A. 34,35,45,58,181,186,
  198,228,233
Gazdar, C. 166,203
Gelehrter, T.D. 166,204
Gerdemann 155
Gestetner, B. 219,233
Gilbert, W. 164,198
Giles, N.H. 167,195
Gill, R.H. 267,269
Glaser, C. 223,234,352,356,443,450
Glazer, A.N. 19,25,311,339,340
Glimelius, K. 361,365
Gooddell, E.W. 78,176,198
Goldstein, I.J. 347,356
Goldstein, J.L. 220,233
Goodenough, P.W. 169,179,187,188,
  189,198
Goodman, R.N. 125,270,295,423,424,
  426,437
Gottelli, I.B. 93,94,100
Gottesman, M. 164,165,197,203
Gottlieb, D. 282,286
Gotz, G. 361,364
Graham, P.H. 397
Grainger, J. 188,198
Granner, D. 166,204
Grassmann, W. 311,339
Gratzner, H. 177,198
Green, R.J. 35,38,60,190,202
Green, T.R. 227,233
Griffey, R.T. 192,198
Griffiths, D.A. 334,335,339
Griffiths, E. 23,25
Griffiths 333
Grodzicker, T. 164,195
Gros, F. 165,194
Gross, S.R. 166,198
Guérin, C. 349,356
Gupta, D.P. 37,38,58,64,168,171,
  180,181,182,186,198,239,282,286

Haass 353
Habig, W.H. 19,24,311,312,329,338
Haggerty, D.M. 166,199
Hahn, M. 134

Hahnel, E. 311,340
Hall 256
Hall, J.A. 34,58
Hall, R. 169,199
Ham, G.E. 408,412
Hamblin, J. 12,25,361,365,373,375,
  377,380,391,394
Hamp, S. 76,77,78
Hanchey, P. 219,238
Hancock, J.G. 45,58,185,199,222,
  223,228,233
Hankin, L. 168,170,182,206,221,238
Harman, G.E. 186,199
Hartley, R. 117
Hathway, D.E. 220,221,233
Hautala, E. 282,287
Hawthorne, D.C. 167,197
Haywood, P.L. 350,357
Heale, J.B. 37,38,58,64,168,171,180,
  181,182,186,198,239
Heath, M.C. 180,199
Heath, M.F. 105,106
Heberlein, G.T. 439,450
Heitefuss, R. 226,233
Hemming, F.W. 87,102
Henderson, S.J. 277,279
Henis, Y. 181,201,215,235
Henkart, P. 11,25
Henze, R.E. 281,286
de Herrera, E.C. 168,197
Heslop-Harrison, J. 26,334,339,340,
  342
Heslop-Harrison, Y. 26,334,340
Higashi, S. 408,409,412
Higgins, V. 149
Hijwegen, T. 221,233,277,279
Hiloick-Smith, G. 151,153
Hiroi, T. 73,78
Hiron 296
Hirotani, M. 169,173,177,205
Hiruki, C. 267,268,269
Hislop, E.C. 214,215,234,241
Holzer, H. 206
Horiuchi, T. 164,203
Horsfall, J.G. 43,58,188,199
Horton, J.C. 42,43,58,59,168,169,
  170,171,177,180,187,199,200
Hough, L. 121
Houldsworth, M.A. 177,196
Housten, D.R. 325,342
Howard, J.B. 19,25,311,339,340
Howell, C.R. 36,37,59,60
Howlett, B. 334,340,342
Hsu, E.J. 168,173,200
Huang, P.Y. 424,426,437

44